Crop Science and Production in Warm Climates

J.Y. Yayock, G. Lombin and J.J. Owonubi

Macmillan Intermediate Agriculture Series

General Editor:
O.C. Onazi B.Sc., M.Sc., Ph.D.

MACMILLAN
PUBLISHERS

First published 1988

Published by *Macmillan Publishers Ltd*
London and Basingstoke
Associated companies and representatives in Accra,
Auckland, Delhi, Dublin, Gaborone, Hamburg, Harare,
Hong Kong, Kuala Lumpur, Lagos, Manzini, Melbourne,
Mexico City, Nairobi, New York, Singapore, Tokyo

Printed in Hong Kong

British Library Cataloguing in Publication Data
Yayock, J.Y.
  Crop science and production in warm
  climates.
  1. Subtropical regions & tropical regions.
  Crops. Production
  I. Title II. Lombin, G. III. Owonubi, J.J.
  631.5'0913

ISBN 0-333-44539-2

Contents

Acknowledgements

The time, research and teaching experience that we have put into the writing of this book result from the academic atmosphere created jointly by the Institute for Agricultural Research and the Faculty of Agriculture of the Ahmadu Bello University, Zaria, for which we are grateful. Several colleagues, too many to list individually, have come to mean a great deal to us in terms of mutual encouragement. We nonetheless wish to specifically acknowledge the untiring support and cooperation of the staff of the Departments of Agronomy and Soil Science of Ahmadu Bello University.

Photographic acknowledgements

The authors and publishers wish to acknowledge, with thanks, the following photographic sources:
G.D. Bull pp 13; 18; 62; 207
Cadbury Schweppes p 267
FAO pp 141 (photograph F. Mattioli); 175 left
R.J. Hart pp 128; 150; 225 top; 228; 232; 263; 276
Hutchison Photograph Library pp 147; 281
Jeffries Ltd Suffolk p 69 top
Rex Parry pp 60; 97; 108; 115 top; 120 top; 132 top; 138; 142; 145; 184; 195; 202; 209; 212; 238; 244; 271
Alan Thomas p 199
Professor Don Tindall pp 111 right; 154; 179; 188; 215; 230; 235; 240; 242; 286
TROPIX Photograph Library pp 19; 66 top; 66 bottom; 70 top; 70 bottom; 71; 93
US Department of Agriculture pp 173; 205
World Bank Photograph Library pp 117; 270
All other photographs are courtesy of the author.

The cover photograph courtesy of COMPIX Photograph Library.

The publishers have made every effort to trace the copyright holders, but if they have inadvertently overlooked any, they will be pleased to make the necessary arrangement at the first opportunity.

Preface to the series

One of the most important problems in developing countries is the level of awareness and importance given to agricultural development. This problem is compounded because of the relatively weak institutional framework for promoting development of the agricultural sector. Countries determined to make rapid progress in agriculture and rural development have started to create appropriate training resources.

The role of agricultural education in this development process is vital. It is directly related to the effectiveness of the economy in providing the requirement of trained manpower for the development process. The general strategy for development relies on the intensive utilization of well trained individuals, particularly at the intermediate level.

One of the major problems facing institutions concerned with the training of middle- and top-level agricultural workers is the availability of suitable textbooks that are relevant to the special problems of agriculture in Third World countries. It is for this reason that we welcome the initiative of Macmillan Publishers in introducing the new Macmillan Intermediate Agriculture Series (MIAS). This series is designed to provide up-to-date textbooks that are geared to meet the needs of agricultural education at schools and colleges of agriculture; for agricultural courses at polytechnics and colleges of science and technology, advanced teachers' training colleges; and even for first year courses in university faculties of agriculture. This series will also be suitable for use in secondary schools where agriculture courses are taught.

The objective of the Macmillan Intermediate Agriculture Series is to provide suitable material with adequate illustration based on local problems and issues that affect tropical agriculture. The authors in the series have used their wide experience in tropical agriculture to considerable advantage. The texts have been based on sound scientific premises and therefore have international relevance for agricultural education in the tropics.

I commend the series for use in schools and colleges where agriculture is given due emphasis as a technical course. I trust that the series will be a valuable addition to the available materials in this field.

Ochapa C. Onazi
Zaria

April 1983

Preface

Increased agricultural production has been widely identified as the key to sustained socio-economic advancement in most of the developing nations of the world, nearly all of which lie within the tropics. Several factors have combined to place agriculture in its present position in tropical and subtropical countries, but the three main considerations are:

1. Over 60% of the work-force in the tropics is engaged in agriculture and allied fields, which provide important employment opportunities.
2. The rapidly expanding populations in the developing countries need to be adequately fed and this requires sustained increases in agricultural production.
3. Farmers need to produce substantially more than their immediate subsistence requirements to allow them to exchange any surplus for manufactured goods and services, both nationally and internationally. In many of the developing countries of the world, agricultural produce provides the only means of procuring the foreign exchange needed for development.

Agricultural production normally comprises two major aspects; crop production and livestock production. This book deals only with crop production. Increased crop production may be achieved by increasing the hectarage in use or the productivity per unit area of land under cultivation. Increasing the hectarage can be achieved by opening up new lands and by multiple cropping. Only marginal increases have occurred from the opening up of new lands across the tropics, while multiple cropping can only be recommended for areas where adequate climatic resources for production are available for a period of the year long enough to accommodate more than one crop.

The books presently available on tropical agriculture are clearly inadequate, and do not meet the demands posed by the continuous efforts to increase crop yields. The inevitable shift from the traditional, non-intensive methods of farming to the semi-intensive and continuous cultivation techniques has to be matched with appropriate production technologies. This book is intended to serve agricultural students at the diploma level, those studying for a National Certificate in Education or for the Interim Joint Matriculation Board, as well as those in the initial years of a first degree programme in Agriculture. In addition, the book is recommended as a reference text in tropical agriculture, particularly for people involved in agricultural services and development.

This volume is divided into three parts. The first part deals with the resources of the crop environment, including climate and soils. The second part covers the general cropping systems used in the tropics and the basic principles of crop production. The final part covers the specific production techniques for various crops grown in the tropics. As far as possible, use has been made of the latest research findings and most examples cited are taken from recent publications and research data.

PART ONE

Resources of crop production

Crop production as an industry depends on the availability of primary requirements such as seed, land, labour, and a suitable environment for growth and development. In this context, the environment means the immediate surroundings of the seed, seedling and the final crop, and includes the soil with which the roots and tubers are in contact, as well as the air that envelops the stem and leaves.

Being a living thing, the crop needs air, water and optimal temperatures (Figure 1.1). Air contains the oxygen required for respiration. Water is a constituent of protoplasm and is also the solvent in which nutrients are carried from the soil to the leaves and food manufactured by the leaves is transported to all parts of the plant. Warm temperatures are necessary to maintain optimal rates of all biochemical reactions within crop plants. In addition to these basic requirements of living things, crops need carbon dioxide and light for photosynthesis. Also a pest, disease and weed-free environment is necessary for optimal growth and development.

While it is recognised that socio-economic factors, such as capital formation, land tenure and marketing policies are important contributors to the development of any organised agricultural

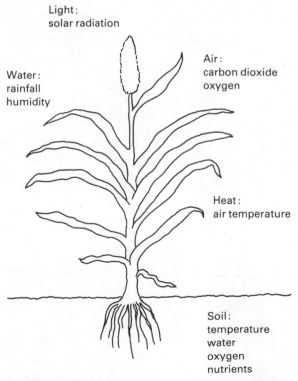

Light: solar radiation

Water: rainfall humidity

Air: carbon dioxide oxygen

Heat: air temperature

Soil: temperature water oxygen nutrients

Figure 1.1   Environmental requirements of a crop plant.

production system, this section, and indeed the volume, is deliberate in focussing largely on those exploitable environmental factors which are directly related to crop adaptation and culture and the maintenance of soil fertility. In this first section on the resources of crop production, it is therefore assumed that land and the labour to work it are available, and that what may be missing, in terms of limiting the yield, is an understanding of the crop and its environment. The section, therefore, focusses attention on climate and soil as resources for crop production in the tropics. In the first chapter on climate, the availability of water and the variability of solar radiation and temperature as the main components of the climatic input to cropping in the tropics are discussed. The second chapter examines the physical and chemical properties characteristic of tropical soils. The objective is to lay a foundation which leads to an evaluation of the nutrient supplying status of soils across the tropics.

CHAPTER 1

Climate

The behaviour of the earth-atmosphere system at any given point and time may be described in terms of physical parameters such as temperature, sunlight, humidity, wind speed and wind direction. This description, which is relevant only over a small geographical area, is referred to as the 'weather' of the area at that given time. When such quantitative data are pooled over a period of at least 30 years, a stable definition of the point emerges which may be true over a large land area.

This averaged weather is known as the 'climate' of the area. The climate of an area depends on its location (latitude and longitude), its elevation above sea level, its proximity to a mass of water and the overall effect of the continuous movement of the earth.

Climate affects crop production through the cumulative influence of various weather elements. This influence, along with the effect of each weather element, is summarised in Table 1.1.

Table 1.1. Main influences of weather elements on crop plants

Weather element	Form of influence	Effect on crops and production	Weather element	Form of influence	Effect on crops and production
Air	CO_2 content	Alters rate of photosynthesis	Temperature	Seasonal fluctuations	Necessitates matching of cropping times
	Density of pollutants	Alters penetration of sunlight; chemical injury		Daily fluctuations	Controls relative photosynthate accumulation
	Ozone	Alters level of ultra-violet injury to protoplasm		Extremely low	Causes possibility of cold injury; vernalization; stratification
Light	Daylength	Controls photoperiodic response		Extremely high	Causes possibility of heat injury; water deficit
	Intensity	Controls rate of photosynthesis	Wind	Air mass movement	Rainfall induction; lodging
	Spectral composition	Controls photosynthetic capability		Pest or spore movement	Pest or disease incidences
Rainfall	Amount	Causes flooding or drought conditions	Humidity	Spore germination	Disease incidences; pest reproduction cycles
	Distribution	Alters length of growing season		Air saturation	Controls precipitation
	Influence on humidity	Alters rate of evapo-transpiration		Air water deficit	Alters rate of evapo-transpiration

Light

The sun provides almost all the energy used on the earth's surface. This energy is received in the form of electromagnetic radiation of varying wavelengths transmitted through space and the earth's atmosphere. Every surface at a temperature above absolute zero ($-273°C$) emits some form of radiation. The various types of light making up solar radiation, along with their effects on crop plants, are given in Table 1.2. Of the visible spectrum, only the blue (400–510 nm) and the red (610–700 nm) wavelengths are used in photosynthesis.

A general picture of the global distribution of sunlight is presented in Figure 1.2. It should be noted that average cloud cover can reflect as much as 25% of total in-coming radiation. Similarly, normal atmospheric absorption (2%) and reflection (7%) can be increased by the presence of large amounts of pollutants, such as dust, in the air.

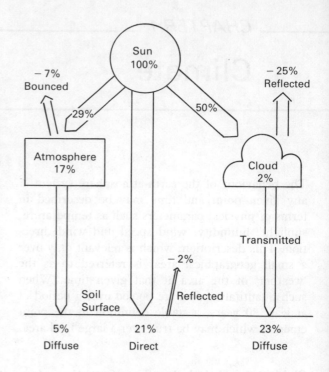

Figure 1.2 Global distribution of sunlight.

Table 1.2. The components of solar radiation and their effects on plants

Light types	Range of wavelength (nm)	Effect on plants
Ultraviolet	10–350	Detrimental
Visible	350–700	Photosynthetic
Near infra-red	700–2000	Morphogenetic
Far red	2000–10 500	Energy balance

The distribution of sunlight across the tropics on a monthly basis is given in Table 1.3. The intensity of solar radiation decreases as one moves away from the equator. Variations from month to month are reflective of the seasons. In the northern

Table 1.3. Mid-monthly intensity of solar radiation (W/m²) on a horizontal surface above the earth's atmosphere (assuming 12 hours daylight throughout)

Month	Northern hemisphere				Southern hemisphere		
	30°	20°	10°	0°	10°	20°	30°
January	518	657	779	882	962	1023	1054
February	640	749	846	913	956	974	962
March	773	846	901	926	920	889	828
April	901	826	826	895	840	762	657
May	974	956	913	846	755	651	529
June	1002	962	901	816	706	584	451
July	987	956	901	821	724	609	475
August	932	932	913	865	792	701	584
September	822	877	907	907	877	822	737
October	688	785	859	913	932	932	901
November	554	682	798	889	956	999	1017
December	481	627	755	871	962	1029	1071
Mean	773	830	867	879	865	831	772

Source: Modified from; List R.J., 1958, Smithsonian Meteorological Tables, Smithsonian Institution, Washington.

hemisphere, maximum solar radiation is received around June while the least is in December. These times are reversed in the southern hemisphere. Sunlight is almost uniformly intense throughout the year at the equator. These observed differences are caused by the rotation and revolution of the earth around the sun as well as the fact that the earth is not only irregularly shaped, but also has its axis of rotation inclined about 66.5 degrees from the plane of ecliptic.

Other sources of local variation in the amount of radiation received at any particular point and time include the amount of cloud cover, the concentration of pollutants in the atmosphere and the angle of inclination of the sun. The effect of the concentration of pollutants in the atmosphere on the incident solar radiation is shown in Figure 1.3.

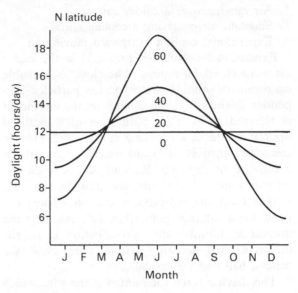

Figure 1.4 Monthly trend in daylength at the equator and at latitudes 20, 40 and 60°N.

The wide variation in diurnal solar radiation on the three dates was caused by different levels of dust concentration in the air.

Differences in daylength is another important crop-related light phenomenon which results from the movement of the earth. The variability of day-length in the northern hemisphere is presented in Figure 1.4. In line with the variability in the intensity of solar radiation, daylength is almost constant, at about 12 hours per day, only at the equator.

Rainfall

Rainfall is the most significant climatic element affecting crop production in the tropics. As much as 90% of cropping is rainfed. The rains are generally seasonal, having distinct wet and dry periods. The time of onset of the rainy season and the total amount, distribution and duration of rainfall, as well as the time of its cessation, all contribute to the type and number of crops grown per season and their expected yield.

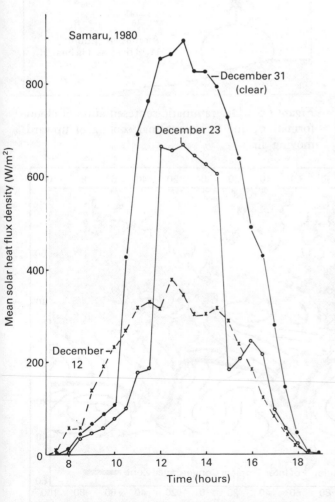

Figure 1.3 Depletion of incoming solar radiation by dust in the atmosphere during the dry season at Samaru, Nigeria.

Rains normally fall when moist, warm air is forced to cool to a temperature below its dew point. The cooling process can be one of the following:

1. Air moving over a colder surface.
2. Stagnant air overlying a cooling surface.
3. Expansional cooling in upward moving air.

Expansional cooling (Figure 1.5) is the major cause of cloud formation. The cloud is a visible aggregate of minute water or ice particles suspended in the air. If the cloud is on the ground, it is referred to as 'fog'. Saturation of a localised updraft produces a towering cloud which may be composed entirely of liquid water, ice crystals or a mixture of the two. Rainfall occurs when the minute water or ice particles grow in size and weight until the atmosphere can no longer suspend them; at that point they fall, reaching the ground as liquid water. Precipitation is an all-inclusive term denoting drizzle, rain, snow, ice pellets, hail and ice crystals.

The driving force for rainfall is the sun, which provides the energy required for evaporating water into the air and creates the surface-energy imbalance that generates winds. Expansional cooling can arise either through convection or when warm, wet air converges with dry air. In West

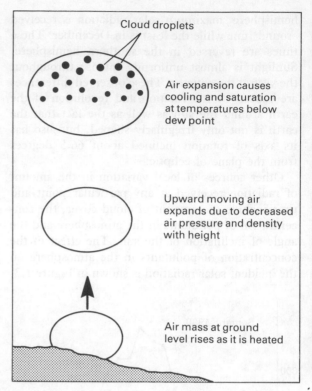

Figure 1.5 Diagrammatic representation of cloud formation, due to expansional cooling of upward moving air.

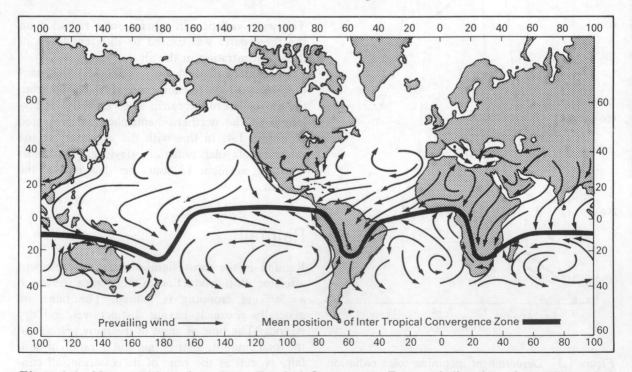

Figure 1.6 Mean position of the Inter Tropical Convergence Zone and direction of prevailing winds throughout the tropics during the month of January.

6

Africa, the moist, warm air is the southwest trade wind while the dry and dust-laden air is the northeast Harmattan; the line of convergence is the Inter Tropical Convergence Zone (ITCZ). The position of the ITCZ and the prevailing winds throughout the tropics are given (Figures 1.6 and 1.7) for the months of January and July, respectively.

Expansional cooling of air is a process involving energy exchange. As mentioned before, the sun is the source of virtually all energy available for work on earth. The intensity of solar radiation is highest at the equator and decreases towards the poles. This causes air near the ground to move from the tropics towards the equator. This air then rises at the equator, moves towards the poles at a higher level and then sinks at the poles (Figure 1.8). However, because the earth is spherically shaped and continuously rotates, a force referred to as the 'Coriolis force' diverts the expected path of moving particles on earth. The Coriolis force deflects air to the right in the northern hemisphere and to the left in the southern hemisphere (Figure 1.9). The mean circulation emerging at the tropics with the influence of the Coriolis force averages eight kilometres per hour and is called the 'Hadley cell' (Figure 1.10). This circulation cell, with the rising

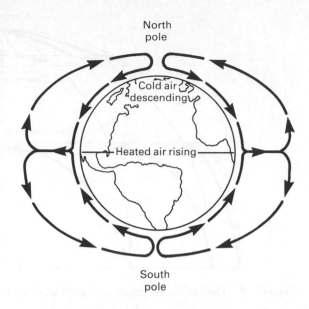

Figure 1.8 Circulation of air in the atomosphere as would be expected with a non-rotating earth.

warm air at the equator, explains why the region receives so much rainfall.

Around latitudes 30° north and south, the air sinks and warms up. There is therefore no rain and most of the world's deserts are found in these

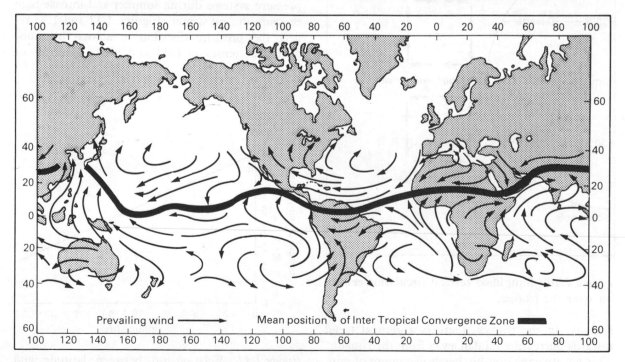

Figure 1.7 Mean position of the Inter Tropical Convergence Zone and direction of prevailing winds throughout the tropics during the month of July.

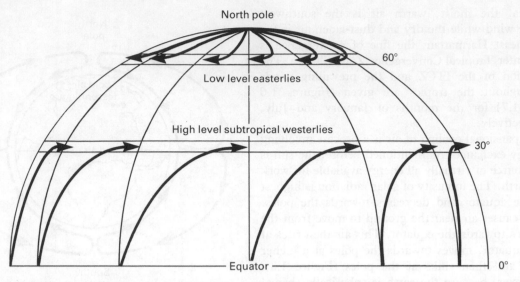

Figure 1.9 Deflection of mean air circulation in the northern hemisphere due to the Coriolis force.

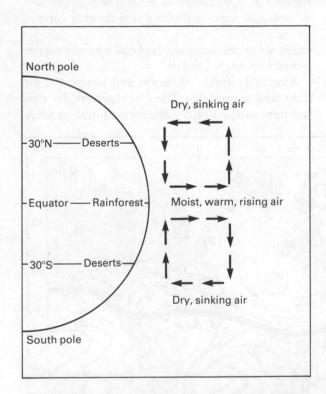

Figure 1.10 Simplified vertical circulation cells of air over the tropics.

regions. In the tropics, there is a general trend towards decreasing rainfall away from the equator. This negative relationship between amount of rainfall and latitude is clearly demonstrated in Figure 1.11 for the savanna ecological zone of Nigeria.

The simplified Hadley cell circulation applies only over the oceans. Over the continents, especially in the tropics, monsoon circulations and the effects of mountain ranges dominate. The large land masses of Asia, Australia and Central Africa warm up and are covered by well developed low pressure systems during summer and intense high pressure during winter. Figures 1.12 and 1.13 reflect the pressure distribution across the globe during extremes of cold and warm (January or

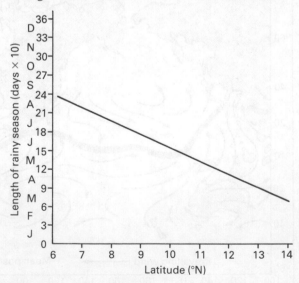

Figure 1.11 Relationship between latitude and length of the rainy season in the savanna ecological zone of Nigeria.

8

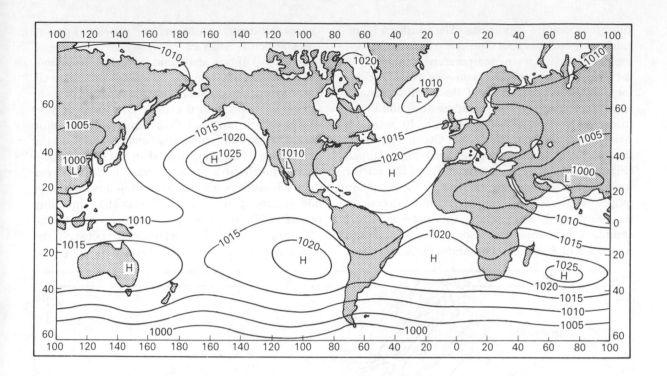

Figure 1.12 Mean world-wide surface pressure distribution in July.

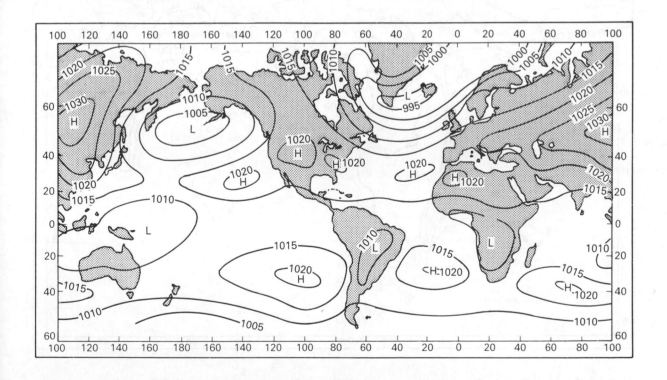

Figure 1.13 Mean world-wide surface pressure distribution in January.

July) temperature conditions. The southern hemisphere experiences cool temperatures (winter) during the times of warm temperatures (summer) in the northern hemisphere, and vice versa. Therefore, in the northern half of the tropics, summer winds are mostly southwest, as the humid air from the oceans, flows over land. In southern Asia, rising air from the Indian Ocean flows over the Western Ghats, causing very high rainfall (Figure 1.14). The reduction in humidity on the western slopes leaves the other side of the mountain ranges relatively dry. Similar conditions occur in the Southern hemisphere as the easterlies bring abundant rain into the eastern shores of South America and Australia. The Andes mountain range causes the desert conditions in the western part of tropical South America.

Rains fall almost daily and virtually throughout the year at the equator. Moving progressively away from the equator, both the amount of rainfall and its duration decrease. The time of year during which rains fall then constitutes the rainy season, while the remaining rainless period is the dry season. The rainy period in West Africa decreases northwards by an average of 19 days per degree latitude, such that at the 350 mm isohyet (a line joining points of equal rainfall), the rainy period is 55 days, compared with 220 days at the 150 mm isohyet. The relationship between the times of onset and retreat of rains, the length of

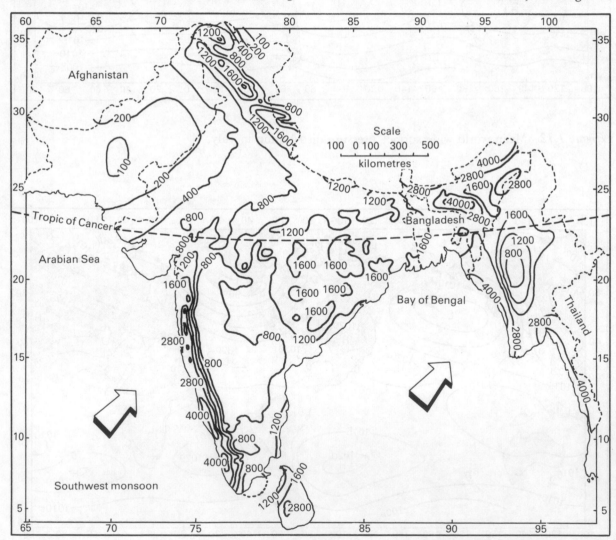

Figure 1.14 Combined effect of relief and wind direction on mean annual rainfall (mm) over southern Asia (adapted from World Survey of Climatology, Volume 9, 1981).

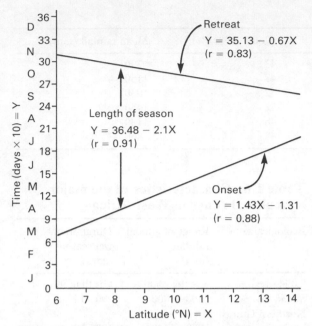

Figure 1.15 Relationship between the time of onset and retreat of rainfall, the length of the growing season and latitude, in Nigeria.

While most parts of the tropics have only one rainy season in a year (uni-modal rainfall), areas close to the equator generally have a long rainy season, lasting almost through the year, with two peak rainfall periods (bi-modal rainfall pattern). The bi-modal rainfall pattern, which is seen within latitudes 9° North and South of the equator, is caused by the relatively long period separating the two occasions a year when the sun is overhead at these low latitudes.

The distribution of the rains and the consequent ecological sub-divisions across the tropics are summarised in Table 1.4. The rainy season lasts for more than four and a half months in three-quarters of the land area in the tropics. The mean annual growing season and the latitude, is shown in Figure 1.15 for Nigeria.

The rainy season may be further subdivided on the basis of moisture availability as in Figure 1.16. The preparatory period (PP) represents the time when rainfall is between 0.1 (10%) and 0.5 (50%) of the full potential evapo-transpiration (PET). The intermediate period (IP) refers to the intervals before and after the wet season when rainfall exceeds or is equal to 50% PET, but remains less than PET. The humid period (HP) refers to the time when rainfall exceeds PET, while the growing season length combines the times designated humid, intermediate and residual moisture depletion (PRM).

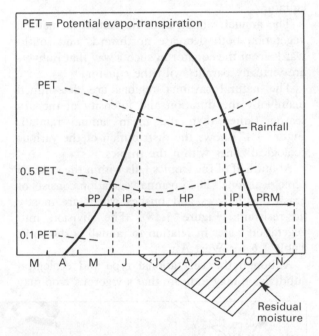

Figure 1.16 Periods of moisture availability for cropping as derived from rainfall and potential evapo-transpiration patterns.

Table 1.4. Distribution of major climatic regions in the tropics

Climatic region	Number of humid months	Predominant vegetation	Tropical America (M ha)	Tropical Africa (M ha)	Tropical Asia (M ha)	Total (M ha)	% of total
Rainy	9.5–12	Rainforest and forest	646	197	348	1191	24
Seasonal	4.5–9.5	Savanna or deciduous forest	802	1144	484	2430	49
Dry	2–4.5	Thorny shrubs and trees	84	486	201	771	16
Desert	0–2	Desert and semi-desert	25	304	229	558	11
Total			1557	2131	1262	4950	100

Source: Sanchez, P.A., 1976, *Properties and Management of Soils in the Tropics*, John Wiley, New York.

Table 1.5. Rainfall distribution in the tropics

Latitude	Land (%)	Ocean (%)	Cloud cover (%)	Mean rainfall (mm)
20–30°N	37.6	62.4	43	790
10–20°N	26.4	73.6	47	1150
0–10°N	22.8	77.2	52	1930
0–10°S	23.6	76.4	52	1450
10–20°S	22.0	78.0	48	1130
20–30°S	23.1	76.9	48	860

Source: Sellers, 1965.

rainfall and the ocean and cloud coverage across the tropics are given in Table 1.5. The range of rainfall is from almost zero at latitudes 20° North and South of the equator to over 2500 mm at the equator.

The annual rainfall and the density of natural vegetation both decrease northwards and southwards from the equator in such a way that isohyets are virtually parallel with the equator.

The natural savanna regions are determined mainly by the duration and intensity of the dry season rather than the mean annual rainfall. Figure 1.17 shows the distribution of the various ecological zones within the tropics.

About half of the tropics falls within the savanna ecological zone. The savanna vegetation consists of grasses with trees and bushes which are mostly fire-resistant (Figure 1.18). The division into vegetation zones in relation to rainfall is shown in Table 1.6 for West Africa.

Although the vegetational type and ecological subdivision may indicate that a vigorous crop may

Table 1.6. Characteristics of the major agricultural zones in West Africa

Ecological zone	Range of annual rainfall (mm)	Duration of wet season (days)
Sahel savanna	less than 500	less than 90
Sudan savanna	500–1000	90–130
Northern Guinea savanna	900–1400	130–190
Southern Guinea savanna	1000–1650	190–250
Forest	1550–2550	more than 250
Coastal swamps	2300–4100	more than 250

thrive if grown in a particular location, the actual boundaries for arable farming in the tropics depend on the following criteria:

1. Minimum rainfall providing the water requirements of a 75-day crop such as millet.

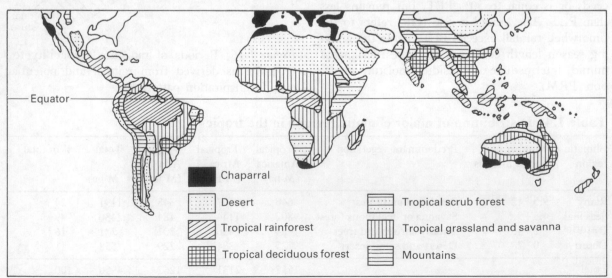

Figure 1.17 Ecological subdivisions of the tropics.

Figure 1.18 Savanna vegetation consists of grasses with trees and bushes which are mostly fire-resistant.

2. At least 55-day rainy period, assuming soil water storage lasts 20 days.
3. Reliability of rainfall and length of rainy period falling within acceptable limits for the farmer.

Areas that do not satisfy these limits are often referred to as arid lands. Available information on crop water requirements (Table 1.7) and losses from seepage and run-off, as well as pre-planting water requirements, show that major crop production would appear limited to within latitudes 12° North and South of the equator.

Table 1.7. Water requirements from planting to harvesting of various crops grown in the savanna zones of West Africa

Crop	Water requirement (mm)
Cotton	480
Cowpea	470
Groundnut	410
Maize	490
Millet	330
Sorghum	670
Wheat	640

Once the rainy season has begun in the zone bordered by latitudes 12° North and South of the equator, crops hardly suffer due to shortage of water. The exceptions to this are the zones between latitudes 6 and 9° North and South of the equator where a transition exists from uni-modal to bi-modal patterns of rainfall; the intervening period may often experience prolonged dry spells. Indeed, during the major part of the rainy season, farm management is often presented with the problems of excessive water accumulation. Pronounced leaching, high surface run-off, persistent high water tables, flooding and soil erosion are common problems of the Coastal Swamps and Forest vegetational zones. The need to profitably manage excess water, especially in the wetter parts of the tropics, has necessitated the use of multiple cropping. The cropping system may take the form of sole, intercropping, inter-planting or relay, depending on the length of the rainy season.

Air temperature

Heat is a form of energy. When a substance contains heat, it exhibits the property measured as temperature, that is its degree of 'hotness' or 'coldness'. It is a measure of the total energy available to a body and its energy distribution characteristics (heat capacity and thermal conductivity). The heat capacity refers to the energy-retaining property of a body or substance, while the thermal conductivity is a measure of the rate of passage of heat through the material.

Temperature, being a quantitative expression of the energy level of any substance or surface, is closely related to the quantity of heat available to that body or surface. It follows, therefore, that either temperature or energy emission from any body or surface can be estimated if the other parameter is known. Similarly, temperature distribution across the globe is primarily dictated by the intensity of solar radiation incident at any particular location. The mean distribution of temperature across the tropics, therefore, follows the same trend as that of incident solar radiation, with mean temperature decreasing away from the equator (Table 1.8).

Table 1.8. Temperature characteristics of the tropics

Latitude	Mean temperature (°C)
20–30°N	20.6
10–20°N	25.3
0–10°N	25.7
0–10°S	25.0
10–20°S	23.5
20–30°S	19.0

Source: Sellers, 1965.

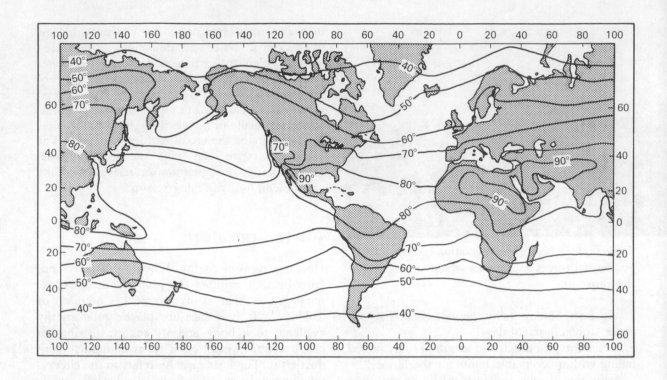

Figure 1.19 World-wide average surface temperatures in July.

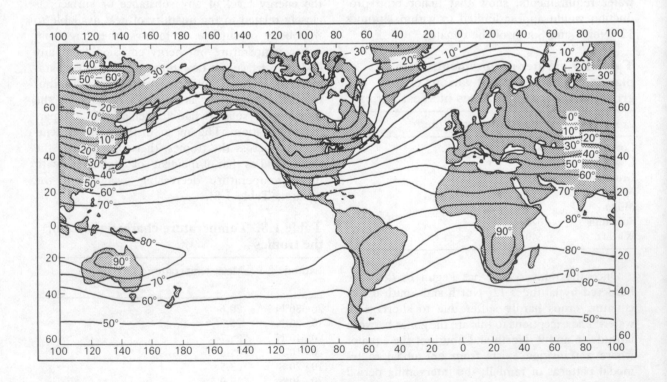

Figure 1.20 World-wide average surface temperatures in January.

14

The high incident radiation which is relatively evenly distributed throughout the year makes the tropics a warm environment, only seasonally modified by changes in water regimes and surface conditions. There is no 'winter' as experienced in the temperate regions and the growth of vegetation or crops is not limited by low temperatures. Worldwide average surface temperatures for the months of July and January are shown in Figures 1.19 and 1.20 respectively.

The mean seasonal or monthly temperature values tend to smooth out daily and diurnal variations. On average, temperatures at a given location and time of year vary relatively little. The pattern of thermal fluctuations during the year depends on location. Figure 1.21 demonstrates the extreme conditions that are found in tropical Africa as represented by Sokoto (13°01′N, 4°35′E), northern Nigeria.

The diurnal range of temperature is generally lowest during the humid periods, as maximum and minimum temperatures are moderated by the presence of clouds. During the dry season, the prevailing winds are characterised by cool night and hot day temperatures. For example, Samaru (11°11′N, 7°38′E), located in the middle of the tropics, has an annual mean temperature of 22°C, but often records daily maximum and minimum temperatures of 39° and 6°C respectively.

In the arid areas of the tropics, air temperatures frequently exceed 40°C, while at the soil surface even higher temperatures (60°C) may be experienced. Soil temperatures as high as 68°C have been reported in the arid parts of Botswana. But in the same location, minimum night temperatures may drop to as low as 6°C during parts of the year. The lowest daily minimum temperatures seem to occur about one to two months after the reversal of winds, signifying the end of the rainy season. The reduction in temperature may be quite pronounced over highlands and minimum air temperatures frequently fall below 0°C in parts of East Africa, Mexico and Cameroon.

Cropping must, therefore, take into consideration seasonal variations in rainfall and temperature. In practice, where water is available either through rainfall or irrigation, cropping can be done throughout the year in the tropics. However, the types of crop, and the time of year they are grown, will be dictated not only by the amount of water available, but also by the prevailing temperatures.

Soil temperature

Soil temperature affects the rate of mineralization of soil organic matter, the rate of evaporation of water from the soil and, more directly, the physiology of crop plants. The average soil temperature in the tropics is much higher than in temperate climates. Since there is no cold season, exposed soil surfaces often become extremely hot, especially during the dry season. The mean monthly soil temperatures at 5 cm depth at Samaru, Nigeria, range from 26 to 35°C, but extreme temperatures frequently rise to 46°C. Consequently all chemical and many biological processes, including degradation of organic residues, are more rapid, often two to four times faster, than in temperate climates.

There is generally a scarcity of soil temperature data across the tropics. However, there is a direct relationship between air and soil temperatures which is moderated by the availability of water in the soil as shown in Figure 1.22. The type and density of material covering the surface also affects the relationship between air and soil temperatures.

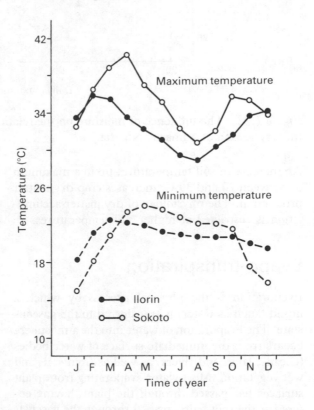

Figure 1.21 Annual trends of maximum and minimum temperatures at Ilorin and Sokoto, Nigeria.

15

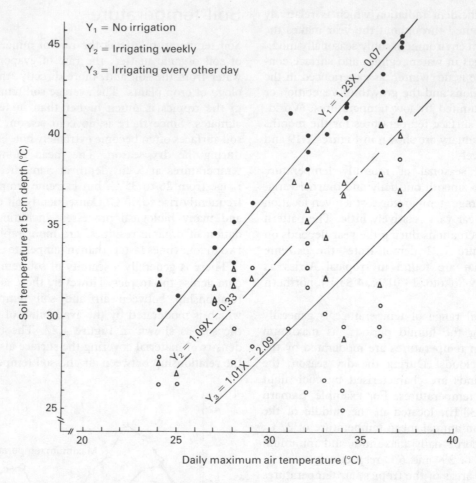

Figure 1.22 The influence of moisture on the relationship between air and bare soil temperatures during the dry season at Samaru, Nigeria.

An increase in soil temperature, up to a maximum of between 20 and 30°C, increases crop dry matter production. The physiology of dry matter accumulation is restricted at higher soil temperatures.

Evapo-transpiration

Evaporation is the physical process by which a liquid (such as water) is transferred to the gaseous state. The evaporation of water into the atmosphere occurs from the immediate surface of water bodies (oceans, lakes and rivers) as well as from soils and wet vegetation. Most water evaporating from plant surfaces has passed through the plant, having entered at the root hairs, passed through the vascular tissue to the leaves or other organs and left the plant into the surrounding air via the stomata or cuticules. The process of evaporation of water

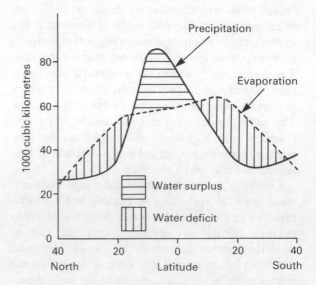

Figure 1.23 Latitudinal variation of evaporation and precipitation in the tropics.

Table 1.9. Mean annual relative humidity and evaporation (or evapo-transpiration for values in parentheses) at various locations across the tropics

Country	Location	Latitude	Longitude	Relative humidity (%)	Evaporation/ evapo-transpiration (cm)
Gambia	Georgetown	14°46′N	13°32′E	56	270
Kenya	Eldoret	0°32′N	35°17′E	60	156
	Nanyuki	0°31′N	37°04′E	61	169
Nigeria	Kano	12°03′N	08°32′E	41	(177)
	Samaru	11°11′N	07°38′E	43	(146)
	Mokwa	09°18′N	05°04′E	63	(159)
Sao Tome and Principe	Rio de Coro	00°23′N	6°43′E	75	92
Seychelles	International airport	04°40′S	55°31′E	81	223
Sudan	Juba	04°52′N	31°36′N	60	234
	Wau	07°42′N	28°01′E	53	263

that has passed through the plant is referred to as 'transpiration'. Since soil evaporation and plant transpiration occur simultaneously in nature, the term 'evapo-transpiration' is used to describe the total process of water transfer into the atmosphere from vegetated land surfaces.

Evapo-transpiration represents the highest fraction of energy loss in any cropped land. Evapo-transpiration rates depend on the amount of water available to be vapourised, the relative humidity, the temperature and the wind speed of the environment. A latitudinal variation in annual evaporation and precipitation across the tropics is given in Figure 1.23. The mean annual relative humidity, the evaporation and evapo-transpiration at some specific locations across the tropics are compared in Table 1.9.

CHAPTER 2

Tropical soils

The land resource

Of all agricultural resources, land is undoubtedly the most critical. To the farmer, land is essentially the soil under his feet, the materials in that soil and the slope that determines the ease of cultivation of his crop plants. To be suitable for crop production, land must have fertile soil and adequate water, and the slopes must not be too steep. To a town planner, land together with the landscape, is basically the space or place to build. To the highway engineer, land is the space for road construction. Thus, in the economic sense, land refers to man's entire natural environment with the forces, opportunities and resources that exist therein. Land is an immovable resource, the supply of which is limited and cannot easily be increased.

Soil, one of the components of land, is a naturally synthesised product of rock weathering and decaying organic matter. It constitutes a thin layer, up to several metres deep in places, on the earth's surface, and it provides the medium for plant growth, thus making possible man's existence. More often than not man tends to forget that land resources are not inexhaustible, and yet one of his greatest challenges is to use land resources with the necessary skill, care and consideration for the good of present and future generations (Figure 2.1).

Characteristics of tropical soils

Tropical soils are diverse, ranging from the arid and semi-arid soils which are characteristically low in organic matter, cation exchange capacity and natural fertility, to the heavily leached soils of the humid tropics. Tropical soils differ from those of temperate regions in several aspects, including differences controlled by temperature and rainfall. As explained in the first chapter, temperatures are generally high in the tropics, and with high humidity and rainfall, the result is intensive chemical weathering and leaching. Thus, soil forming processes and degradation are greatly accelerated and generally result in low levels of weatherable minerals, and soils that are predominantly acidic in nature (pH 5.0−5.8 in water).

In the drier semi-arid tropics, physical weathering assumes greater importance than chemical weathering. High temperatures are also conducive to intense microbial activity, resulting in rapid organic matter turnover. However, organic matter build-up also occurs in the cooler regions under anaerobic, seasonally waterlogged conditions.

Many desirable soil properties tend to deteriorate rapidly with cultivation. Prolonged cultivation generally leads to compaction of the soil rendering it less permeable and thus encouraging surface erosion (Figure 2.2). The ability of soils to retain water and supply it to crops thus constitutes one of the major limitations to crop production in many of the agriculturally important areas of the tropics.

Figure 2.1 The estuary of a large river in the tropics showing a high level of suspended material resulting from soil erosion and surface runoff, which substantially increase on newly cleared agricultural land.

Figure 2.2 A low yield cotton cash crop growing on eroded soil, Shinyanga, Tanzania.

Soil physical properties

The productivity of a soil is determined by both its physical and chemical properties, including nutrient availability, the utilisation of nutrients being dependent on the suitability of the soil as a rooting medium. In many areas of the tropics, the mean annual precipitation is less than the potential evapo-transpiration, with the consequence that water supply becomes the most critical factor limiting crop growth and development. The moisture storage characteristics of a soil thus become a key factor in its productivity. The two most important physical properties of soil are texture and structure. The structure influences aggregate stability and the ability of the surface soil to withstand the impact of rain drops. The infiltration rate, which is also dependent on the texture and structure of the surface soil, invariably determines the extent of surface run-off and erosion.

Soil texture

Soil texture refers to the relative proportions of sand, silt and clay in the soil. Feeling the soil usually gives some indication of these proportions. For instance a very sandy soil (generally coarse-grained, loose, gritty) is often described simply as 'sand'. A soil with a high proportion of clay is described as 'clay'. An intermediate soil, not exhibiting the predominant characteristics of either sand or clay, is termed a 'loam' (Figure 2.3). A loamy soil is usually considered the best medium for crop production since it combines the good physical qualities of sand with the good chemical qualities of clay, as well as often containing a good percentage of humus. The main systems used for soil textural classification are summarised in Table 2.1.

Fine-textured soils containing a considerable amount of clay, particularly of the montmorillonite type, normally become sticky and difficult to work when wet. When dry, such soils crack and form hard clods which, if broken with tillage implements, result in the fine powdery material that can easily get washed away by water or blown away by wind. Soils in which kaolinite is the predominant clay type are less sticky and easier to work when wet and more friable when dry.

Soil texture greatly influences the amount of water a soil can hold at any given time. In general,

Table 2.1. Comparative systems for classifiying soil texture

Soil fraction	International scale (mm)	US Department of Agriculture (mm)	British Standards Institute (mm)
Coarse sand	2.0–0.2	2.0–0.05	2.0–0.06
Fine sand	0.2–0.02	–	–
Silt	0.02–0.002	0.05–0.002	0.06–0.002
Clay	Less than 0.002	Less than 0.002	Less than 0.002

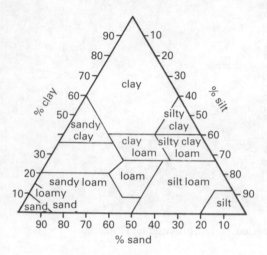

Figure 2.3 Soil classification according to texture.

the higher the clay content the greater the water retention capacity. The large surface area provided by clay holds a large amount of water by forces of adhesion and cohesion. Such soils contain little air because of the narrow pore spaces which tend to retard drainage and encourage capillary uptake and water retention. Soils high in clay are often described as 'heavy' because they are usually difficult to work and cultivate. In general, the higher the amount of clay in a soil, the less water it has available to plants. In contrast, the higher the amount of silt the greater the amount of water available to plants. A sandy soil is often well drained because percolation is easy. However, it heats up readily in the day and cools rapidly at night. The heat tends to destroy soil microorganisms, which break down organic material and thereby help in humus accumulation. This effect of heat, coupled with the high rate of leaching, tends to impoverish sandy soils. From the agricultural standpoint, the silt and clay fractions are the most important. They largely determine the

nutrient availability and, to a large extent, the productive potential of the soil. This is because, in general, the smaller the mineral particles the greater the chemical activity, due to the large total surface area. Clay also provides a strong anchorage system for the roots.

Many tropical soils, and especially those in the semi-arid regions, are predominantly sandy, at least at the surface. The small percentage of clay is mostly of the kaolinite type. Two reasons advanced for the sandy nature of the soils relate to the widespread occurrence of sedimentary and granitic parent materials and the downward translocation of clay within the profile. The general increase in the proportion of clay with depth is often related to rainfall. Thus, a pronounced subsoil clay accumulation is often found in the heavily leached humid tropical soils. Although topsoils are usually sandy, fine sand often predominates. It is this feature, particularly where little silt is present, that is largely responsible for the widespread tendency of surface soils to form a crust. Because of their sandy nature, tropical soils generally present minimal problems for root penetration and development.

Soil density and porosity

Soil density is of two types:

1. **Particle density**. This refers to the solid particles, with roughly constant values of between 2.60 and 2.75, because of the presence of quartz, feldspars and colloidal silicates.
2. **Bulk density**. This concerns the entire soil, including the total pore space.

From the point of view of crop production, bulk density is a more important soil property than particle density. Loose and porous soils usually have lower bulk densities than more compact soils. Sandy soils, which are low in organic matter, usually have higher bulk densities than

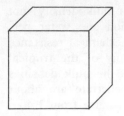

A cubic centimetre of soil weighing 1.33 g is made up of both solids and pore spaces

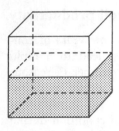

If all the solids were compressed to the bottom, the cube would look like this – ½ pore spaces and ½ solids

Bulk density (BD)

$$BD = \frac{\text{Weight of soil}}{\text{Volume of soil}}$$

Volume = 1 cm³;
Weight = 1.33 g

Therefore $BD = \frac{1.33}{1}$

= 1.33 g/cm³

Particle density (PD)

$$PD = \frac{\text{Weight of solids}}{\text{Volume of solids}}$$

Volume = 0.5 cm³;
Weight = 1.33 g

Therefore $PD = \frac{1.33}{1}$

= 2.66 g/cm³

Figure 2.4 Bulk density and particle density of soil.

loams, silt and clays or soils high in organic matter. For example, the bulk densities of clay, clay loams and silt loams often range from 1.0−1.6 g/cm³ compared with 1.2−1.8 g/cm² for sands (Figure 2.4).

The pore space of a soil is that portion occupied by air and water and its amount is determined largely by the arrangement of the solid particles within the soil. When these particles lie close to each other (as in sands), the total porosity is low. However, when they are arranged in porous aggregates, as is the case in medium-textured soils high in organic matter, the porosity is high. This infers that the finer the soil texture the greater the surface area and total pore space. Thus, clays have the highest percentage of pore space at 50−60%, while sands are lowest at 20−30% and loams are intermediate at 30−50%.

Sandy surface soils have higher bulk densities than clayey soils, mainly because of the lower pore space of sandy soils. Yet it is often observed that

water moves much faster in sandy soils than it does in clayey soils. The reason for this is that although the total pore space in a sandy soil may be low, a large proportion of the space is made up of large pores (macro-pores) which permit efficient water movement and air circulation. The percentage of small pores (micro-pores) in a coarse sandy soil is very low, hence its lower water-holding capacity. Fine textured soils, on the other hand, are high in total pore space (low bulk density) with a relatively large proportion of the space being micro-pores. Such soils have a high water retention capacity and are well aerated; water movement may, therefore, be impeded and may be inadequate for satisfactory root development in the subsoil. In a moist and well-drained soil, the macro-pores are usually filled with air and are therefore sometimes referred to as aeration pores. Micro-pores, on the other hand, are usually filled with water and are sometimes referred to as capillary pores.

Soil structure

The sand, silt and clay particles in a soil do not occur completely separately from each other. They are often cemented together to form natural units or aggregates, separated by pores, cracks or planes of weakness. The term 'structure' refers to the arrangement, size and stability of the aggregates which are usually described as fine, medium or coarse, depending on their size. The degree of soil aggregation largely determines the magnitude and distribution of pore spaces. Along with texture and organic matter content, this is responsible for aeration and the water-retention properties of soils.

A soil profile with no clearly observable evidence of regular aggregates is said to be structureless. Many of the soils of the West African savanna are in this category, generally containing weak aggregates. The condition of a soil making it suitable for effective root growth and development depends on the size and distribution of its pores which, in turn determine its permeability to rain water and to roots. A well aggregated soil is described as having good crumb structure. In the tropics, the maintenance of crumb structure is difficult under fairly continuous and intensive arable cropping. This is because soil crumbs on the surface tend to break up, due to the impact of falling rain drops, and the fine materials formed clog the large pores, thus forming a surface skin largely impermeable to water. Soil organic matter plays a major role in soil aggregation and one of reasons for the poor structure of many tropical soils is their low organic matter content.

Cultivation also breaks down aggregates and accelerates the decomposition of organic matter, the presence of which aids in the formation of aggregates. Stable aggregate formation caused by hydrated iron oxides occurs in certain red soils (latisols), lending them to fairly intensive cultivation without total destruction of their soil structure.

Effects of cultivation

In general, the immediate effect of cultivation is the loosening of the soil, resulting in increased soil aeration and water infiltration (Figure 2.5). The long-term effect, such as occurs under continuous cultivation, is reduced soil aggregation and increased compaction. Under these conditions, the total pore space decreases while the bulk density increases; there is thus reduced aeration and water circulation.

The soil physical parameters most closely linked with the development of crop roots are the interrelated properties of porosity and bulk density.

Increase in porosity (or lower bulk density) not only enhances efficient aeration and water permeability, but also reduces mechanical resistance to root penetration. For many of the tropical savanna soils of West Africa, the bulk densities above which roots fail to penetrate are about 1.75 g/cm^3 for sandy soils and range from $1.46-1.63$ g/cm^3 for clays.

Physical properties and soil productivity

In most areas in the tropics, a few years of continuous or intensive cultivation lead to a rapid decline in soil productivity. A number of reasons may be adduced for this, but the two most important are related to declining soil fertility and the deterioration of soil physical condition during crop production. The deterioration in physical condition is due largely to a degradation of soil structure primarily in terms of an alteration in pore size distribution, leading to reduced porosity and increased bulk density. Most tropical soils are low in organic matter and generally high in bulk densities, ranging from $1.5-1.7$ g/cm^3, implying porosities of only about $36-40\%$. Yield losses of up to 55% have been reported for groundnuts grown on soils with bulk densities between 1.5 and 1.60 g/cm^3 (Figure 2.6).

A predominance of fine sand particles in a soil

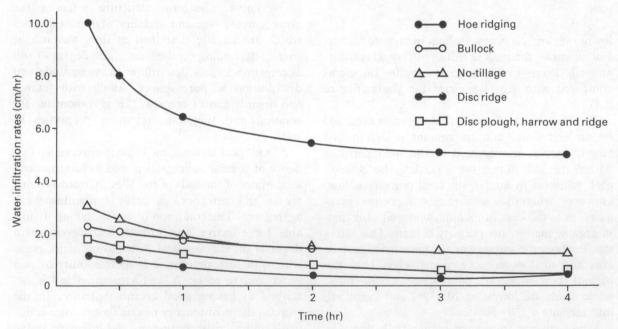

Figure 2.5 The effect of tillage practice on water infiltration into soil eight weeks after tillage operation (after Ike, 1986).

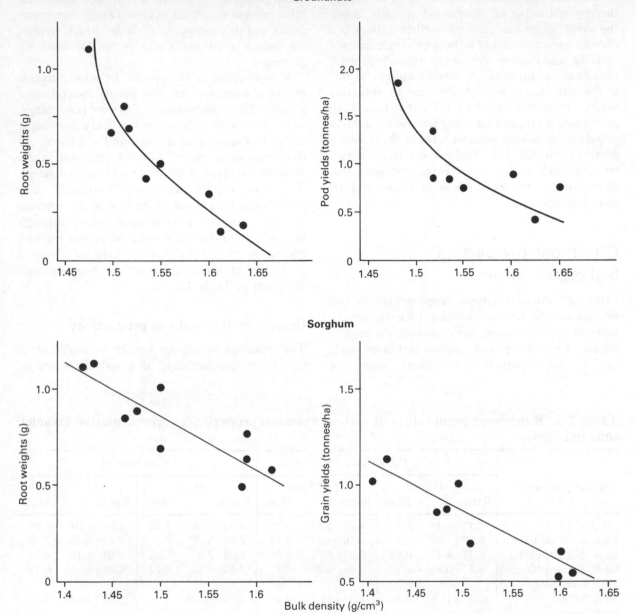

Figure 2.6 Relationship between bulk density, root development and yields of groundnut and sorghum in sandy alfisols, Bambey, Senegal.

partly account for a tendency towards compaction, capping and crust-formation, which in turn result in increased surface run-off, erosion and reduced water permeability. All these invariably adversely affect seedling emergence. In general, most tropical soils contain only weak aggregates, probably because of their low clay content and the predominance of kaolinite in the clay fraction. The

formation of clay pans in the subsoil leads to decreased water infiltration, and constitutes a severe limitation to root development. Surface soil under natural vegetation, especially in the more humid tropics, tends to have a good porous structure, but this is rapidly lost on cultivation. Tillage, in most cases, produces only some unstable clods, which are easily broken after a few heavy rains; the smooth soil surface dries out and becomes crusty, thereby restricting infiltration and aeration. When the crust is broken through further cultivation, the soil structure tends to become single grained and the aggregation even more short-lived with continuous cultivation. A classic example of this is the pale dusty soil, mainly fine to very fine sands, commonly found on fields that have been continuously cropped for many years in the densely populated semi-arid savanna areas of West Africa north of latitude 9°N. Some of the farming systems, as well as the soil management practices, employed to counteract these problems will be discussed later.

Chemical properties
Soil organic matter

Plant and animal residues, supplemented by soil microflora and fauna, including bacteria, actinomycetes and fungi, together constitute soil organic matter. Chemically, soil organic matter is made up of carbohydrates (in various stages of decomposition), proteins, fats, resins, waxes and other compounds. Various types of micro-organism attack and decompose the residues, which serve as the source of nutrients and energy for their life processes.

Mineralisation is the process by which organic matter is broken down into simple mineral compounds. The main products of this are heat energy, water, carbon dioxide, ammonia, and a few simple mineral salts such as nitrates and sulphates. All dead organic materials added to the soil finally break down through the process of mineralisation. The more readily decomposable organic compounds can be utilised as fresh residues, whereas the more resistent organic compounds, although also subject to mineralisation, can persist for long periods as part of soil humus. Selected chemical properties of some representative Nigerian soils are given in Table 2.2.

Organic matter and soil productivity

The influence of organic matter on the fertility, and hence productivity, of a soil takes several

Table 2.2. Range and mean values of various chemical properties of representative savanna soils in Nigeria

Soil characteristics	Fallow soils				Cultivated soils			
	0−15 cm		15−30 cm		0−15 cm		15−30 cm	
	Range	Mean	Range	Mean	Range	Mean	Range	Mean
pH (CaCl$_2$)	4.45−6.40	5.38	4.20−6.30	5.20	4.30−5.60	5.12	4.06−6.04	4.98
Exch. Ca (meq/100 g)	0.85−7.35	3.10	1.16−9.25	3.24	0.57−6.75	2.38	0.43−9.35	2.58
Exch. Mg (meq/100 g)	0.33−4.12	0.88	0.23−4.77	0.96	0.08−3.37	0.61	0.08−4.12	0.73
Exch. K (meq/100 g)	0.12−0.86	0.33	0.09−0.65	0.27	0.08−0.56	0.22	0.08−0.42	0.19
Total exch. bases (meq/100 g)	1.46−11.61	4.30	1.45−14.18	4.46	0.75−10.28	3.21	0.63−13.61	3.49
Exch. acidity (meq/100 g)	0.00−0.15	0.01	0.00−0.30	0.04	0.00−0.28	0.04	0.00−0.30	0.08
Effective CEC (meq/100 g)	1.61−11.61	4.31	1.70−14.18	4.51	0.75−10.28	3.25	0.78−13.60	3.57
Base saturation (%)	90.0−100.0	99.7	85.4−100.0	99.0	91.5−100.0	98.8	77.1−100.0	97.7
Organic C (%)	0.098−1.518	0.441	0.085−1.537	0.365	0.059−0.987	0.284	0.098−1.071	0.259
Total N (%)	0.024−0.92	0.049	0.017−0.057	0.038	0.017−0.068	0.038	0.014−0.083	0.035
C/N ratio	3.0−20.0	8.3	2.7−27.4	9.4	1.3−24.1	7.5	2.4−18.2	7.8
Resin extractable P (kg/ha)	0.4−26.3	5.6	0.1−24.4	5.6	0.2−60.4	18.9	1.1−60.4	18.8
Water soluble P (kg/ha)	0.07−10.30	1.50	0.07−10.30	1.72	0.08−38.50	7.03	0.08−36.40	7.03
Clay (%)	4−24	10.4	4.30	14.4	4−24	9.9	2−30	13.2
Silt (%)	4−40	18.2	2−40	16.8	0−42	17.4	2−38	15.1

forms. Organic matter contains a wide variety of mineral nutrients, including nitrogen, phosphorus, sulphur, calcium, magnesium, potassium, boron as well as many other inorganic elements. All of these are present in complex organic forms and are therefore not always immediately available to crop plants. However, as the organic materials undergo decomposition, the mineral nutrients are released as inorganic ions which become available to plants. Relatively large proportions of the nitrogen, phosphorus and sulphur that are used by crops in subsistence agriculture come directly from the oxidation of organic materials. In areas where intensive agriculture is practised, it is common to supplement the soil supply of these and other nutrients by means of commercial (inorganic) fertiliser applications. Under such conditions, the added nutrients from fertilisers are of increasing importance, although they do not diminish the importance of soil organic matter.

Organic matter has a high capacity for absorbing water. Consequently, increasing the organic matter content in a soil improves its water holding capacity. As a colloidal material, organic matter also has a high cation exchange capacity. Thus, it tends to hold nutrient cations against the leaching action of percolating water. Soil organic matter also has an influence on the development of soil aggregates. Soils which contain large amounts of organic matter are usually well-structured. Although organic matter is colloidal in nature and has high absorption properties, it does not exhibit high plasticity. Indeed, organic matter can to some extent ameliorate the plasticity that is commonly found in soils that are high in clay. In coarse textured soils, on the other hand, organic matter tends to bind the soil particles together, thus improving the physical condition of the soil. Many of the useful soil micro-organisms, including those which transform inorganic materials into available nutrients or fix atmospheric nitrogen, need a separate source of energy. In most instances, the energy is derived from the decomposition of organic matter. When organic residue decomposition commences, the less complex compounds are oxidised first to release energy, while simple products are cycled through living organisms to form new organic compounds which, on the death of the organism, are also added to the soil. As the more easily decomposed material is broken down, the more resistant forms tend to remain. With time, the organic matter that remains in the soil gradually becomes more resistant to further decomposition, although it never becomes completely stable. This partially decomposed and relatively resistant material is an amorphous, dark brown and odourless colloidal matter referred to as 'humus'. It is light, so helps to lower soil bulk density, and has a high water absorption capacity, about five to seven times that of clay, but does not exhibit the properties of adhesion and cohesion characteristic of most clays.

Soil organic matter under cultivation

Under normal soil conditions, there exists an equilibrium between soil organic matter accumulation and its decomposition. Cultivation disturbs this natural equilibrium in two ways. Firstly, relatively little organic matter is usually returned to the soil through unused crop residues. Secondly, decomposition processes are accelerated by such farming operations as cultivation, which usually lead to better aeration as well as enhanced microbiological activity, surface erosion and leaching. The greatest losses in organic matter normally occur immediately after fallow land is opened up and farming is started. Thereafter, the rate of decline lessens as cultivation continues, until the organic matter content of the soil reaches a new equilibrium. This process may take anything from 5 to 20 years, depending on the prevailing ecological conditions.

Bush fallowing which, until fairly recently, was the predominant practice in many tropical areas, is a traditional farming system that affords some restocking of soil organic matter. The length of fallow period necessary for the restoration of organic matter to about 75% of the initial equilibrium has been estimated at 3 to 6, 9 to 20 and 9 to 22 years for every year of cultivation in the rainforest, humid tropical and semi-arid tropical savanna, respectively. During fallowing, nutrients extracted from the lower soil horizons by plants are returned to the surface through organic matter accumulation. The relatively undisturbed vegetative cover also reduces the amount of water passing beyond the root zone, thereby minimising leaching losses. The amount and type of fresh organic material added to the soil will depend on the type of vegetation involved and how quickly it establishes itself. The amount of plant residue added to the soil annually varies widely with climate and local conditions and is usually severely limited by burning. Since it is unlikely, in practice, to exceed about 1 tonne of residue per hectare per year, only about 10% of the carbon in the added organic matter is retained as humus in the soil.

Although the bush fallowing system is still in use in many subsistence farming communities in the tropics and particularly in parts of Africa, the practice is rapidly being replaced by continuous and intensive cultivation, especially in West Africa, India and the far East, due primarily to population growth and other socio-economic pressures on limited land resources. In such situations, the organic matter level may drop by a further 40−50% from the initial level (when the fallow is newly opened up) before a new equilibriun level is attained. At equilibrium, the rate of organic matter addition equals the rate of mineralisation. Most soils in the densely populated areas of West Africa fall into this category.

As indicated earlier, the high temperatures characteristic of the lowland humid tropical regions are more conducive to the mineralisation of soil organic residues than to soil humification. In semi-arid areas, the situation is accentuated by the inherently low rate of organic residue accumulation and the high solar radiation with minimimal cloud cover.

One of the key soil management problems in tropical agriculture is that of sustaining sufficient levels of soil organic matter to ensure satisfactory soil structure. The problem is more serious in the savanna areas where the surface soils have very weak aggregates which are easily destroyed under cultivation. The incorporation of crop residues, the addition of farmyard manure where available and the practice of crop rotation involving legumes, are some of the comparatively inexpensive ways of increasing soil organic matter. The use of inorganic fertilisers, particularly nitrogen and phosphate, normally results in increased dry matter production and consequently a greater return of organic residue to the soil. Green manuring and legume leys with no immediate economic return are unfortunately unlikely to be readily acceptable to the majority of farmers in the tropics.

Cation exchange capacity

The cation exchange capacity (CEC) is a measure of the exchangeable bases retained on soil colloids for plant use. The CEC of most tropical soils is low, often less than 5 meq/100 g of soil. As mentioned earlier, the organic matter content is critically important in determining the actual exchange capacity because the clay content is often low, averaging about 10−15%. Furthermore, because the clay is predominantly kaolinite, its contribution to the CEC is often less than 10%. The low exchange capacities imply that many soils are low in exchangeable cations normally required as crop nutrients. This infers that under intensive cultivation the soils are rapidly depleted of their nutrients. A further implication is that nutrient imbalances may occur from the continuous use of conventional fertilisers, especially those containing only one or two nutrient elements. For instance, nutrient cations such as ammonium, potassium, magnesium and certain trace elements may be weakly held by soil colloids and so are easily leached. Rapid soil acidification may also occur when ammonium sulphate or urea are used. These problems are general for most soils throughout the world, it is their extent that differs. The problems are obviously more serious where values of CEC are low, as is the situation found in many tropical soils.

In summary many tropical soils, and especially those of the savanna areas, are characteristically sandy, with their small clay fraction being dominated by kaolinite. The contribution of clay to CEC is, therefore, minimal and organic matter becomes the single most important factor in this regard. The effective management of soil organic matter is, therefore, synonymous with the maintenance of CEC.

Soil reaction and buffering capacity

Soil reaction (pH), a measure of the acidity or alkalinity of a soil, and the soil's buffering capacity are fundamental chemical barometers used in approximating a soil's potential productivity and several other properties. For instance, strongly acid soils (pH 4.0−5.0, in Ca Cl_2) usually have high and toxic concentrations of iron, aluminium and manganese. Most nitrogen-fixing bacteria in legumes are not normally active in strongly acid soils. Bacteria that decompose soil organic matter and thus release nitrates, sulphates, phosphates and other nutrients for plant use are also hindered by high soil acidity. Phosphate availability from fertilisers is greatly impeded under highly acid or alkaline conditions. For these and other reasons, efforts are often made to reduce the effects of acidity by liming. Most field crops do best in slightly acid soils (pH 6.5) and most liming efforts usually aim at this value. High soil alkalinity (pH 7.5 and above) is more difficult to correct than high soil acidity. High soil alkalinity reduces the solubility and availability of phosphates and all micronutrients, except molybdenum.

Most tropical soils are slightly to moderately acid, (pH 6.0–6.8). There are, however, some soils in the humid tropics which are quite strongly acid, having pH values of 4.5–5.0 (Figure 2.7). Factors which help to mitigate the development of acidity in traditional farming practices are the low anion content of soils, the low rate of removal of bases by cropping (because of generally low yields), the return to the soil of crop residues and the practice of burning crop residues on the field.

Cases of dramatic crop responses to liming in the tropics are not widespread and are usually restricted to soils in the humid tropics. Many well-drained tropical soils may have low pH, as well as a low content of exchangeable aluminium, because of their low exchange capacities. In such soils, the pH may need to fall to about 4.5 before soluble aluminium becomes an important problem. In latosols iron hydroxide may perform the role of calcium sources in stabilising pH. Furthermore, many tropical crop varieties have been developed under acid soil conditions such that they do not readily respond to liming. However, liming can be useful in tropical agriculture. In fact, very impressive crop responses to liming have been reported in the heavily leached soils of the humid tropics, notably in Central and Latin America. Results of similar studies in the semi-arid tropics of Africa and Asia have, in general, been much less striking.

In view of the profound influence of soil pH on the biological and chemical properties of a soil, especially as they relate to nutrient availability and uptake, it is normally desirable that pH should not fluctuate too widely. When it does, both higher plants and soil micro-organisms are adversely affected before they are able to adequately adjust. The ability of soils to resist wide changes in pH following the addition of acid or alkaline materials, such as fertilisers and soil amendments, is termed the 'buffering capacity' of the soil. The higher the amount of soil organic matter or clay, the greater the CEC; the higher the buffering capacity, the greater the amount of lime required to effect a given change in pH. One of the major consequences of the high sand and the low clay and organic matter content of most tropical soils is poor buffering capacities. Injudicious liming or fertiliser use can easily lead to cation imbalances, especially on lighter soils.

Nutrient supply and fertilisers

Fertility potential of tropical soils

Tropical soils are generally low in nutrient reserves. While parent materials are the basic determinants of potential fertility, in many areas of the tropics climate is an important modifying factor. For instance, the low fertility of soils in the humid tropics is, to a large extent, a reflection of the leaching resulting from high rainfall. On the other hand, the situation in savanna areas is generally attributed to the low levels of organic matter, which

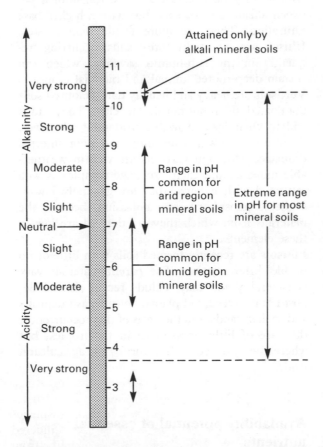

Figure 2.7 Extreme range of pH for most mineral soils.

are also related to rainfall. There are several major soil classes which commonly occur in the tropics, and without the modifying effects of climate and vegetation, the fertility characteristics of these various soil classes would be fairly predictable. Consistent with the scope of this text, only broad generalisations are made here about soil fertility in the following descriptions of the various soil classes.

Alfisols (ferruginous tropical soils) constitute the largest proportion of tropical soils and occur widely between the 500 and 1200 mm isohyets in areas with distinct wet and dry seasons. They are usually formed from aeolian (wind-blown) deposits overlying basement complexes and in many cases from sedimentary deposits. The topsoils are notably sandy and low in organic matter. The clay fraction is low (rarely exceeding 20%) and predominantly kaolinitic. Cation exchange capacities are low (rarely exceeding 4 meq/100 g) and depend mainly on the soil organic matter status which is in turn related to rainfall, clay content and previous land use. The agricultural value of alfisols is usually rated average; their nutrient status, except for phosphates, is moderate in the better soils formed on a basement complex. Structural stability is generally poor and the consequent surface compaction tends to increase surface run-off and erosion, which is mitigated by the low rainfall. Alfisols can support most annual crops but are not well suited to tree crops.

Inceptisols (brown and reddish-brown soils of the arid and semi-arid regions) are essentially young soils with limited profile development. They occur principally where the mean annual rainfall is less than 500 mm. They are derived mainly from aeolian material. Weathering and leaching are light and so the soils may contain fair amounts of montmorillonite clays and, depending on the nature of the parent material, some reserves of weatherable minerals. Organic matter content is often low, but the CEC values, although also generally low, are usually higher than in the alfisols. The base saturation is also higher and the structure is better, although the latter deteriorates rapidly on cultivation so that the problem of surface run-off may also be evident, but to a much lower extent. Inceptisols are most suited to annual crops such as cereals, groundnuts, and cotton.

Oxisols (ferrallitic soils) are typically forest soils which are excessively weathered and heavily leached, with little or no weatherable minerals

remaining. The sub-humid savanna oxisols are generally much less severely leached than those occurring in areas with higher rainfall. The clay fraction is predominantly kaolinitic with iron and aluminium. The subsoils tend to show low base saturation (a measure of the relative proportion of basic cations on the exchange complex, expressed as a percentage of cation exchange capacity). However, the higher organic matter in the topsoil provides more exchange sites. These surface soils are, therefore, moderately well saturated. Nutrient supply, except for nitrogen and phosphates, may therefore be adequate; however, these are easily depleted under continuous cultivation unless they are supplemented by additions of mineral fertilisers. In general, oxisols are considered better suited to forestry and tree crops than to annual crops.

Ultisols (ferrisols) are the leached and acidic soils of the humid and sub-humid tropics which are often less than 35% saturated with metallic cations. Ultisols may be considered to be transitional between alfisols and oxisols, but with a higher base saturation. They are more fertile than oxisols. Ultisols are essentially forest soils, occurring frequently in the sub-humid savanna, which can sustain deep-rooted annual and biennial crops.

Vertisols are heavy and usually dark-coloured soils dominated by montmorillonite clay. They crack widely when dry and swell considerably when wet. They are of low permeability and poor internal drainage. Although vertisols are rich in weatherable minerals and their exchange complex is well saturated with cations, they are generally low in available phosphorus and potassium because the materials from which they are derived are low in these elements.

Entisols are recently formed soils with little or no profile differentiation. The parent materials vary considerably and may include recent alluvial deposition in river flood plains, volcanic ash deposits and aeolian sands. On the basis of their occurrence, they are of little importance in the tropics, but where they do occur they form rich agricultural lands.

Availability potential of essential nutrients

Of the 16 elements known to be essential for plant growth and development, seven are normally

Table 2.3. Essential nutrient elements for plant growth and development

Nutrient element	Chemical symbol	Ionic forms commonly absorbed by plants
Carbon	C	CO_2
Hydrogen	H	H_2O
Oxygen	O	O_2, H_2O, CO_2
Nitrogen	N	NO_3^-, NH_4^+
Phosphorus	P	$H_2PO_4^-$, HPO_4^{2-}
Potassium	K	K^+
Calcium	Ca	Ca^{2+}
Magnesium	Mg	Mg^{2+}
Sulphur	S	SO_4^{2-}
Manganese *	Mn	Mn^{2+}
Iron *	Fe	Fe^{2+}
Boron *	B	BO_3^{3-}
Zinc	Zn	Zn^{2+}
Copper *	Cu	Cu^{2+}
Molybdenum *	Mo	MoO_4^{2-}
Chlorine *	Cl	Cl^-

* Micronutrients; all others are macronutrients.

required only in very minute quantities (Table 2.3). The seven trace (or micronutrient) elements are boron, copper, chlorine, molybdenum, iron, managanese and zinc. All other elements are termed macronutrients because they are needed in relatively large quantities by plants. Calcium, magnesium and sulphur are sometimes referred to as secondary elements because they fall between micronutrients and macronutrients in terms of the relative amounts required by plants. Apart from carbon, hydrogen and oxygen, all the remaining 13 essential nutrients are acquired by plants either entirely or mostly through the soil.

Soil nitrogen

Of all the essential nutrients, nitrogen appears to have the most pronounced effect on plant growth and development. Nitrogen promotes vegetative growth and imparts the characteristic deep green colour to foliage, because it is a component of the chlorophyll which is essential for photosynthesis. It is the major ingredient of proteins, enzymes, amino-acids, amides and nucleic acids. It also regulates, to a considerable extent, the utilisation of phosphorus and is responsible for the degree of succulence of most vegetable crops. When in short supply in the soil, the deficiency symptoms of

nitrogen appear first on the older leaves, since nitrogen reserves are transported to young active leaves. The young plants become stunted and develop restricted root systems. The older leaves turn yellow or yellowish-green, and there is considerable leaf shedding.

Where nitrogen is over supplied through fertiliser application, the leaves become dark-green, soft and sappy. This is because carbohydrates are utilised to form more protoplasm and more cells. Crop maturation may be delayed because of excessive vegetative growth and this can result in crop lodging because the stems become tall and weak (Figure 2.8). Over-supply of nitrogen can also affect the keeping quality of certain fruits and such grains as malting barley.

The soil nitrogen cycle is made up of two major facets: the additions and the losses. Soil nitrogen is lost in several ways, including leaching (chiefly in the nitrate form); surface erosion; removal by crops; burning of crop residues or their complete removal from the field; denitrification (conversion of nitrogen in the form of nitrates in the soil into gaseous nitrogen); ammonia volatilisation as well as immobilisation (incorporation of nitrate nitrogen by soil micro-organisms in their body tissues). In flooded soils, much nitrogen can be lost through denitrification involving obligate and facultative anaerobes. Ammonia volatilisation takes place

Figure 2.8 Over supply of nitrogenous fertilisers in crops such as wheat can cause excessive vegetative growth and the development of tall weak stems, resulting in crop lodging.

when ammonium fertilisers or urea are used or during mineralisation of organic matter.

Unlike other essential plant nutrients, nitrogen can be used by plants in either the cation form (ammonium, NH_4^+) or the anion form (nitrate, NO_3^-). However, only a small fraction of soil nitrogen occurs in these two forms at any one time. As already mentioned, nitrogen in nitrates is easily leached and both the nitrate and ammonium may be consumed by soil micro-organisms. In addition, the ammonium can also be fixed by soil clays while nitrates may be converted to gaseous nitrous oxide or nitrogen and the NH_4^+ lost to the atmosphere as gaseous ammonia (NH_3). Because of the relatively high temperatures in the tropics, problems of denitrification and ammonia volatilisation attain greater significance than in more temperate regions.

Nitrogen is added to the soil in various ways including atmospheric accession and fixation; organic matter mineralisation; and through applications of mineral and organic fertilisers. About 10 kg/ha of soil nitrogen comes from rainfall each year. This atmospheric accession is greater in the vicinity of industrial centres and near the sea. Within the soil, elemental nitrogen is fixed by both symbiotic and non-symbiotic micro-organisms. Symbiotic fixation is by bacteria in the roots of legumes, the best known being the *Rhizobium* species. The plant uses some of the nitrogen fixed and, in turn, supplies the bacteria with energy and other nutrients through the root nodules.

Non-symbiotic fixation is carried out by *Azotobacter* and *Beijerincka* species, both of which are widely distributed in the tropics. Some species of *Clostridium* and certain blue-green algae common in flooded rice soils and some forest soils also fix nitrogen. *Azotobacter* species are more abundant in well-drained soils which have a pH near neutral. In contrast, *Clostridium* species are anaerobic and develop best in poorly-drained acid soils.

The decomposition of organic matter (mineralisation) results first in ammonia production by a process referred to as ammonification. The factors affecting the release of ammonium are those influencing decomposition, including the nature and age of the plant material, its state of subdivision as well as the moisture, aeration, and temperature of the soil. The biological process by which the ammonium is converted into the nitrate form is referred to as nitrification and it is a two-way enzymic oxidation process. The oxidation of ammonium to nitrite is carried out by *Nitrosomonas* bacteria, while the ultimate conversion of nitrite to nitrate is accomplished by *Nitrobacter*. Nitrification, like most other biochemical processes, is influenced by such factors as temperature and moisture, as well as the availability of nutrients. The high temperature and low moisture content of tropical soils during the dry season tend to have a partial sterilising effect. As a result, the amount of nitrogen available for nitrification rises during the dry season. As the soil moisture content rises with the onset of rains, there is an upsurge in bacterial activity and the nitrate level rises rapidly, but soon falls as the rains continue.

Fertiliser use and management are discussed in a subsequent chapter. The use of farmyard manure and nitrogenous fertilisers is the best way of increasing the amount of land available for cultivation.

Phosphorus

Next to nitrogen, phosphorus is often the most important limiting nutrient element in many tropical soils. Phosphorus is a constituent of many compounds in plants, including phospholipids, nucleic acids, nucleotides and co-enzymes. It is important in most plant metabolic processes.

Inadequate supply of phosphorus in the soil invariably leads to young plants becoming stunted, leaves becoming bluish-green or purple with red pigmentation along the veins. Crop maturation (ripening) is often also delayed. Because of the ease and rapidity with which applied soluble phosphates in soil solution are fixed (precipitated), the problem of over-supply of phosphate fertiliser is rare in practice. Where such a situation occurs, the uptake of zinc is likely to be impeded.

The total phosphorus content of tropical soils is generally low, although there is considerable variation depending on the parent material, the degree of weathering and the organic matter content. Soils derived from basic igneous rocks or volcanic ash are often very high in total phosphorus; by contrast, those derived from granites and sedimentary sandstones are characteristically low in total phosphorus. However, the total phosphorus content of a soil does not indicate its fertility; what is important is the amount of available phosphorus which can be extracted with standard extractants.

In the absence of inorganic fertilisers, organic phosphorus is the source of phosphorus for crop plants, especially under traditional farming systems in the tropics. Organic phosphorus forms about 20–30% of total phosphorus in savanna soils and 60–80% in the more highly weathered Alfisols, Ultisols and Oxisols of the humid tropics. The most important solid inorganic phosphate compounds in soils fall into two main groups, the calcium phosphates and the iron and aluminium phosphates. In general, the transformation of one form into another is largely controlled by soil pH. As soils become more acidic, the activity of iron and aluminium increases and the relatively soluble calcium phosphates are converted into the less soluble iron and aluminium phosphates. In practice, the processes are slow enough to permit the presence of considerable amounts of calcium phosphate in strongly acid soils. Phosphate fixation is the most important single factor governing the availability of the nutrient to crop plants. Phosphate fixation is caused either by soluble iron, aluminium and manganese (under acid conditions) or by elemental calcium and calcium carbonate in alkaline or calcium-rich soils. Available phosphates, when applied to such soils in the form of fertilisers, will normally react with either soluble iron, aluminium and manganese (if the soil is acidic) or calcium and calcium carbonate if the soil is alkaline. In either case, the phosphate is rendered unavailable to plants. Soil phosphates may also be fixed by the hydrous oxides of iron and aluminium, the compounds formed as a result of this fixation being hydroxy-phosphates, as is the case with the chemical precipitation discussed earlier. The higher the iron and aluminium oxide content, the higher the phosphorus-fixing capacity of the soil. This is why the highly weathered Oxisols and Ultisols of the tropics usually have high phosphorus-fixing capacities.

Because of the problem of phosphorus fixation, the efficiency of phosphorus recovery by crops is generally low, ranging from less than 5% to about 18–20% for most tropical arable crops such as maize, soyabeans and potatoes.

When acid soils are limed to keep the pH between 6.0 and 6.5, phosphorus fixation is kept to a minimum. The general practice of liming soils to neutrality is not effective and, in fact, not desirable in most of the highly weathered soils of the tropics. In order to prevent rapid reaction of phosphate fertilizers with the soil, the fertilisers are usually banded. Phosphate fertilisers are almost always pelleted or aggregated to further retard their release into the soil.

Potassium

Potassium is required as a co-factor (enzyme-activator) for over 40 different enzymes and it helps to maintain electroneutrality in plant cells (anions such as organic acid groups must be balanced by cations). It helps to maintain cell turgidity and to regulate the opening and closing of stomata. The latter is caused by an influx of ions, mostly K^+ and organic acid groups, into the stomatal guard cells, inducing osmosis and causing water to pass into the guard cells from neighbouring cells. The guard cells increase in turgor and become enlarged, increasing the aperture of the stomata.

Where potassium is deficient in the soil, plant leaves appear dry and scorched at the edges; the leaf surfaces become irregularly chlorotic and photosynthesis is impaired. Over-supply of potassium through fertiliser application normally leads to an excessive intake by the crop. In this way, there is a disproportionately high intake of potassium which is not reflected in increased crop yield. Care must be exercised to guard against such luxury consumption of potassium when considering field fertiliser management.

Of all nutrient elements, potassium may be considered the most abundant in tropical soils. Soils derived from sandy parent materials, such as sedimentary sandstones, are low in total potassium. Potassium is derived from such minerals as feldspars and micas. The problem of potash deficiency in the tropics is more localised than that of either nitrogen or phosphorus.

Soil potassium is usually lost through leaching, but crop removal is also particularly important in view of the excess (luxury) uptake already referred to. For this reason, and partly because of leaching, potash fertilisers have low residual effects, especially in coarse-textured soils.

Potassium is added to the soil either through the return of crop residues (direct incorporation or burning of the residue and returning the ash) or through inorganic fertilisers and organic manures. Under continuous and intensive cultivation, fertilisers appear to be the most dependable source of potassium for optimum crop production.

Calcium and magnesium

Calcium is a critical structural component of cell walls and its deficiency is often associated with dead plant tissue, particularly at the growing points with active meristematic activity. Magnesium is an important component of the chlorophyll molecule. Magnesium deficiency often results in a reduction in green coloration, commencing with chlorosis between the veins which gradually extends over the entire leaf. Deficiency symptoms of magnesium first show on the older leaves.

The well-leached soils of the humid tropics are generally low in exchangeable calcium, but others, such as the Inceptisols of most of the savanna zones, are fairly high in both calcium and magnesium. However, calcium deficiency is not usually a serious problem in humid areas because, as in Latin America and Brazil, liming takes care of the calcium requirements of most cultivated crops. The use of dolomitic limestone ($CaCO_3$ $MgCO_3$) also supplies adequate magnesium to the soil. In soils where liming is not required, calcium is often supplied as an incidental addition in fertilisers such as single superphosphate. However, a deficiency of calcium in groundnuts, resulting in poorly filled pods, has been reported in some sandy soils in the Sudan zone of northern Nigeria, where rainfall is also often inadequate. Only a fraction of the calcium and magnesium taken up by the crop is actually used in the economic yield, except perhaps in the case of groundnuts where both haulms (tops) and pods are of value. The traditional practice of returning the crop residue to the soil, either by direct incorporation or as ash, is therefore an important way of minimising the loss of these two nutrients. Gypsum is one of the most widely recommended sources of calcium, especially where soil acidity is not a problem.

Sulphur

Widespread sulphur deficiencies have been reported all over the tropics. In sub-Saharan Africa, the deficiencies are particulary pronounced in savanna zones with a mean annual rainfall of less than 700 mm, as well as in inland areas where the atmosphere is relatively free of industrial pollution and low in gaseous sulphur. Soils which are subjected to repeated annual burning are often sulphur deficient, since about 75% of soil sulphur is volatilised by fire. Coarse-textured soils, low in organic matter or high in iron and aluminium oxides, are also often low or deficient in sulphur. The deficiency is characterised by light-green to yellow leaves, commencing with the young ones, since sulphur is not easily transported within the plant from older tissues. Groundnuts and grain legumes, such as soyabeans and cowpeas, seem particularly sensitive to sulphur deficiency, partly because of the relatively high sulphur content of legume tissues and partly because of the influence of sulphur on root nodulation.

Most of the sulphur in unfertilised tropical soils is in the organic form, averaging 85–95% in medium or long-term fallows. However, only a small fraction is in the inorganic (sulphate) form. The availability of organic sulphur depends on the rate of mineralisation of organic matter. Mineralisation of organic sulphur is dependent on the same environmental factors that control the mineralisation of organic nitrogen, including moisture, aeration, temperature and soil pH.

Inorganic sulphur is often associated with calcium, potassium or magnesium in tropical soils which are susceptible to leaching. Under anaerobic conditions such as occur in waterlogged soils, sulphur is mainly in the sulphide form which later oxidises to the sulphate form. Besides mineralisation and the use of fertilisers, sulphur is also added to the soil in gaseous form from the atmosphere.

Sulphur is depleted from the soil principally through crop removal, leaching and sulphate adsorption by clays.

Micronutrients

In traditional farming systems in the tropics, involving bush fallowing or semi-intensive cultivation, micronutrient deficiencies are rarely a problem in crop production. However, with the increasing use of such modern agricultural technology as the introduction of high-yielding crop varieties under a system of intensive cultivation as well as the increasing use of the major nutrient elements, micronutrient deficiencies have become a common feature of tropical agriculture. For example, boron, molybdenum and zinc responses have been recorded in India, Latin America and Africa for both tree crops (cocoa, rubber, oil palms, citrus) and annual crops (cotton, groundnuts and maize). Micronutrient deficiency is usually a result of either low inherent amount, a highly acidic soil (reduces molybdenum availability) or a highly alkaline soil (reduces zinc, manganese, boron and iron availability).

A major factor controlling the micronutrient status of soils in the tropics is the soil organic matter. In general, soils high in clay are usually high in micronutrients; where clay content is low, organic matter becomes the crucial factor. In practice, the application rate for micronutrients in relation to response curves is not critical as one rate of application is often sufficient for a wide range of annual crops.

Inorganic fertilisers

Conventionally, the term 'fertiliser' refers to chemically synthesised plant nutrient compounds which may be added to the soil to supplement its natural fertility. Fertilisers may thus contain one, two or all the three major nutrients, nitrogen, phosphorus and potassium. Fertilisers containing only one of these elements are described as single, simple or straight fertilisers; those with two or more elements are called mixed, compound, or complex fertilisers. On the basis of the major nutrient contents, the straight fertilisers fall into three categories: nitrogenous (nitrogen), potassic or potash (potassium) and phosphatic or phosphate (phosphorus) fertilisers.

Nitrogen fertilisers

Ammonium sulphate is a fine crystalline salt, generally white in colour, but sometimes bluish or yellowish. It is readily soluble in water and very quick-acting, only slightly slower than nitrogen from nitrates. When in soil solution, the NH_4^+ ion may be taken up by plants or it may be absorbed by soil colloids (humus and clay). When absorbed, it replaces calcium on the exchange complex, so that calcium sulphate and calcium nitrate are formed in the soil. These calcium salts are easily leached out, the soil thus becoming more acidic. A major drawback for the use of ammonium sulphate as a fertiliser, therefore, relates to this soil acidifying property. The extent of acidification depends on the quantity of fertiliser applied annually, the texture of the soil and the amount of rainfall.

Calcium ammonium nitrate (or nitrochalk) is a mixture of ammonium nitrate and finely precipitated carbonate of lime. One half of the nitrogen is in the ammonium form while the other half is in the nitrate form. This means that a portion of the nitrogen, the nitrate form, is immediately available, while the rest becomes available a little more slowly. Nitrochalk may be applied at planting or as a top-dressing. Because of its calcium carbonate content (40%), its acidifying power is much less than that of ammonium sulphate. A further advantage of the calcium carbonate content of nitrochalk is that it reduces the powerful oxidising property of the nitrate radical (NO_3^-). This oxidising property makes ammonium nitrate fertiliser explosive. The calcium carbonate also reduces the high hygroscopicity of the ammonium nitrate. Calcium ammonium nitrate is being used on an increasing scale in the tropics and has, in West Africa, virtually replaced ammonium sulphate.

Urea is a white and finely crystalline material which is readily soluble in water. In fact, its most serious limitations are its high deliquescence and poor handling properties. Its relatively high nitrogen content reduces transportation costs. Urea readily undergoes hydrolysis in the soil to produce ammonium carbonate, which is then converted to nitrate by nitrifying bacteria. Thus, urea ultimately produces both NH_4^+ and NO_3^- ions for plant absorption, thereby acting in a similar way as ammonium fertilisers.

Ammonium nitrate is a concentrated, quick-acting nitrogenous fertiliser. When in soil solution, both the NH_4^+ and NO_3^- ions become available to plants. The NO_3^- radical is a powerful oxidising agent, so the fertiliser may explode under pressure or heat. To prevent this, the granules are often coated with an inert powder (kaolin) which also prevents them from forming a compacted mass in the bags. The fertiliser tends to acidify the soil in the same way as ammonium sulphate, although somewhat more slowly. Because of its inherent danger, ammonium nitrate is not widely used in the tropics where many farm operations are carried out manually.

Slow-release nitrogen fertilisers. In certain situations, fertilisers that allow gradual and extended use of nitrogen are preferred to conventional nitrogen sources, which tend to make nitrogen too readily available. From the point of view of plant physiology, such slow-release sources should supply nitrogen to the soil solution at a rate and

Figure 2.9 Farmyard manure spread on ploughed land, prior to harrowing and ridging, at Samuru, Nigeria.

concentration which allows the growing plant to maintain maximum expression of its genetic capability. Such fertilisers are developed mainly to increase the efficiency of nutrient use by plants. Improvement of fertiliser nutrient recovery through control of nutrient transformations in the soil may require only slow or delayed release. Most slow-release materials have been designed to reduce or slow down the rate of nutrient delivery to the soil solution. Some of the advantages of slow-release fertilisers include a reduction in leaching of applied nitrogen, less chemical and biological immobilisation, a reduction in denitrification and ammonia volatilisation, less seed and seedling damage from highly localised concentrations of salts, increases in residual value of applied nitrogen, better economy of use and improved storage and handling properties.

Phosphate fertilisers

It is customary to speak of the phosphorus nutrient both in soils and fertilisers in terms of phosphoric acid or oxide, represented by the same formula, P_2O_5. All phosphorus specifications in inorganic fertilisers are given in this form. As an example, single superphosphate $(0-18-0)$ is 18% P_2O_5. This refers to the available phosphorus, including both the water and citric acid-soluble fractions.

Single superphosphate is the principal phosphate fertiliser in use in many parts of the tropics. It is a greyish-white granular material with an acidic odour. Single superphosphate is prepared from concentrated sulphuric acid and finely ground rock phosphate. Both monocalcium and dicalcium phosphates are contained in single superphosphate. These, together with gypsum, are the most important components of single superphosphate. It contains 7–9% P, 12% S and 18% Ca. However, in the superphosphate trade, only the water-soluble fraction (the monocalcium phosphate) is taken into account.

A complicated series of chemical changes occur in water soluble phosphates (apart from possible absorption by colloids) when phosphate fertiliser is applied to the soil. Although the phosphate nutrient in the fertiliser is water-soluble, no significant leaching occurs in soil because the monocalcium phosphate is soon converted to the insoluble form. Therefore, application of superphosphate must be at planting or preferably about two weeks before planting. In contrast to nitrogen, phosphate is generally not mobile in soils and remains largely in the region of application. At best, monocalcium phosphate moves only a few centimetres before it is completely precipitated. This means that where possible superphosphate should be worked well into the soil. All phosphatic fertilisers are unsuited to top-dressing, since it may be difficult to adequately work them into the soil.

In view of its quick reversion to insoluble forms, it is not economically advisable to use ordinary (single) superphosphate on markedly acidic soils (pH less than 5 in $CaCl_2$) without first correcting the soil acidity.

In favourable conditions only about 10–20% of the phosphate supplied as fertiliser is absorbed by plants during the year of application. This is because of the difficulty of distributing phosphatic fertilisers in the soil and the relative insolubility of the phosphate precipitates. Therefore, several times the amount of single superphosphate required by a given crop should be applied to compensate for the availability factor. In general, the residual effect of phosphorus on a neutral or slightly calcareous soil lasts for several years relative to an acid soil where the residual effect may be negligible even in the second year after application.

Concentrated (triple) superphosphate contains 18–20% available phosphorus (47% P_2O_5). It is manufactured from rock phosphate using a higher ratio of sulphuric acid than in the manufacture of single superphosphate. This high-analysis superphosphate differs from ordinary superphosphate in that it contains more phosphorus and virtually no gypsum (calcium and sulphur).

Ammonium phosphate. Mono and di-ammonium phosphates are the two important forms prepared by neutralising phosphoric acid with liquid ammonia. Mono-ammonium phosphate (12% N, 50–54% P_2O_5) is more common. Di-ammonium phosphate is a more recently developed material containing up to 21% N and 53% P_2O_5. Ammoniated superphosphate contains from 3–4% N and 16–18% P_2O_5, and is usually prepared by treating superphosphate with liquid ammonia or nitrogen solutions.

Potash fertilisers

Potash fertilisers are manufactured from natural deposits of potassium salts found in various parts of the world, notably in France, Germany, the USA and Canada. The crude potash minerals are dissolved in water and the various salts separated by fractional distillation.

All potash salts used as fertilisers are water-soluble and the potassium content is considered readily available to plants. Unlike nitrogen fertilisers, most potash fertilisers have no effect on soil pH. The two most important potash fertiliser materials available are muriate and sulphate of potash.

Muriate of potash is a coarse or fine salt closely resembling ordinary domestic salt, but with a bitter and somewhat acidic taste. It contains 80–90% potassium chloride (KCl), the remainder being mainly sodium chloride. The fertiliser therefore contains about 42% K (60% K_2O). It is very soluble and quick-acting and so is best applied at the time of planting.

Sulphate of potash is a fine yellowish salt containing 75–80% potassium sulphate (K_2SO_4) and 33–50% K. Potassium sulphate has an advantage over muriate of potash in that it also supplies adequate sulphur. It is a preferred potassium source for tobacco because a large dosage of the chloride may lower the leaf quality.

The fate of applied soluble potassium in the soil has already been considered. Whatever is not taken up by plants and not absorbed onto soil colloids is lost through leaching. In the humid tropics, potassium loss through leaching is quite a serious problem.

Compound fertilisers

A compound or mixed fertiliser contains at least two of the major fertiliser elements: nitrogen, phosphorus and potassium. A fertiliser containing all the three elements is described as complete. Compound fertilisers are usually formulated to solve particular soil fertility problems and to meet specific crop needs.

Advantages associated with the use of a compound fertiliser are as follows:

1. The mixture is usually dry, fine and well mixed, and can therefore be applied by hand satisfactorily as well as through mechanical fertiliser drills.
2. The mixture does not usually consolidate, form lumps or deteriorate in any way if not used immediately.
3. The mixture usually contains two or all three

major plant nutrients so there is no fear of one-sided fertiliser application.

4. For a farmer who does not know how to make his own mix, the ready-made mixtures are straightforward to use.

5. A compound fertiliser saves time and labour when applied in the required amount instead of using several separate fertilisers.

However, compound fertilisers are not without disadvantages. The prime consideration is that, except where the formulation is made to order, the mixture may be unsuitable for many soils or specific crop needs. For instance, a particular soil may be high in potassium and low in nitrogen, with the result that the application of a compound fertiliser may render the potash component redundant while the nitrogen may be too low to satisfy the crop's needs.

A farmer who is familiar with the fertility and manurial requirements of his field and who is conversant with the properties of different fertilisers may decide to buy straight fertilisers which can then be physically mixed as necessary. In doing this, it is important to ensure that the fertilisers are mixed thoroughly and preferably on a hard cement floor. However, care should be taken to avoid mixing certain fertilisers because some of the nutrients may volatilise, they may become less available or the mixture may consolidate and become hard and lumpy. For example, if single superphosphate, ammonium sulphate and a potassium fertiliser are mixed, the resultant mixture becomes hard if not used immediately.

Farmyard manure

The term farmyard manure (FYM) refers essentially to animal dung, including poultry droppings, plus any liquid wastes as well as straw and materials used as bedding (Figure 2.9). In general, the solid and liquid components of FYM are in a ratio of approximately 3:1. Chemically, there is considerable variation in the composition of FYM, depending on the source, the type of livestock, the food consumed by the livestock and the way the manure is diluted, handled and stored. In general, dung has a relatively high moisture content (about 30%) and contains approximately 0.5% N, 0.25% P_2O_5 and 0.5% K_2O. On average, 30% of the N and P_2O_5 and 50% of the K_2O in FYM are available to crop plants in the first season after application under field conditions.

Table 2.4. Mean crop yields (kg/ha) using farmyard manure and a fertiliser dressing with an equivalent nutrient content at Matuga, Kenya

Crop	No manure	Fertiliser annually	7.5t/ha farmyard manure annuallly	22.5t/ha farmyard manure annually
Maize grain, average of six crops	584	970	875	944
Sorghum grain, average of six crops	1058	1929	1649	1586
Cassava roots, average of six crops	6206	10228	9600	9270
Sweet potatoes, average of four crops	2585	5118	5432	4687

Source: Grimes and Clarke, 1962.

As a source of nutrients, FYM is a source of nitrogen and potash and to a lesser extent certain trace elements. Its phosphorus content is invariably low. When applied with a phosphate fertiliser, the use of FYM is likely to sustain good crop yields almost indefinitely. In the short-term the use of farmyard manure provides no significant benefits over the use of fertilisers containing an equivalent ratio of nutrients (Table 2.4). However, in the long-term the use of FYM helps to maintain good soil structure as well as maintaining the level of soil organic matter, the buffering capacity and the nutrient retention capacity of the soil by improving the cation exchange capacity.

The value of FYM has long been recognised in tropical agriculture but its widespread use has been limited by its bulky nature and local availability. Apart from transportation costs, it is difficult to procure FYM in sufficient quantities to meet the complete nutrient requirements of crops. This practical limitation notwithstanding, it has been clearly demonstrated experimentally that FYM can effectively maintain soil productivity under continuous cultivation (Table 2.5).

Table 2.5. Effect of long-term use of FYM, nitrogen, phosphate and potash on grain yields of maize (kg/ha) at Samaru, Nigeria

Treatment	1971	1973	1975	1977	1979	1981	1983
FYM (dung)							
Nil dung	891	1695	1534	1685	1186	1541	1267
2.5 t/ha	2493	2824	2567	2959	1991	1728	2321
5.0 t/ha	3549	3461	3368	3519	2902	2527	3055
Nitrogen							
Nil N	1593	1041	1252	1762	1271	952	1183
68 kg/ha N	2410	3136	2693	3018	2252	1536	2522
136 kg/ha N	2930	3803	3525	3382	2556	2309	2937
Phosphate							
Nil P	1998	1985	1530	1982	1212	938	1854
9.5 kg/ha P	2427	3090	2988	3172	2375	1924	2585
13.5 kg/ha P	2507	2906	2952	3009	2492	1935	2203
Potash							
Nil K	2391	2631	2479	2662	2028	1645	2147
25 kg/ha K	2350	2721	2529	2855	2042	1527	2281
50 kg/ha K	2192	2627	2462	2646	2009	1625	2215
LSD (0.05)	241	352	325	456	295	528	511

PART TWO

Principles of crop production

The development and effective implementation of any meaningful agricultural programme is not only dependent on a proper appreciation of the farming system of the area in question, but is also governed by the successful application of those principles which are basic to the various production practices. In addition to considering cropping systems as they relate to tropical environments, this second section also covers the various physiological and agronomic foundations of crop production.

CHAPTER 3

Physiological aspects of crop production

The resources available for crop production have been presented in the first two chapters. Obviously, the individual parameters examined under climate and soils do not affect crop production in isolation; rather, they influence each other and hence their individual contributions are modified. In this third chapter, the mechanisms by which climate and land resources are interrelated and incorporated into crop production are discussed. Included in the discussions are those concepts which help to explain variations in yield.

Photosynthesis

The basis of crop production is photosynthesis. This is the process by which the plant pigments present in chloroplasts absorb light and thus produce chemical energy for the subsequent conversion of carbon dioxide into carbohydrates. The process can be represented by the following simplified equation:

$$nCO_2 + 2nH_2O + light \xrightarrow{chloroplasts} (CH_2O)n + nH_2O + nO_2$$

The chloroplasts, which are mainly greenish in colour, contain chlorophylls 'a' and 'b', the pigments facilitating the transfer of energy from light for the oxidation of water and reduction of carbon dioxide to form organic compounds.

Details of the structure, pigment content and formation of chloroplasts, and the biochemistry of carbon fixation, are provided in standard texts of biology. In this chapter the type, availability and role of light, as well as various other factors that influence photosynthesis, are briefly discussed.

Selective absorption of light

Gases present in the atmosphere act as a screen for removing fractions, if not all, of the wavelengths

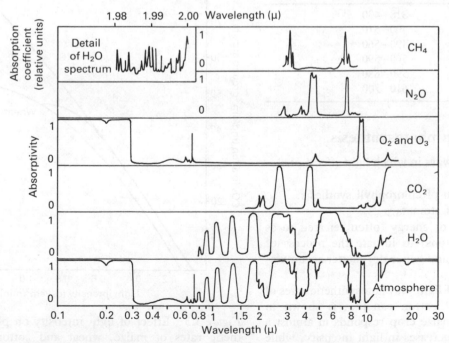

Figure 3.1 Absorption spectra for various atmospheric gases (Rosenberg, 1974).

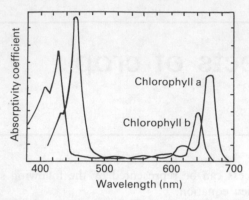

Figure 3.2 Absorption spectra of chorophylls 'a' and 'b' (Salisbury and Ross, 1978).

outside the visible spectrum (Figure 3.1). The visible component thus accounts for about 50% of the total solar radiation incident at ground level. Visible radiation is further divisible into its component colours (Table 3.1). Only the blue and red wavelengths of visible light are used in photosynthesis. The absorption spectra of chlorophylls 'a' and 'b' show that very little green and yellow light between 500 and 600nm is absorbed by green plants (Figure 3.2).

Table 3.1. Wavelength characteristics of visible light

Type of radiation	Range of wavelength (nm)
Violet	318−430
Blue	400−510
Green	490−560
Yellow	560−590
Orange	590−630
Red	610−700

Role of light in photosynthesis

The fundamental ways in which light affects photosynthesis relate to:
1. The stimulation of chlorophyll synthesis.
2. The opening of stomata.
3. The provision of energy (often referred to as quanta or photons) to initiate the process of photosynthesis and hence dry matter accumulation.

The influence of light on photosynthetic rates of maize, wheat and cotton canopies is shown in Figure 3.3. The maize crop responds in almost a linear fashion to increases in light intensity, while cotton and wheat appear to exhibit a distinct tend-

ency towards light saturation. The observed difference in the response of these crops is related to the basic differences in the physiology and anatomy of crops originating in the tropics (C4), as exemplified by maize, and those originating in temperate climates (C3), as exemplified by wheat and cotton. Important distinguishing characteristics between the two groups of crops are summarised in Table 3.2. Selected examples of C4 (tropical) and C3 (temperate) crops are also listed in Table 3.3.

Other uses of light in crop canopies

Any fraction of light intercepted by crop canopies but not utilised in photosynthesis may be involved in thermal and photomorphogenic interactions. The term thermal interaction refers to the conversion of light into heat, which is needed for transpiration and convectional exchange with the air above the canopy. This form of light usage involves the longer wavelengths, the near infra-red and a fraction of the photosynthetically active wavelengths, and accounts for about 70% of light interception. It also determines the temperature of leaves and other plant parts.

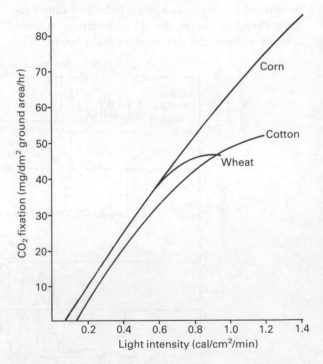

Figure 3.3 Effect of light intensity on photosynthetic rates of maize, wheat and cotton plants (Salisbury and Ross, 1978).

Table 3.2. Important distinguishing characteristics between C3 and C4 plants

Characteristic	C3 plants	C4 plants
Leaf anatomy	No distinct bundle sheath of photosynthetic cells	Well organised bundle sheath
Transpiration ratio (H_2O/g dry wt. increase)	450–950	250–350
Requirement for sodium as a micronutrient	No	Yes
Ratio of leaf chlorophyll 'a' to 'b'	2.8	3.9
CO_2 compensation point (ppm)	30–70	0–10
Photosynthesis inhibited by 21% O_2?	Yes	No
Photo-respiration detectable?	Yes	Only in bundle sheath
Optimum temperature for photosynthesis (°C)	15–25	30–40
Dry matter production (t/ha/yr)	22	39

Source: Modified from; Black C.C., 1973. Photosynthetic carbon fixation in relation to net CO_2 uptake. Ann. Rev. Plant Physiol. 24: 253–286.

Table 3.3. Selected examples of crop plants of tropical (C4) and temperate (C3) origins

C3 crops	C4 crops
Barley	Sugarcane
Beet	Millet
Bean	Sorghum
Carrot	Maize
Cotton	
Groundnut	
Oats	
Rice	
Wheat	
Soyabean	

The photomorphogenic effects of light refer to the control and regulation of growth and developmental processes in plants. The photosynthetically active radiation, the near infra-red and the ultraviolet portions of sunlight are believed to exercise some photomorphogenic effects on plants, although the exact nature of the influence is presently not completely understood.

Availability of carbon dioxide

Photosynthetic rates are not only enhanced by increased light intensities, but also by higher carbon dioxide concentrations, unless the stomata are closed by shortage of water. Figures 3.4 and 3.5 illustrate the mechanism by which increased carbon dioxide levels in the air stimulate photosynthesis in selected plants.

Other factors relating to photosynthesis

Basically there are four processes involved in the conversion of a carbon dioxide molecule to a carbohydrate molecule. These are the interception of photosynthetically useful radiation; the diffusion of carbon dioxide in to the leaf; the photochemical reactions supplying energy for carbon dioxide fixation, and the biochemical reactions for the conversion of carbon dioxide to carbohydrates. Any factor that influences one or more of these processes can affect the rate of photosynthesis.

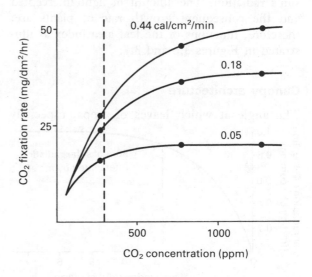

Figure 3.4 Atmospheric carbon dioxide enrichment and fixation in sugar beet (Salisbury and Ross, 1978).

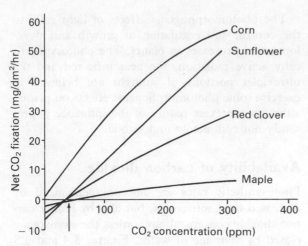

Figure 3.5 The effect of the atmospheric concentration of carbon dioxide on the rates of photosynthesis in various crops (Salisbury and Ross, 1978).

Such plant related and environmental factors include the total area of green leaf surface, the canopy architecture and temperature, the level of photorespiration and the incidence of disease or pest attack.

Leaf area

Plant leaves are the main site for photosynthesis. The leaf is the plant organ directly exposed to sunlight and is used for the interception of the sun's radiation. The amount of light intercepted and the consequent growth rate of plants are, therefore, functions of the leaf area index as illustrated in Figures 3.6 and 3.7.

Canopy architecture

The angle at which leaves of crops, especially

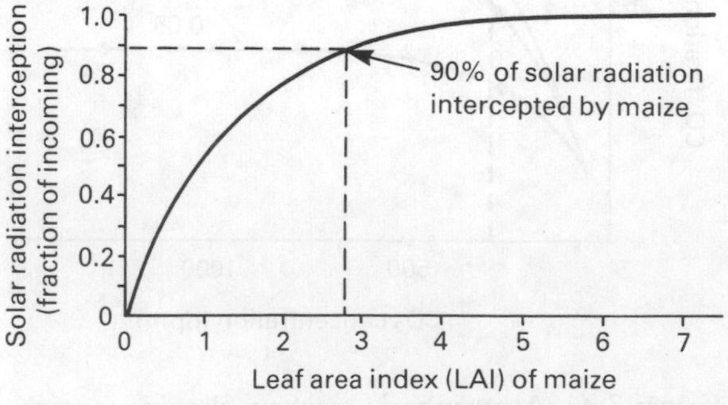

Figure 3.6 Relationship of the leaf area index of maize to solar radiation interception by the canopy.

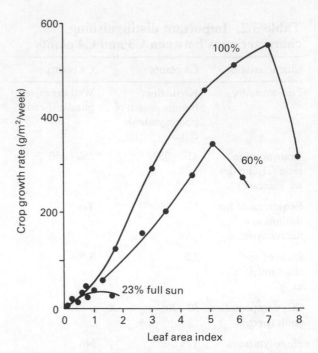

Figure 3.7 Growth as related to leaf area index and light intensity (given as per cent of full sunlight) in a sunflower canopy (Salisbury and Ross, 1978).

cereals, are held, influences the depth of light penetration through canopies. As more light reaches the lower leaves, a greater leaf area is involved in photosynthesis. Most crops with upright leaf orientation, therefore, tend to intercept less solar energy per unit leaf area. However, because of the enhanced photosynthesis, they produce greater dry matter and grain yields on a canopy basis at leaf area indices greater than 4.0.

Canopy temperature

Crop growth is very sensitive to temperature, often a difference of a few degrees leads to a noticeable change in the growth rate (Figure 3.8). Each plant species or crop variety has, at any given stage of its life cycle and set of conditions, a minimum temperature below which it will not grow. Each species or variety also has an optimum temperature (or temperature range) at which it grows best, as well as a maximum above which it will not grow and may even die. These temperature limits, often referred to as cardinal temperatures, are listed for the critical growth stages of some major crops cultivated in the tropics (Table 3.4). The growth

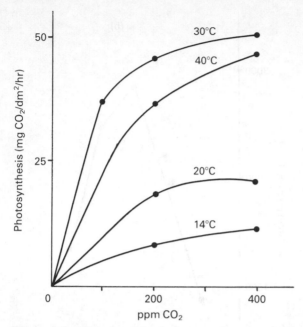

Figure 3.8 Relationship between photosynthesis of maize and carbon concentration at different temperatures (Rosenberg, 1974).

of various species is typically adapted to their natural environmental temperature. Temperate crops have relatively low minima, optima and maxima, while tropical species have much higher cardinal temperatures. In general, the minimum temperature for growth of most tropical crops is about 10°C, the optimum is between 20 and 30°C while the maximum is about 40°C. The temperature response curve at various stages of the growth of millet is illustrated in Figure 3.9.

Temperature affects crop production by controlling the rate of physiochemical and biochemical reactions, as well as the rate of evaporation of water from crop and soil surfaces. Excessively high temperatures cause a large deficit between the atmospheric water demand and the water extraction capability of plant roots. This deficit may result in stomatal closure. The closure of stomata restricts photosynthesis and evaporative cooling. Cell respiration rises along with the temperature of the plant, such that in very hot environments the plant effectively progressively burns itself up.

Low temperatures, on the other hand, desiccate tissues and increase protoplasmic permeability. They can also cause an accumulation of toxic compounds in plants (for example, cyanides in Sudan grass and sorghum) and accelerate respiration over photosynthesis. Low soil temperatures limit the rate of seed germination and determine the start and end of seed dormancy, especially in temperate climates.

Low temperatures are not always deterimental to crop production. When a normally cool season crop is grown in a warm environment, a period of low temperatures must be present to induce flowering in the crop. This low temperature

Table 3.4. Cardinal temperatures (°C) at various growth stages of some major crops grown in the tropics

Crop	Germination			Vegetative growth			Flowering and grain development		
	Maximum	Minimum	Optimum	Maximum	Minimum	Optimum	Maximum	Minimum	Optimum
Wheat	32	5	20−24	41	5	18−24	33	10	18−24
Barley	30	3	15−22	40	6	16−18	35	8	18−20
Potato	30	12	20−28	32	8	20−25	na	na	18−22
Tobacco	35	13	28	na	na	na	na	na	18−22
Maize	44	10	32−35	38	10	28−32	35	10	25−28
Millet	45	12	32−35	42	18	30	35	18	25−28
Rice	38	12	30−32	40	22	28−30	40	20	30−32
Cowpeas	42	12	25−35	40	10	25−35	35	18	25−30
Soyabean	40	10	25−30	35	12	25−32	32	19	25−28
Groundnut	35	12	24−30	38	15	28	33	18	25−30
Cotton	39	13	25−32	40	13	24−30	38	18	22−25
Sorghum	40	10	32−35	35	16	27−30	28	15	24−26
Sugarcane	46	21	32−37	41	13	26−29	na	na	24

Note: na = data not available

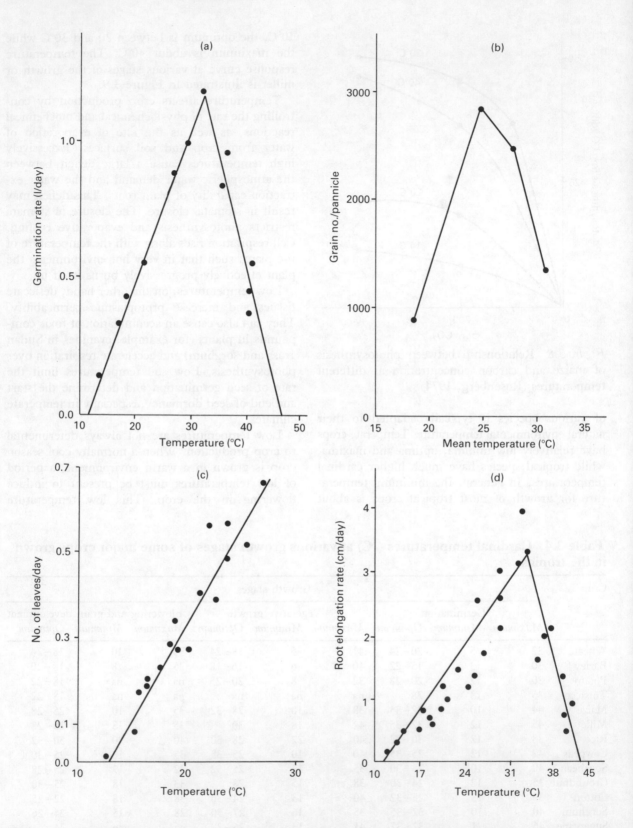

Figure 3.9 Temperature effect on (a) germination rate (b) number of grains per panicle (c) rate of leaf appearance and (d) rate of root elongation, in pearl millet.

promotion of flowering is referred to as 'vernalisation'. Low temperature treatment is also required for the germination of some seeds, including peach, plum and cherry seeds. Germination of such seeds requires low temperatures, moist conditions and the presence of oxygen, often for weeks or even months. The low temperature promotion of germination is known as 'stratification'. Figures 3.10 and 3.11 illustrate the effects of low temperature on flowering and germination, respectively.

In the tropics, temperatures of exposed soil surfaces often measure as high as 55°C at midday. Excessively high soil temperatures, especially at the peak of the dry season, may therefore cause more damage to crops than low soil temperatures. The extent of cold or heat damage depends on the intensity of the hazard, the rate of recovery and the frequency and length of crop exposure.

Photorespiration

Respiration is the inverse of photosynthesis. It is the oxidation of carbon compounds to release energy for use in the various biological and chemical processes in the plant. Photosynthesis begins at sunrise, but the rate is low and it is usually sometime after sunrise before the fixation of carbon dioxide exceeds the rate of respiratory release. The light intensity at which photosynthesis equals respiration is known as the light compensation point.

All plants respire throughout the day and night.

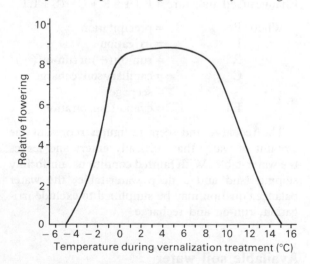

Figure 3.10 Flowering response of rye as a function of a six-week temperature treatment during vernalization (Salisbury and Ross, 1978).

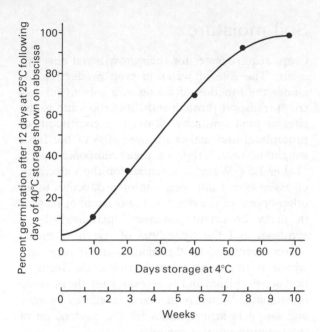

Figure 3.11 Germination of apple seeds as a function of storage time at 4°C (Salisbury and Ross, 1978).

Tropical species appear to respire through the same chemical pathway during both day and night. In comparison temperate species have, in addition to their dark respiration, a second light-driven respiratory pathway that operates only during the day and is known as photorespiration. Photorespiration of C3 plants is partly responsible for their high light compensation point and the relative inefficiency of their net photosynthesis as compared with C4 plants (Table 3.2). This is because C3 plants loose as much as 50% of assimilated carbon through photorespiration.

Diseases and pests

Pests and diseases affect photosynthesis specifically and crop production in general. The affects are achieved in the following ways:
1. By reducing the effective leaf area of crops through defoliation, discolouration and senescence.
2. By restricting the uptake of water and nutrients from the soil.
3. By diverting some of the material synthesised by the crops into the tissues of the pests or parasites.

Soil moisture

Crops require water for their growth and development. The role of water in crop production includes the function of acting as a solvent for nutrient transport through soils, into roots and to the sites of food synthesis. Water is a constituent of protoplasm and makes up over 80% of the fresh weight of some crops or their component parts (Table 3.5). Water is essential for the movement of assimilates from their site of production to all other organs of the plant. It is a reactant or reagent in many important processes, including photosynthesis and the hydrolysis of starch to sugar. Water is required for the maintenance of turgidity, which is important in maintaining the form of herbaceous plants and is necessary for the opening of stomata. Water is involved in energy balance and also acts as a medium for the moderation of temperature in crop canopies.

Because over 90% of agricultural activities in the tropics take place under rainfed conditions, it is important to know whether the amount of rainfall is an adequate index from which to determine the availability of water in the soil for crop growth and development at any given location.

The hydrologic cycle

The total amount of rainfall measured at a particular location is not all available for crop production. The falling rain may be partly intercepted by vegetation. Interception of as much as 10% of rainfall has been recorded over maize. Most of the rain, however, reaches the ground and may go into soil surface run-off (run-on in the case of depressions) or it may infiltrate the soil. Rainwater that infiltrates soil may be totally absorbed by the topsoil if the soil is dry. Otherwise it may wet the complete profile and percolate down to the water-table. Ground water, and water from surface run-off, eventually reaches streams or large open water areas where, along with exposed wet soil and leaf surfaces, it contributes to evapotranspiration. The circulation of water from evapotranspiration, through rainfall and back to evapotranspiration, makes up the hydrologic cycle (Figure 3.12).

Table 3.5. Water content of various plant tissues expressed as a percentage of fresh weight

Plant part and crop	Water content (%)
Roots	
Barley	93.0
Carrot	88.2
Sunflower	71.0
Stems	
Asparagus	88.3
Sunflower	87.5
Pine	50–60
Leaves	
Maize	77.0
Cabbage	86.0
Lettuce	94.8
Fruits	
Tomato	94.1
Watermelon	92.1
Seeds	
Barley	10.2
Groundnut	5.1
Maize	11.0

Source: Kramer, 1969.

Soil water balance

The water balance over a given land area and time may be estimated from components of the hydrologic cycle as follows:

$$\text{Change in soil moisture} = P + I \pm R + C - (S + ET)$$

Where	P	$=$ precipitation
	I	$=$ irrigation
	R	$=$ run-on $(+)$ or run-off $(-)$
	C	$=$ capillary soil recharge
	S	$=$ seepage
	ET	$=$ evapo-transpiration

The recharge and seepage figures represent the amount of water that vertically enters and leaves the water-table. With rainfed conditions, uniformly sloping land and a deep water-table, the water balance equation may be simplified to exclude irrigation, run-on and recharge.

Available soil water

The soil is a complex system made up of varying proportions of mineral or rock particles, non-living organic matter, soil solution and air. The soil

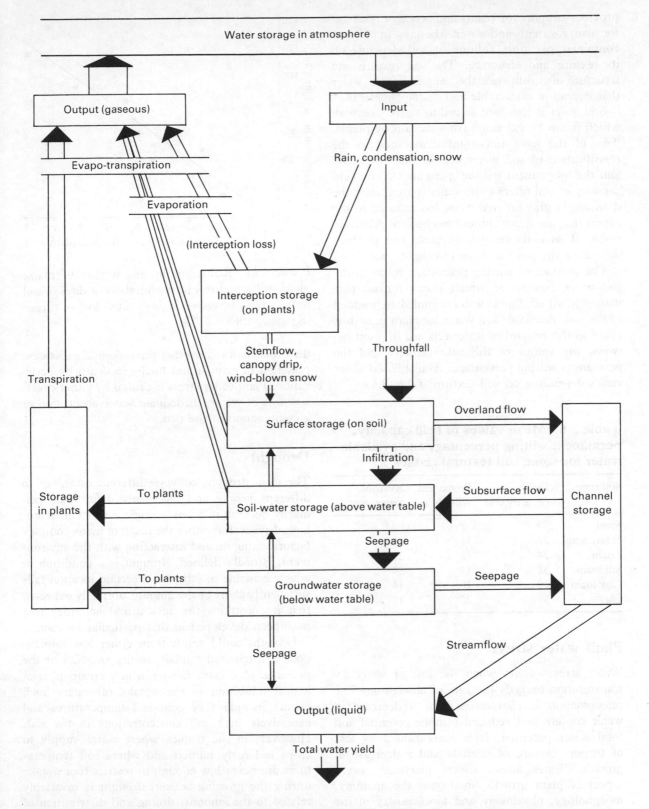

Figure 3.12 The hydrologic cycle consists of a system of water-storage compartments and the solid, liquid or gaseous flows of water within and between the storage points (Anderson *et al*. 1976).

provides support for plants and acts as a reservoir for plant nutrients and water. The amount of water contained per unit volume of soil depends on its texture and structure. The soil texture and structure also influence the amount of soil water that is readily extractable and usable by plants.

Soil water is classified according to the ease with which it can be extracted from the soil by plants. Two of the most important terms used in the classification of soil water are the 'field capacity' and the 'permanent wilting' percentage. The field capacity of soil refers to its water content after the drainage of gravitational water has reduced to the extent that the water content has become relatively stable. This situation usually exists one to three days after the soil has been thoroughly wetted.

The permanent wilting percentage refers to the soil water content at which plants remain permanently wilted, unless water is immediately added to the soil. Available soil water for plant growth is taken as the amount of water retained in a soil between the values of the field capacity and the permanent wilting percentage. Available soil water varies depending on soil texture (Table 3.6).

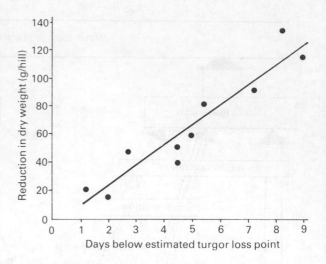

Figure 3.13 Reduction in dry weight of maize plants subjected to various numbers of days of soil water stress, severe enough to cause loss of turgor (Kramer, 1969).

Table 3.6. Mean values of field capacity, permanent wilting percentage and available water for some soil textural classes

Soil type	Field capacity (%)	Permanent wilting %	Available water (mm)
Sand	9	2	7
Sandy loam	27	11	16
Loam	34	13	21
Silt loam	38	14	24
Clay loam	30	16	14
Clay	39	22	17

Plant water stress

Water stress occurs when the loss of water by transpiration exceeds the rate of absorption. The phenomenon is characterised by a decrease in water content and reduced osmotic potential and total water potential. It is accompanied by loss of turgor, closure of stomata and a decrease in growth. Water stress affects practically every aspect of plant growth, modifying the anatomy, morphology, physiology and biochemistry of the plant. If severe, water stress results in a drastic reduction in photosynthesis (Figure 3.13),

disturbance of many other physiological processes, cessation of growth and finally in death by desiccation. Plant water stress is caused by either excessive loss of water, inadequate water absorption or a combination of the two.

Drought

The term drought conveys different meanings to different people and has many definitions, primarily because it is not a single distinct meteorological event, but rather the result of many complex factors acting on and interacting with the environment. Broadly defined, drought is a condition in which available moisture at a certain location falls sufficiently short of the amount normally expected that it constrains the agricultural activities that have been developed in that particular location.

Drought could result from either low rainfall, poorly distributed rainfall, scanty snowfall or the presence of certain factors which create physiological inhibition to the uptake of water. Such factors include very cold soil temperatures and excessively high salt concentrations in the soil. However, in the tropics where water supply to crops is largely rainfed and where soil temperatures are never low enough to restrict root uptake during the growing season, drought is invariably related to the amount, timing and distribution of rainfall. Figure 3.14 shows total rainfall and its distribution (into pre-moist and post-moist periods)

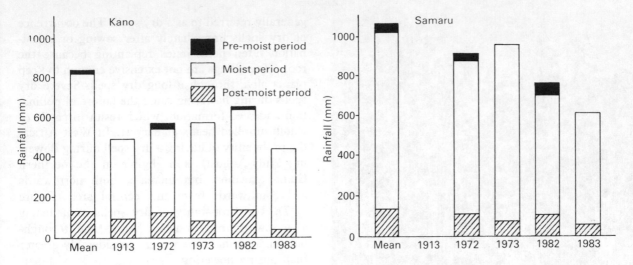

Figure 3.14 Comparison between total rainfall during various droughty years and long-term mean annual rainfall at Kano and Samaru, Nigeria.

during droughty and normal years at two locations in Nigeria. Figure 3.15 illustrates how the drought in the savanna zone of West Africa in 1973 was caused by a reduction in the length of the rainy season from latitude 11°N northwards, in addition to low total rainfall. Figure 3.16 is a picture of an irreversibly drought-stressed crop of sorghum in the savanna zone of Nigeria in 1983.

Drought arising from the uneven distribution of rains within the rainy season in the tropics is

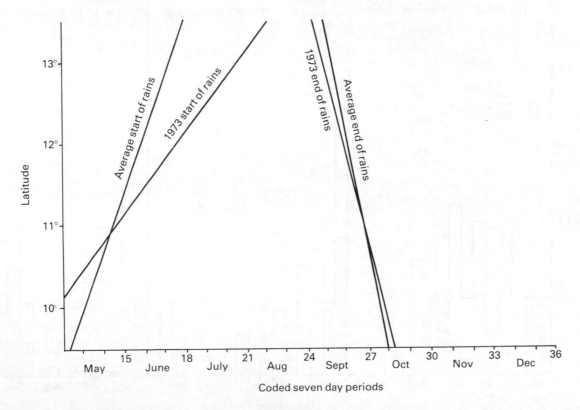

Figure 3.15 Relationship between the start and end of the rains and latitude for an average year and for a droughty year (1973).

Figure 3.16 Drought-afflicted crop of sorghum at Samaru in 1983.

generally referred to as a dry spell. The occurrence of dry spells immediately after sowing or germination often necessitates replanting because the roots of seedlings are not extensive enough to keep plants alive through a long dry spell. Severe dry spells during flowering cause the failure of pollination and seed formation, which result in partly or wholly unfilled heads at harvest. In West Africa, the probability of having a dry spell during flowering (July-August) is negligible in the northern Guinea savanna, but increases both northwards and southwards from this central area (Figure 3.17). The probability of dry spells immediately after sowing or germination (April-June) throughout the savanna is, however, considerably greater than during flowering.

Although dry spells cause a false start to the growing season and can significantly reduce grain yields, they are a local phenomenon, as distinct from the generally widespread rainfall fluctuations following world-wide droughts. Available literature indicate that world-wide droughts occurred at the

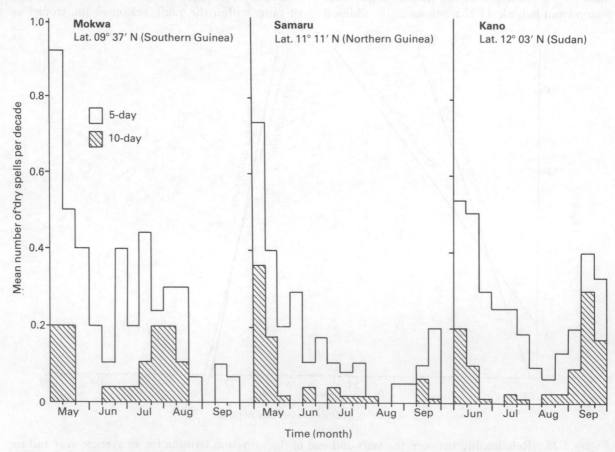

Figure 3.17 Frequency of dry spells during the rainy season at selected locations in Nigeria.

beginning of the nineteenth century, in the late thirties and the early forties, and in the late sixties and the early seventies, with the climax years being 1913, 1942, 1972 and 1983.

Drought resistance

Drought resistance refers to the various means by which plants survive periods of environmental water stress. Basically, plants are drought resistant either because their protoplasm is able to withstand dehydration without permanent injury or because they possess structural or physiological characteristics which result in avoidance or postponement of a lethal level of desiccation. Most drought resistant crops avoid desiccation through the development of an extensive root system (as in sorghum). Others control the rate of transpiration by leaf shedding, by closure of stomata as soon as water stress begins or with deposits of cutin in their cell walls.

Growth analysis

Carbon synthesis and dry matter accumulation result from the combination of water and nutrients, absorbed by plants from the soil via the roots, with carbon dioxide diffused into the leaves through stomata in the presence of chloroplasts, and energy supplied by sunlight. The continuous synthesis of large, complex molecules from the smaller ions and molecules that are the raw materials for growth, leads not only to larger cells, but often to more complex ones. Furthermore, not all cells grow and develop in the same way, so that a mature plant consists of numerous cell types. The process by which cells become specialised is called differentiation, and the process of growth and differentiation of individual cells into recognisable tissues, organs and organisms is often called development.

Measurement of growth

Growth is an increase in size resulting from an increase in weight, cell number, amount of protoplasm and complexity. Measurement of growth, therefore, involves the determination of either the increase in volume or weight. Volume or size increases are often approximated by measuring ex-

pansion in only one or two directions, such as length, height, width, diameter or area. Increases in weight can be determined by harvesting the entire plant, or a particular plant organ, and weighing it immediately to obtain the fresh weight, or drying it for 24−48 hours at 70−80°C to give the dry weight. Measurement of dry weight is normally considered to be more reliable than fresh weight, as the latter varies depending on the environmental conditions and the time of harvest.

Growth indices

Plant growth may be expressed in various ways. The most common indices for measuring growth include:
1. **Net assimilation rate (NAR)**. This is defined as the dry matter produced per unit leaf area per unit time.
2. **Crop growth rate (CGR)**. This is the increase in dry matter per unit soil area per unit time.
3. **Relative growth rate (RGR)**. This is the increase in dry matter per unit dry weight of plant material per unit time.

The values of the NAR, CGR or RGR depend on the ratio of leaf area per unit soil area (LAI). The relationship of CGR and NAR to LAI, measured for a canopy of sweet potatoes, is illustrated in Figures 3.18 and 3.19. The response of RGR to LAI is similar to that shown in Figure 3.18.

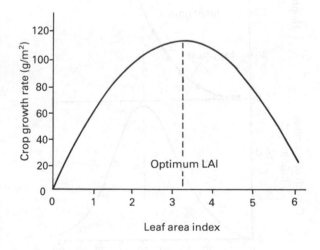

Figure 3.18 Relationship between leaf area index and crop growth rate in sweet potatoes.

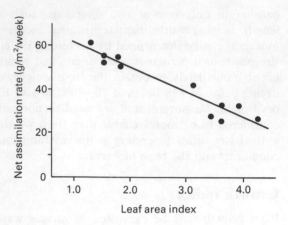

Figure 3.19 Relationship between leaf area index and net assimilation rate in sweet potatoes.

Growth curve

The changes in the size of plants as a function of time are referred to as a growth curve. Most annual plants and individual plant parts exhibit an S-shaped (sigmoid) growth curve as shown in Figure 3.20.

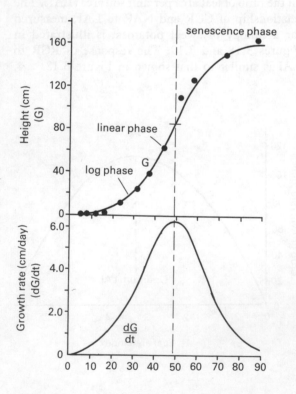

Figure 3.20 Idealised growth and growth rate curves.

Dry matter partitioning

There is competition for nutrients between the vegetative and reproductive organs of plants. Developing flowers and fruits, especially young fruits, have a relatively large requirement for mineral salts, sugars and amino acids as compared with vegetative plant structures. Studies with radioactive tracers show that nutrient accumulation in developing flowers, fruits or tubers occurs mainly at the expense of materials in nearby leaves. Similarly, competition for nutrients between individual fruits of the same plant results in a decrease in fruit size as the number of fruits per plant increases. The mechanism by which plants can divert nutrients from leaves to fruits, sometimes against apparent concentration gradients, is presently not well understood. Various plant hormones, especially cytokinins, are probably involved.

Modifying the crop environment

The limiting climatic factor for crop production in the tropics is usually water. The humid tropics (Figure 3.21) has a rainy season longer than 200 days in a year and a mean annual rainfall in excess of 1500 mm (Table 1.5). The high cloud cover throughout the rainy season means that much of the direct sunlight is reflected in the upper atmosphere. This effectively reduces the amount of solar radiation reaching the crop and lowers air temperature in the crop environment. The high rainfall also causes enhanced soil erosion and leaching.

The dry tropics refers to those areas whose rainy season lasts for less than 200 days and where rainfall exceeds potential evapo-transpiration for less than 100 days. This makes it possible to grow only one crop per year. Dry spells and shortages of water are evident at the beginning and towards the end of the rainy season. The dry tropics have a higher potential evapo-transpiration, maximum temperature and level of solar radiation as compared with the humid areas. The dry season features large diurnal temperature ranges, with low minimum temperatures while still maintaining high maximum temperatures. The general objective of modifying the climate in the tropics must, therefore, be looked at in the context of the prevailing weather conditions at specific locations. Any

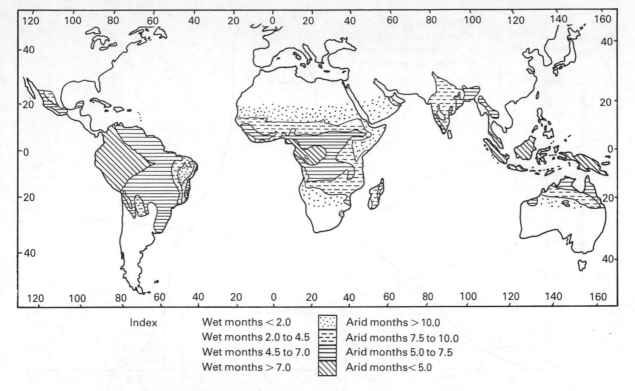

Index	Wet months < 2.0	Arid months > 10.0
Wet months 2.0 to 4.5	Arid months 7.5 to 10.0
Wet months 4.5 to 7.0	Arid months 5.0 to 7.5
Wet months > 7.0	Arid months < 5.0

Figure 3.21 Rainfall distribution in the tropics.

solution must include strategies for making water available all the year round, in moderate quantities to avoid erosion and excessive leaching, as well as sunlight and temperature depression. Specific approaches involve attempts to alter the weather, changing the crop's response to the weather, taking advantage of probable weather conditions or amelioration of the crop environment.

Alteration of weather

Alteration of the weather is not an easy task and only moderate success has been reported in the area of cloud seeding. The method commonly used involves the injection of substances that can act as condensation nuclei for the growth of water droplets in the cloud. Current estimates of expected increase in rainfall due to cloud seeding average about 10%. While this addition might be important for agriculture in marginal rainfall areas, its usefulness is limited because clouds must exist before they can be seeded. The seasonal rainfall distribution in the tropics would remain unchanged even if the technology of cloud seeding were available for use.

Alteration of crop responses to weather

The possibility of changing plant responses to the environment, especially to such factors as temperature, daylength and water supply, would be expected to significantly affect crop production. Possible approaches include the application of growth regulators and the breeding of crops with the objective of incorporating specific plant characters.

Growth regulators are organic compounds which are normally synthesised by plants and translocated to areas where, in low concentrations, they cause physiological responses and changes. Present levels of technology for tropical crops have confined gains in this area to the progress made in plant breeding. The factors that are most relevant include breeding for drought resistance and erect leaves, especially in cereals. An erect disposition of the leaves permits deep penetration of light into the crop canopy.

Matching crops to probable weather

Taking advantage of the probable weather requires a thorough understanding of the climate at a given

55

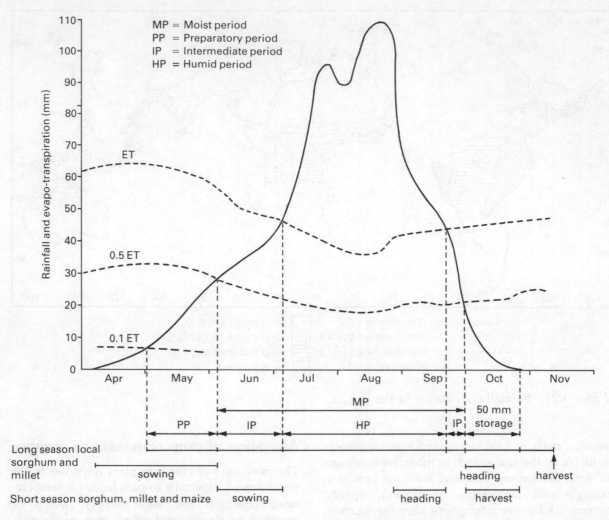

Figure 3.22 Suggested cropping times to match the mean rainfall distribution at Kano in the Sudan savanna zone of Nigeria.

location so that crops can be matched to the climatic resources. The rainfall data for Kano in the Sudan zone of northern Nigeria has, for example, been used to illustrate how cropping could be planned to take advantage of this crucial resource (Figure 3.22). In the humid tropics, the emphasis for the agriculturist would be on minimising the damage to crops during the periods when rainfall exceeds potential evapo-transpiration.

The low minimum temperatures during the dry season should not be disadvantageous to crop production if water is available through irrigation. Temperate (C3) crops such as wheat and barley can be substituted for the regular tropical (C4) crops in areas where the minimum temperature often falls below 10°C.

Amelioration of the crop environment

Amelioration of the crop environment includes all management programmes designed to favourably modify the micro-climate of the crop. Methods that can be used include heating to protect high value crops such as orchards from frost, carbon dioxide enrichment, shelterbelts, windbreaks and irrigation. While an attempt to warm crops in the tropics might appear ridiculous, it should be noted that some vegetable crops, such as tomatoes, often suffer frost damage in the East African highlands and during cold Harmattan nights in the Sudan and Sahel savanna of West Africa. The difficulty and cost of making and keeping dry ice make carbon dioxide enrichment impracticable under

Figure 3.23　Sprinkler irrigation.

field conditions. Elsewhere, there have been reports of increased carbon fixation (and hence dry matter production) with supplementary carbon dioxide under controlled environmental conditions.

Windbreaks generally reduce windspeed, and consequently evapo-transpiration, over a crop. A well designed windbreak (50% porosity) reduces windspeed to less than half of that in the open, but its effectiveness downwind is limited to a distance equal to or less than about 25 times the height of the windbreak.

Irrigation is the provision of supplementary water to support crops through part or all of their growth (Figure 3.23). In most of the tropics, even though there is a long dry season lasting up to seven months, the rainy seasons are often so wet that much of the rain-water can be conserved and stored for cropping during the dry season. In certain areas it has been found beneficial to make such irrigation water available to carry rainfed crops through droughts and dry spells (Table 3.7).

Because the climatic problems are different between the dry and humid tropics, the approach to adapting land and crop management must of necessity differ. The general principles for land and crop management to match the features of both the dry and humid tropics are discussed in a subsequent chapter.

Table 3.7. Effect of supplementary irrigation on yields of rainfed groundnuts at Kano

Year	Total rainfall (mm)	Rainfed yield (kg/ha)	Supplementary irrigation (mm)	Irrigated yield (kg/ha)	Yield increase over rainfed (%)
1982	630	1614	160	3107	93
1983	432	580	110	3145	442
1984	507	950	115	1580	66
Mean		1048		2614	150

CHAPTER 4

Cropping systems

The term cropping system may be defined to include the scheduling and cultivation of the various crops within a farm enterprise in a given agricultural year. The cropping system refers to the way and manner in which the farmer actually organises the growing of various crops and how he arranges them in his field. In contrast, the term farming system is a broader concept of which cropping system is but a part. The farmer operates in a system which is influenced by environmental, technical, economic and human factors. The farming system that results is determined by how he produces, uses, markets or consumes his farm products, both crops and livestock. Therefore, the concept of a farming system considers the farmer, his farm operations and the biological and socioeconomic environment in which he operates as a whole.

Cropping systems, and indeed farming systems in general, have often evolved in response to climatic conditions (Figure 4.1), soil resources, available crop species, socio-economic priorities and proximity to urban centres. Two important and interrelated factors that influence the evolution of a cropping system are the availability of suitable land for cultivation and the productivity of the land with respect to the farmer's requirement for total production in any given year. Indigenous cropping systems in the tropics have thus been adapted to maximise the use of the available growing season which is invariably limited by the duration and amount of rainfall. The need to minimise the risk of crop failure is also of prime importance. For instance double cropping is practised where the rainy season exceeds six months. However, a more common practice is that of relay cropping, where a short duration crop, such as groundnut, maize or a cucurbit, is sown following the first rains, and a longer duration crop, such as sorghum or cassava, is interplanted one to two months later. In this way, the later crop makes little growth until the early crop begins to mature, and then fully utilises the soil and air environment after the early crop has been harvested. In the mixed multiple system, a fairly sparse stand of the long duration crop is intersown with the first crop and a second short duration crop is intersown after the harvest of the first. In areas with less than five wet months, the sowing dates of the crops are generally closely synchronised. Here also the mixtures invariably include as their two principal components, one short duration crop (millet, early sorghum, maize or groundnut) and one of a longer duration (long-season sorghum, cotton or pigeon pea). The techniques may differ, but the goal is to maximise use of the growing season. The ingenuity of these indigenous adaptations to different ecological circumstances is remarkable, particularly with respect to their potential for maximising photosynthetic productivity and balanced nutrient removal from the soil.

In recent times, mounting population pressure, improvements in agricultural technology and other changing economic circumstances have caused certain changes in the traditional cropping systems of the tropics.

Shifting cultivation

Many soils in the tropics are low in potential fertility and are usually not suitable for continuous cultivation, unless sustained by heavy mineral fertilization and organic manuring. The acquisition of fertilisers still remains difficult for many tropical farmers. Even where fertilisers are available their relatively high cost limits their use. Until recently, shifting cultivation was the cheapest way of maintaining soil fertility. In this practice the land is cleared and cultivated for a few years, usually three to four years, and thereafter allowed to revert to bush for five to ten years, before it is again

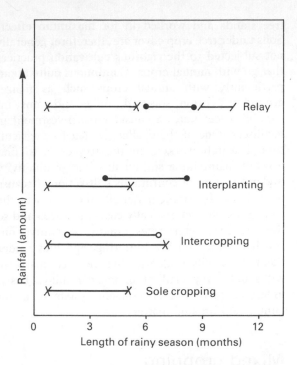

Rainfall (amount) →

Length of rainy season (months)

Figure 4.1 Types of cropping system in relation to the length of the rainy season in the tropics.

opened up for cultivation. If sufficient land is available and the population is low enough to allow adequate intervals between successive cultivations, the system can sustain reasonable yields almost indefinitely. If, through population pressure or other socio-economic demands for land, the intervals are unduly shortened, sufficient natural fertility regeneration becomes impossible. This is one of the strongest factors against the practice. Population growth in most tropical countries of the world, where agricultural development is still in its infancy, has been rapid. This, coupled with rapid urbanisation and the demands made by industrial expansion, has placed considerable pressure on land. As a result, less and less land is available for bush fallowing. Shifting cultivation is, nevertheless, an effective and cheap method of maintaining soil productivity. Returning the land to fallow restores organic matter and nutrient levels in the soil.

In general two types of shifting cultivation are recognised under subsistence farming systems in the tropics. The first type is that in which the people build temporary villages and practice shifting cultivation in the immediate vicinity for several

years until crop yields fall significantly. The whole community then migrates elsewhere to build a new village and open up new land. This practice used to be a common feature of agriculture in Africa and Malaysia. Where such movement occurs, land is usually reopened only after a prolonged period of fallow.

The second type of shifting cultivation is that in which the people live in permanent villages or towns with their cultivated land covering a large area. The prolonged use of a relatively limited amount of land naturally results in a more rapid rotation of the cultivated farms and so fallow periods tend to become gradually shorter and as the productivity of the immediate vicinity of the village declines, the distance from dwellings to the farms may continue to increase.

The fertility restoration during the resting period is dependent on the length of the fallow, the nature of the vegetation and the rate at which soil nutrients are taken up by the fallow vegetation from the subsoil. Some nutrients are stored in the vegetation and are eventually returned to the surface soil through fallen leaves and in the form of dead roots. The decomposition of these organic residues results in a build-up of soil organic matter which in turn leads to an improvement in soil structure and fertility.

In forest areas, the practice of leaving tree roots and stumps undisturbed after felling and the prevailing high rainfall as well as the hot and humid temperatures all make for very rapid soil fertility regeneration. Subsoil feeding by the fallow vegetation has the important effect of transferring nutrients via the vegetation and litter fall, and in ash after burning, to the surface soil. This process counteracts leaching losses and crop removal of nutrients.

In the savanna zones, the amounts of nutrients stored in the fallow vegetation, and the organic matter content of the soil after fallow, are much less than in the forest zones. A number of reasons are responsible for this. Most savanna areas have comparatively low rainfall and experience a marked dry season. Consequently, after a few years of cultivation, when the land is reverted to fallow the grass vegetation may be slow in forming a cover and the rate of dry matter production becomes limited by inadequate moisture. Nutrient storage in vegetation is low, except in the tall grass savanna on the fringe of the forest. Growth of trees is slow and yet the amount of nutrients stored in the fallow vegetation depends largely on the extent to which trees and shrubs are present. The transfer of nutrients from the subsoil to the topsoil is less because grass vegetation is generally surface rooting.

The rate of decline in fertility and productivity of a soil depends largely on the cropping period, climate, topography and the crops grown. Although shifting cultivation confers a number of advantages to a soil, the system carries a number of drawbacks which make it impracticable in the present socio-economic climate of which agriculture is part.

Continuous cropping

In many parts of the tropics, population growth, urbanisation and other socio-economic demands on land, have made it impossible to maintain farming systems which include resting fallows. The emphasis is, therefore, shifting from bush fallowing to continuous and intensive cultivation. By continuous (or intensive) cultivation is meant a system of cropping whereby the land is cultivated on a continuous basis, either by rotation monocropping, monoculture or mixed cropping. With tree crops such as rubber, cocoa, coffee or cola grown in plantations, the land has to remain under cultivation for several years, in many cases for more than a decade. During this period, the land is usually subjected to minimum conventional cultivation, although fertiliser has to be applied to the

tree stands and worked in for maximum effect. Soils under tree crop cover are, therefore, generally not subjected to the rigorous cultivation practices that go with annual crops. Continuous cultivation, particularly with annual crops such as maize, groundnut, cotton, sorghum, millet and yams, invariably necessitates a considerable investment in fertilisers. One of the challenges that face agricultural research workers in the tropics is to find ways of maintaining soil fertility and productivity under a system of continuous cultivation. Because of their sandy parent materials, most soils in the tropics are characteristically coarse textured and so the structure deteriorates rapidly on cultivation. Furthermore, their low buffering capacities render them chemically fragile so that they are unable to withstand heavy fertilisation on a continuous basis, unless coupled with an efficient system of maintaining soil organic matter.

Mixed cropping

Except for the cultivation of rice as a monoculture, mixed cropping is the most popular system of cropping in tropical Africa. The practice involves growing two or more crops simultaneously on the same piece of land (Figure 4.2). For example, sorghum and millet are mostly grown as mixed crops in West Africa. Millet is usually the first crop to be planted and, about four weeks later, the sorghum is sown between the millet stands. The

Figure 4.2
(a) Maize and cowpeas grown as a mixed crop.
(b) Maize and cassava grown as a mixed crop.

a)

60

terms mixed cropping and intercropping are often used synonymously.

Mixed cropping has a number of technical and economic advantages which make the system efficient:

1. Mixed cropping makes better use of the environment in terms of space, water and nutrients. It permits a higher plant population than would a single crop and makes effective use of available plant nutrients, since the different crops in the mixture obtain their nutrients at different soil depths.

2. Growing of more than one crop on the same field, at the same time, reduces the risk of total crop failure resulting from pest or disease attack.

3. The more extensive root system produced by two crops tends to hold the soil particles together, thereby giving the soil good structure, which in turn minimises erosion. Erosion is also reduced by the more continuous crop cover usually produced by mixtures.

4. Where legumes are included in the mixture, they may leave some residual nitrogen in the soil, which may benefit a subsequent crop. Recent research in northern Nigeria suggests that the inclusion of a legume in a crop mixture does not give any obvious nutritional advantage to the second crop in the mixture, but any residue left by the legume increases soil fertility and benefits subsequent crops.

5. Economically, the return per unit of labour is higher as a result of greater total yields, and more dependable returns can be secured from year to year.

6. Pests and diseases do not spread as quickly in crop mixtures as they do in monocultures.

There are two major disadvantages of mixed cropping. The system complicates the interpretation of crop performance while making mechanisation difficult. The difficulties inherent in the system, particularly those which relate to mechanisation, constitute a serious drawback in the realisation of increased agricultural productivity per unit of labour. The use of labour saving devices developed for monocrop conditions, particularly those developed for temperate environments, often do not lend themselves to effective use in mixed cropping. Thus, the need for devising appropriate crop production technology for mixed cropping systems cannot be overemphasized.

Indigenous farming systems compatible with mixed cropping have been developed by African farmers who have adopted animal draft. Improved crop varieties, for example hybrid maize, and fertilisers have been incorporated into mixed cropping systems in both East and West Africa. This suggests that mixed cropping is perceived by farmers in the tropics as worthy of adaptation to new technology. There appears to be justification in suggesting that the system merits greater research attention, to develop improvements on present farming methods.

For new technology which is not compatible with mixed cropping, the farmer needs to be convinced that the benefits outweigh the cost of abandoning the traditional practice. For this, the dependability of a new technology may be more important to the tropical farmer than its potential performance.

Current research is centred on finding the best configuration and density for planting the various major crop mixtures. Another area of research activity relates to fertilising mixed crops. Most fertiliser recommendations are based on monocropping. This is because early research on the responses of crops to fertilisers was based on the philosophy of trying to change the practice of the small-scale farmer from mixed cropping to monocropping, without taking into consideration his socio-economic circumstances. The folly of this philosophy has since been recognised by researchers and efforts are now being made to find out the best ways of fertilising mixed crops.

b)

Mixed farming

Whereas mixed cropping refers to a system of cropping on the field, mixed farming is the integration of animal and crop production on the same farm. It provides for the combination of crop production and livestock in a single enterprise, such that the farmer is able to feed his animals or poultry with his own crops. Farmyard manure produced by the livestock is also used on the crops.

In many parts of West Africa where a large number of farmers practice mixed farming, crop farms are used as livestock feeding grounds once the crop has been harvested. Cattle feed on the crop residues and leave their dung in the field, thus increasing the fertility of the soil. Mixed farming can be very demanding in terms of time and energy, but it provides excellent insurance against failure of any one farm enterprise. Besides the farmyard manure produced, the bulls are available for use in the cultivation of crops, thus increasing the total land area available for cropping.

Monocropping

Monocropping (monoculture or sole cropping) is the growing of a single crop on a piece of land within a growing season. Even if the climate, and particularly the rainfall, permits the growing of two or more crops in one year, when only one crop is cultivated at a time, the practice remains that of monocropping.

Continuous monoculture implies the exclusive cultivation of the same crop on the same piece of land year after year. This may apply to both annual and perennial crops. Monocropping, particularly continuous monoculture, has a number of disadvantages, including the following:
1. The practice carries with it the risk that the farmer could lose his entire crop in the event of such natural hazards as drought and pest or disease attack.
2. Monocropping tends to encourage pest and disease build-up.
3. It creates an imbalance in nutrient removal from the soil.

The system has one important advantage in that it encourages specialisation in the techniques of production. Rices (Figure 4.3) and sugarcane are two crops that have been produced in monocropping systems for many years in the tropics. In semi-arid and sub-humid tropical savanna regions, research has shown that cotton and groundnuts, if adequately fertilised and protected from weed, pest and disease infestation, can be successfully grown under continuous monoculture for five to seven years without any significant drop in yield.

Crop rotation

Crop rotation is a system of farming in which the same piece of land is kept under cultivation for years, but the crops grown are changed from one field to another on a rotational basis with a regular sequence. Several advantages are associated with crop rotation, among which are the following:
1. The practice of rotation allows for balanced nutrient removal from the soil. Crops such as maize, yams and cassava are considered to have heavy nutrient demand and, therefore, can cause soil exhaustion. Vegetables such as tomatoes, pumpkins and groundnuts are relatively less demanding, although there are many examples of soil exhaustion in West Africa from attempts to continuously crop groundnuts (see Table 7.7). This means that a soil may be low in nutrients with respect to one crop but ade-

Figure 4.3 Rice grown as a monocrop in Southeast Asia.

quate for another. Legumes generally tend to enrich the soil by adding nitrogen through symbiotic fixation. For this reason, most rotations include a leguminous crop. For example, a typical crop rotation system in the savanna region of Nigeria would be cotton−sorghum−groundnut.

2. Another advantage is that the different plant species are able to extract nutrients at different soil depths. This is as a result of differences in the development of their root systems. It is advantageous to rotate deep-rooted crops with those having shallow root systems.
3. Soil-borne fungal, bacterial and viral diseases are often plant species specific; by rotating crops, there is a tendency to prevent the pathogens from completing their life cycle. Thus, the practice of crop rotation reduces the incidence of disease.
4. A further advantage associated with crop rotation relates to improvements in soil structure with a consequent reduction in erosion.
5. Several socio-economic advantages also exist, including the spread of available labour, the spread of economic risks and the diversification of marketable raw materials which in turn diversifies the locally available diet.

Ley farming

This system alternates pastures with crop production. The pasture usually selected for inclusion in a ley farming system is of sufficient nutritional and morphological quality to enable it to fit into a crop rotation system. After the arable crop (usually a cereal) is harvested, the field is sown to pasture and grazed for one or two seasons before it is ploughed again for arable cropping. The planted pasture is usually a mixture of grasses and legumes.

The problem with this system is that it involves a planted fallow which many farmers are unable to justify in economic terms and are, therefore, reluctant to include in their crop rotation. In West Africa, for instance, it has been difficult convincing subsistence and small-scale farmers to accept this system because they see the fallow period as a waste. This is understandable when it is considered that the farmer's first pirority is usually the production of food crops needed to sustain his family.

Alley cropping system

The restoration of fertility by shifting cultivation, particularly under humid and forest conditions, depends to a large extent on the nutrient recycling ability of deep-rooted trees and shrubs. Until recently, efforts to find ways of maintaining soil fertility with reduced fallows or even permanent cultivation have been devoid of a tree or shrub component. However, the results of recent work pioneered by the International Institute for Tropical Agriculture, Ibadan, Nigeria, have indicated a high potential for developing a stable and productive system, the alley cropping system, with a small tree or shrub, *Leucaena leucocephala*, which recycles plant nutrients and at the same time provides material for mulch. It provides support for such twining crops as yams, green leaf for enriching the soil organic matter and increased nitrogen levels in the soil.

The concept of alley cropping retains the basic features of bush fallowing, but has the following modifications:

1. Selected species of fast-growing small trees and shrubs, usually legumes with the ability to fix nitrogen, are used to replace the variable species of the naturally regenerated bush fallow.
2. The small trees or shrubs are planted in rows with inter-row spacing wide enough to allow the use of mechanised equipment.
3. The trees or shrubs are cut back and kept pruned during the cropping period and the leaves and twigs are applied to the soil as a mulch, providing a source of nutrients and organic matter. The bigger branches are used as stakes or for fire wood. Fire is never used for land clearing or for tree suppression.
4. The height to which the trees or shrubs are cut back depends on the shade tolerance of the associated crops.
5. The land is (periodically) ploughed in order to cut tree roots to reduce competition with crops.
6. The trees or shrubs are allowed to recover during the dry season, when they develop new growth ready to be used on the next crop.

In addition to *Leucaena leucocephala*, several tree or shrub species, including both legumes and non-legumes, have been evaluated for use in alley cropping systems. Among other considerations, desirable species are those that can be established easily and which can be maintained from basal sprouts and coppices when periodically cut back.

Table 4.1. Some characteristics of Apomu surface soil (0−15 cm), before and after alley cropping at Ibadan, Nigeria

Soil characteristic	Before cropping	After one year cropping			
		Nil−N*	Nil−N	40 + 30kgN/ha	80 + 60kgN/ha
pH (H$_2$O)	6.2	5.7	5.7	5.5	5.3
Organic carbon (%)	0.98	0.96	1.47	1.24	1.41
Total N (%)	0.12	0.11	0.12	0.13	0.13
P (ppm)	24.7	19.4	21.5	18.7	22.8
K (meq/100g)	0.25	0.16	0.16	0.14	0.18
Ca (meq/100g)	2.63	5.07	5.33	5.72	4.43
Mg (meq/100g)	1.02	0.35	0.43	0.34	0.28

* *Leucaena* prunings removed from this treatment.
Source: IITA Annual Reports and Research Highlights, IITA, Ibadan, Nigeria.

Field trials conducted at Ibadan, Nigeria, on the effect of nitrogen application in a maize−*Leucaena* alley cropping system gave results which can be summarised as follows:

(a) Substantial *Leucaena* dry matter and nitrogen yields were produced with a total of six prunings.

(b) The annual yield of nitrogen totalled over 180 kg N/ha, without the application of nitrogenous fertilisers.

(c) There was a significant effect on *Leucaena* dry matter production and nitrogen yield of low rates of nitrogenous fertiliser applied to the maize. In contrast, with high rates of applied nitrogen, the effect was only slight.

(d) Despite the high amount of nitrogen produced, there was still a need for the application of low rates of nitrogenous fertiliser to maximise maize yields.

(e) The removal of *Leucaena* prunings reduced the yield of maize to about 45%.

(f) Without the application of nitrogenous fertiliser, but with the prunings left in place, the total grain yield was maintained at about 3.5 t/ha, a good yield for the Ibadan location.

(g) The addition of *Leucaena* prunings had a distinct effect on levels of soil organic matter as evidenced in Table 4.1.

CHAPTER 5

Principles of crop husbandry

The art and science of crop production is a dynamic process. The breeding and selection of new crop varieties; improvements in methods of planting, cultivation, soil and water conservation and management, harvesting, storage and marketing; new techniques for weed, pest and disease control; the formulation of better fertilisers as well as improvements in their efficient use are all changes aimed at achieving improved living standards and a better way of life for mankind and his animals.

The attainment of increased agricultural production is essentially dependent on the successful application of the fundamental principles which govern the various crop management practices. The need for a sound appreciation of the basic principles of crop husbandry becomes particularly important with new and improved production technologies and the challenge of increasing demands for food, fibre and shelter.

Tillage

Irrespective of ecological conditions, the production of all crops demands some form of tillage, an operation which destroys weeds and improves the characteristics of the soil. In general, tillage is an expensive farm practice and, for good crop production, it is to the farmer's advantage not to till the land more than is necessary. In addition to controlling weeds, tillage also achieves other important purposes, including the incorporation of crop residues, loosening of the soil, improvement of soil aeration, destruction of insect pests, preparation of land for surface irrigation, and possible erosion control. With regard to erosion control, it is important to note that tillage is more often responsible for accelerated erosion than for its reduction. However, contour tillage can aid in erosion control.

In the absence of weeds and especially under relatively dry tropical conditions, the practice of not tilling the land or minimally cultivating it has been in existence for centuries. In such circumstances, the farmer simply uses a hoe to scoop the soil where he intends to sow the seeds so that the planting operation is made easy. However, in a field which has been overtaken by weeds, the farmer traditionally makes ridges or mounds with the hoe, such that the weeds are buried and planting is facilitated.

Tillage systems

In traditional methods of farming in the tropics, the tillage system is an integral part of the cropping system, reflecting the requirements of the crops and determining, at least in part, the manner and time of planting. Except on the most friable soils, both hoe and animal farmers in many developing countries in the tropics are unable to expend the additional energy required to cultivate dry soil. In the savanna region, for instance, millet is frequently sown into small heaps of loosened soil at relatively wide spacings, so that only 5–10% of the land is 'prepared' before the first crop is sown. Later, after more rain, the land is considered ready for more thorough preparation, an operation which also kills off the first weed seedlings. In these systems, late season weeding, often including the rebuilding of ridges or mounds, is often considered more as land preparation for the next season than as an operation necessary for the crop occupying the land at the time (Figure 5.1).

In more humid areas, especially where weeds are present at the onset of rains and where probable negative effects from delayed sowing are not so great, land preparation is completed before the first sowing. However, it is not uncommon to sow into undisturbed soil following the burning of fallow vegetation to kill weeds (Figure 5.2). To be effective, special requirements for a system

Figure 5.1   Women hoeing a cooperative maize field, Mbambara Ujamaa village, Tanga Region, Tanzania.

Figure 5.2   Savanna subjected to fire by farmers in Sierra Leone.

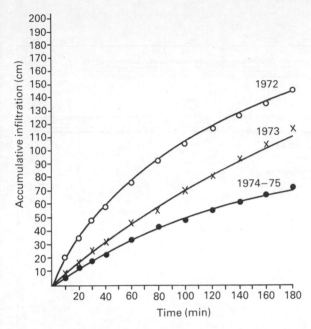

Figure 5.3 Progressive reduction in the infiltration rate of an untilled field from which crop residues have been removed.

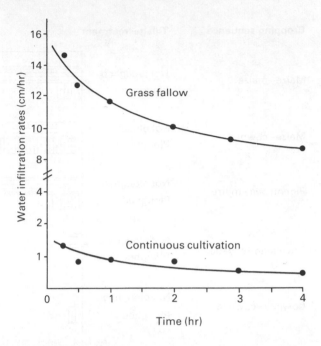

Figure 5.4 Infiltration rates under grass fallow soil and continuously cultivated soil (Ike, 1986).

without tillage include the availability of adequate amounts of crop residue, the effective control of weeds, a minimal dependence on herbicides and adequate crop rotation and cropping systems to eradicate insects, other pests and pathogens. Lack of adequate quantities of plant residues on the soil surface can result in degradation of soil structure (Figures 5.3 and 5.4). For example, the infiltration rate of soil in a plot under a no-tillage system and from which crop residues were deliberately removed declined progressively with time within three hours from 145, 110, 75 and 75 cm in 1972, 1973, 1974 and 1975, respectively. The effects of tillage and crop rotation on earthworm activity are shown in Figure 5.5.

Tillage operations

In large-scale mechanised agriculture, the principle of land clearing follows a sequence of tillage operations under animal or tractor power systems. Tillage practices generally involve primary and secondary operations using appropriate implements (Figures 5.6 to 5.8). In primary tillage, the soil strength is broken and weeds are destroyed. Weed destruction is effected by the complete in-

version of furrows which are slice-cut by the tillage implements used in ploughing. The ploughed field is left for one to two weeks for the weeds to decay; any new weed growth is destroyed in the subsequent secondary tillage operations. One characteristic of primary tillage is that the furrow-slice is left in large clods. Since this leaves the field unsuitable for planting, a secondary tillage operation is necessary for the final seedbed preparation.

Development of appropriate technology

Although hoe-farming systems often show elegant adaptations to minimise drudgery, they cannot be expected to eliminate it. Unfortunately, the search for appropriate technology to minimise drudgery in tropical farming has been slow. The improper use of tractor power for annual crops can be ecologically disastrous, particularly under forest conditions and even in savanna areas. Indeed, the operation of tractors in the tropics is generally more expensive and less effective than in temperate conditions, mostly due to deficiencies in infrastructure and manpower training. While tractors may offer prospects of alleviating drudgery and the constraints of manual labour in large-scale

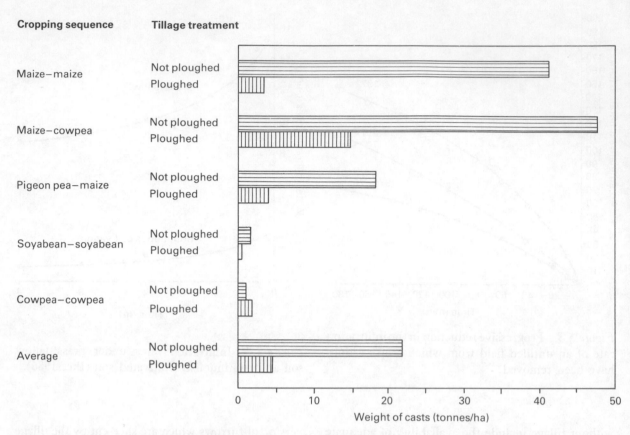

Cropping sequence	Tillage treatment	

Figure 5.5 Effect of tillage and crop rotation on earthworm activity expressed in weight of casts. (No tillage effects on soil properties in Western Nigeria, R. Lal, 1978.)

Figure 5.6 A tractor operated mouldboard plough for general ploughing, especially where complete inversion is needed.

Figure 5.7 A tractor mounted disc plough is preferred in situations where complete inversion is not necessary or where a field has stubble and penetration by a mouldboard type is difficult.

production in the future, immediate relief must consider more comprehensive and appropriate tool systems for animal draft as well as the use of herbicides.

Irrigation

The artificial application of water to a crop from an external source is necessitated where the total rainfall is insufficient to supply the needs of the crop. In dry tropical areas, irrigation is one of the most important operations in the cultivation of crops, especially vegetables. In the humid tropics, seasonal deficits may occur. Whether to irrigate or not depends not only on the extent of the rainfall deficit, but also on the crop. For example, while about two weeks of drought may have serious consequences for young banana plants, a citrus crop normally suffers little harm from two months without rain.

When considering irrigation facilities, it is necessary to assemble and analyse data on the climate and hydrology of the region as well as to consider the social, economic and financial implications. The application of irrigation water may be by gravity or under pressure (Figures 5.9 to 5.11). In general, gravity irrigation is especially useful for land with a gentle slope and smooth topography. The main methods of gravity irrigation include

Figure 5.8 A tractor operated spring tooth harrow is a secondary tillage tool for final seedbed preparation.

Figure 5.9 Gravity-fed irrigation by border flooding on a small-scale farm in Nigeria.

border flooding and furrow irrigation. Both methods involve low capital investment and high labour costs. In contrast, irrigation under pressure is used for installations which are portable, semi-portable or permanent. A pump may be used to lift water from a low level and deliver it under pressure to fields through normal irrigation channels or through portable installations. In portable installations, perforated aluminium pipes are used, whereas sprinklers are fixed to the pipes in the semi-portable and permanent pressure irrigation methods. Labour costs decrease as capital costs increase from the portable, semi-portable to the permanent methods.

In special circumstances, water may also be applied by drip or trickle irrigation under low pressure to the root zone of a crop. This approach achieves great savings in irrigation water while also restricting weed growth.

Young seedlings, especially of vegetable crops, should be irrigated in the early mornings and evenings. Watering in the middle of the day should normally be avoided. However, if this appears essential, water should be applied to the soil rather than to the foliage. Over-watering can be harmful and the practice often encourages the development of disease. In drought conditions, and especially with tree crops, leaves tend to wilt during the hotter hours of the day, but recover their turgor later in the afternoon. If the drought continues, wilting starts earlier and lasts longer each day, thus signifying the need for irrigation.

Figure 5.10 Irrigating agricultural land using a shadoof in Kano, Nigeria.

Figure 5.11 A water engineer with experimental bamboo pipes, Ruvuma Region, Tanzania.

Drainage

Drainage implies the process of removing excess water from a soil by providing open or covered channels through which the water can freely move. The practice of drainage is, therefore, largely dictated by local circumstances of climate, soil, topography and crops as well as by economic considerations.

Problems of drainage normally arise either as the result of a high water-table (as can be experienced with excess irrigation or in low-lying areas that receive water from surrounding higher land), or because the physical condition of the soil impedes the flow of excess surface water to a depth below the root range of crops. While each of these causes normally requires different ameliorative measures, the main effect of providing drainage in both situations is to improve the aeration of the soil so as to enable the development of deeper root systems for effective nutrition as well as withstanding periods of droughts.

Soils having a water-table that is permanently or seasonally high enough to adversely affect crop growth normally require some drainage. The depth to which the water-table should be lowered invariably depends on local conditions, particularly the nature of the soil, the rainfall and evapo-transpiration regimes and also the crops to be grown. For example, tree crops which are characterised by deep root systems to anchor them firmly against high winds, invariably require a lower water-table than various vegetable or annual crops or such surface-rooting perennials as pineapple.

In the presence of abundant rainfall which is distributed throughout the year, it is desirable to ensure that the water-table is maintained below the crop rooting depth. However, considering the unlikely economic benefit, such a thorough practice is often considered unnecessary, especially since short periods of water-logging of part of the root system need not result in significant harmful effects. This is particularly true in dry areas with seasonal rainfall where the chief objective of the farmer is to conserve water. In practice, drainage by open surface drains is usually preferred, despite the disadvantage of taking up land, the inconvenience to mechanised tillage operations and the need for regular maintenance. Apart from being cheaper to construct, open surface drains can be made mechanically, and are easy to check and clear when necessary.

Downward movement of excess surface water may be impeded by several factors, chief amongst which are:

1. A lack of sufficient coarse pores and cracks in the surface soil owing to the breakdown of aggregates by the impact of rainfall, by excessive tillage or by cultivation when the soil is too wet.
2. The presence of a subsurface layer of soil of low permeability, below which the profile is free-draining, as for example in horizons of clay accumulation.
3. Heavy equipment passing over the soil and compressing it to form a compact layer below the surface.

Where the structure of the surface soil has been destroyed, the remedy lies not in drainage but in

71

protecting the soil surface with a crop cover or mulch and avoiding untimely or excessive tillage (Figure 5.12). Because of differences in the time of sowing and harvesting as well as the rate and habit of growth, crops differ in their ability to absorb the kinetic energy of raindrops. As shown in Tables 5.1 and 5.2, there is considerable difference in the rate of accelerated soil erosion, and therefore run-off, in relation to crop cover. The utilisation of crop residues, as a mulch or incorporated in the soil, also drastically reduces soil erosion. Cultivation methods which produce a rough soil surface and the practice of ploughing the land immediately after harvest (provided the soil moisture is not too low for cultivation) reduce erosion and hence surface run-off. In general, flat cultivation results in much less erosion and run-off than cultivation on ridges, except where these are cross-tied.

Table 5.1. Annual rate of soil erosion for various types of vegetative cover at Sefa, Senegal

Vegetative cover	Erosion (kg soil/ha)
Natural vegetation, burnt	200
Natural vegetation, unburnt	100
Fallow, sparse cover	4900
Groundnut	6900
Cotton	7800
Sorghum	8400
Maize	10 300
Millet	10 300

Source: Roose E., 1967. Dix années de mesure de l'erosion et du ruissellement au Senegal. Agron. Tropic., Paris, 22: 123−152.

Table 5.2. Annual soil loss with various land management practices in an area with 0.3% slope at Samaru, Nigeria, 1964−68

Cultivation treatment	Erosion (kg soil/ha)
Minimum cultivation on flat, cropped	9400
Flat cultivated, cropped	8900
Flat cultivated, bare fallow	8300
Flat cultivated, cropped, residue on surface	3000
Flat cultivated, cropped, residue ploughed	2200
Ridged, no cross-ties, cropped	43 100
Ridged, alternate cross-ties, cropped	12 600

Source: Modified from; Kowal and Kassam, 1978.

Table 5.3. Effect of depth of cultivation on soil bulk density and total pore space, determined five weeks after ploughing at Samaru, Nigeria

Depth of cultivation (cm)	Bulk density 0−7 cm	10−20 cm	Porosity (%) 0−7 cm	10−20 cm
0	1.45	1.50	40	42
5	1.23	1.49	53	47
15	1.20	1.21	53	52
30	1.19	1.18	54	54

Source: Adeoye, K.B., 1982. Effect of tillage depth on physical properties of a tropical soil and on yield of maize, sorghum and cotton. Soil and Tillage Research, (2): 225−231.

Where drainage is impeded by a subsurface layer of low permeability, improvements may be achieved by deep ploughing or by sub-soiling, to break up the compacted or structureless layer. As shown in Table 5.3, the bulk density of soil decreases with the depth of cultivation as the total pore space increases.

A well-known method of improving heavy soils with poor drainage is the 'ridge and furrow' system in which the land is formed into broad ridges with a slight gradient on the furrow or ditches between them. A modification of this is the 'cambered bed' system (Figure 5.12) which has been extensively employed on heavy, poorly-drained soils for the cultivation of such plantation crops as sugarcane, cocoa and citrus. The beds are usually constructed by first marking out the lines of the drainage ditches or furrows, 6−10 m apart. Overlap ploughing is then carried out, starting from the centre of the bed and working towards the drain lines. This should be effected as deeply as possible, usually 25−35 cm. Subsoiling is carried out next, to a depth of 45−60 cm. It is essential that the subsoiling be done when the clay is relatively dry as the objective is to open up and shatter the soil in order to provide a deep layer of loosened, well aerated and well drained soil in which roots can develop and grow vigorously.

Crop improvement

The major objectives of crop improvement programmes include the development of varieties that

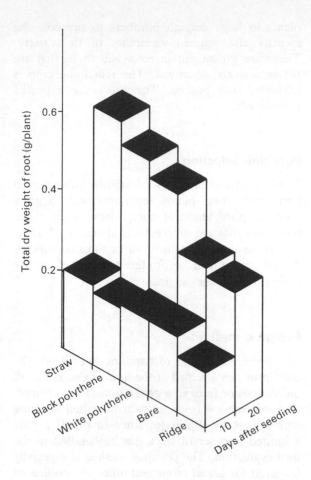

Figure 5.12 Total dry weight of roots of maize seedlings up to three weeks after seeding as affected by different soil surface mulches (Maurya and Lal, 1981).

are adapted to the environment in which they are to be cultivated. Such varieties should be able to yield better than the existing ones under the same soil and environmental conditions. They should also be able to tolerate, if not actually be resistant to, the pests and diseases of the area.

It is not usual for an improved variety to have all these attributes. It may, therefore, be necessary for example, to control pests and diseases with chemicals and to improve the nutrient status of the soil by adding fertilisers. Other crop improvement objectives include earliness in maturity, short plant stucture, non-shattering pods, stiffness of straw and lodging resistance.

Crop improvement can generally be achieved by two approaches; direct plant introduction and plant breeding methods.

Plant introduction

Not all desirable plant characters can be found in a specific locality or within one country. It is, therefore, necessary in any crop improvement programme to systematically collect, assemble and evaluate both local and exotic genotypes for the plant characters desired.

Local collections are normally obtained from farmers' fields, backyard gardens and markets. There are presently institutions in almost every country dealing with the collection and preservation of crop genotypes from which materials can be obtained. The International Board for Plant Genetic Resources (IBPGR), operating under the

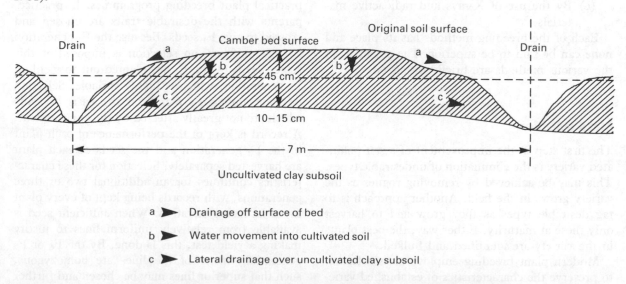

Figure 5.13 A diagrammatic section of a cambered bed (Webster and Wilson, 1980).

aegis of the Food and Agricultural Organization of the United Nations, is charged with co-ordinating the efforts of several institutions throughout the world to conserve the world's plant genetic resources. The IBPGR helps to establish and maintain gene banks where genotypes of crop species are kept and documented regularly. Thus, the IBPGR is an invaluable source for introducing valuable germplasm into crop improvement programmes.

Breeding methods

Several methods of plant breeding exist, including the following:
1. Selection (without hybridisation).
   (a) Pure line method.
   (b) Mass selection method, either by ear-to-row, maternal selection or simple culling.
2. Hybridisation for new varieties.
   (a) Pedigree method.
   (b) Bulk-cross method.
   (c) Backcross method, as applied to either pure lines or Mendelian populations.
   (d) Synthetic combinations.
3. Development of hybrids.
   (a) By the use of inbred pure lines.
   (b) By the selection of clones.
4. Mutation methods.
   (a) Polyploidisation using colchicine.
   (b) By the use of chemical mutagens.
   (c) By the use of X-rays and radioactive materials.

Each of the breeding methods has its place and none can be said to be superior. The essentials of the various methods are briefly discussed.

Mass selection

The first step in the improvement of a self-pollinated variety is the elimination of undesirable types. This may be achieved by removing rogues as the variety grows in the field. Another approach is to tag desirable types as they grow and to harvest only these at maturity. Either way, the best plants in the variety are identified and bulked.

Modern plant breeding employs mass selection to preserve the characteristics of established varieties. Usually this involves the harvest of typical plants in large enough numbers to preserve the identity and original variability of the variety. These are grown out in rows and those that are not typical are destroyed. The remaining crop is harvested and bulked. The process is repeated periodically.

Pure line selection

A refinement of the mass selection technique is to harvest the best plants separately and to grow them as pure lines for comparison. Only those pure lines that are superior and similar are bulked to give an improvement of an established variety. A new variety can be produced if it is obviously different from the original.

Pedigree method

The pedigree method of handling hybrid populations provides a record of the lines of descent of all individuals or lines in a generation. The accumulated information becomes valuable when deciding which lines to eliminate, since in practice only a limited number of lines can be handled in the final evaluation. The pedigree method is especially favoured for cereal crops and other self-pollinated species. It has also been widely used in the development of potatoes, cotton and inbred lines of maize. The method is versatile, relatively rapid and enables genetic studies to be carried out with practical plant breeding programmes. In practice, parents with the desirable traits are chosen and crossed to give F_1 seeds. Because the F_1 generation is very uniform, no selection is imposed at this stage. Seeds of the F_1 are planted out to yield an F_2 generation, which is highly variable. Selection is made among F_2 plants on the basis of traits which are not greatly affected by the environment. A record is kept of the performance of each plant in the F_2 generation and the seeds of each plant are harvested separately. Selection for these characteristics continues for an additional two or three generations, with records being kept of every plant saved in each generation. When sufficient seed is available from relatively uniform lines to justify making a yield test, this is done. By the F_6 or F_7 generation, many of the lines are homozygous, such that superior lines may be chosen and further tested for yielding ability.

Bulk-cross method

The bulk-cross method of handling hybrid populations is similar to the pedigree method described above, except that the early generations are grown as bulk populations rather than as individual plants. The seeds are planted in bulk and harvested in bulk for several generations without any selection being practised. Thereafter, individual plants are chosen, as in the pure line method, and the most promising of the resulting lines are compared with each other and with existing varieties for yielding ability, quality, disease resistance and other desirable characteristics.

Backcross method

If a variety of wheat or another self-pollinated species gives a satisfactory yield of excellent quality, but lacks resistance to rust or another disease, the backcross method is ideally suited for its improvement. In practice, a cross is made between the good variety, referred to as the recurrent parent and the variety or strain which is resistant to the disease. The F_1 plants are then crossed with the desirable variety. If resistance is dominant over susceptibility, the backcross progeny are tested for resistance and all of the susceptible plants are discarded. The plants that are saved are again crossed with the recurrent parent and the progeny are tested for resistance to the disease. This is repeated through several generations of backcrossing and in each case only the resistant plants are saved.

If resistance to the disease is recessive, several backcrosses are made without testing for resistance. A number of plants are then allowed to self-pollinate and their progeny are tested for resistance. The procedure is otherwise the same as when the gene for resistance is dominant. The backcross method of breeding is most widely used with self-pollinated crops, but it can also be used with cross-pollinated crops (Figure 5.14). It is of no value in the breeding of asexually propagated species.

Synthetic varieties

A synthetic variety is a combination of a number of inbred lines, sibbed lines, clones or other populations of cross-pollinated crops. Most new varieties of cross-pollinated species are synthetic varieties. The production of synthetic varieties includes the following features:

Figure 5.14 Covered sorghum heads, illustrating a technique used in plant breeding to prevent cross-fertilisation.

1. They are tested for their combining ability.
2. They can be preserved for future synthesis of the variety.
3. They can be combined so as to permit random crossing.

Because the parental material of a synthetic combination can be maintained indefinitely, one of the advantages of synthetic varieties is that purity can be guaranteed in certified seed by constantly returning to the parental material for foundation stock.

In selecting superior parental material for combination to produce a new synthetic variety, it is important to choose parents that are able to produce superior progeny. Because a synthetic variety is a mixture of many genotypes, it will be subjected to change from selection pressures.

Hybrid varieties

Hybrid varieties are those in which F_1 populations are used to produce the commercial crop. Parents of the F_1 may be inbred lines, clones, varieties or other populations. Hybrid varieties are used where the increased yield resulting from hybrid vigour more than compensates for the development and extra cost of seed. An additional premium in favour of hybrids of inbred lines, as in the case of maize and sorghum, is uniformity.

Apart from the relatively high cost of seed, another disadvantage of using hybrids is that seed from commercial crops cannot be used successfully. This means that parental stock must be maintained from which the hybrid seed is produced each year. In maize, inbred lines are produced by self-pollination of individual plants. When two inbred lines are crossed, the resulting hybrid is usually more vigorous than either parental line and may be more vigorous than the open-pollinated varieties from which the inbred lines were developed. It should be noted that inbred lines are chosen as hybrid parents on the basis of their combining ability, not on performance.

Mutation breeding

A mutation involves a change in the basic structure of a gene or a chromosome, thus causing a change in the phenotype. Agents that are known to cause changes in genes are X-rays, radioactive materials and chemicals. While mutagenic agents have different effects, and some seem to affect certain regions of chromosomes more than others, none has been found that can produce specific desirable gene changes.

The objectives of mutation breeding are as follows:

1. To alter the genotype of a variety only by causing a change in a gene or a few genes to improve the variety. This procedure is suitable for varieties which suffer only a slight deficiency, such as lack of resistance to a specific disease. Mutation breeding is particularly useful to breeders of vegetatively propagated crops where, because of the presence of only one genotype, neither backcrossing nor other conventional breeding methods can alter the variety.
2. To combine a desirable character induced by a mutagen with other desirable characters. This is done using conventional breeding methods.

3. To alter the chromosome structure, to allow translocation of desirable segments.
4. To increase the frequency of crossing-over in some regions of the chromosome.

Propagation of fruit crops

In fruit crops, propagation is done either sexually by seed or vegetatively (asexually) by stems, buds, roots, tubers, leaves or other plant parts. Many fruit crops are grown from seed using the techniques of plant breeding already outlined.

Sexual methods

Before raising fruit crops from seed, care must be taken in the selection of fruits for seed extraction. In particular, it is essential that the fruits selected should possess the desired characteristics, including good size, shape, colour, flavour, taste and keeping quality. Additionally, the fruits should be fully mature and borne on vigorous, healthy trees. After extraction, the seeds are washed and dried in shade for a number of days. Seeds of certain fruit crops, such as pawpaw, mango and citrus, should generally be sown within a week of their extraction.

Sexual methods of propagating fruit crops have the following advantages:

1. Seedling trees are generally longer-lived, bear more heavily and are comparatively hardier than vegetatively propagated trees.
2. Propagation from seed is the only means of reproduction where vegetative propagation is not easy or economical, as in pawpaw.
3. When breeding new fruit varieties the hybrids are first raised from seed, thus employing the sexual method of propagation.
4. Propagation from seed has been responsible for the production of some chance seedlings of great merit which have been of great benefit to the fruit industry. As an example, the 'Washington Navel' variety of sweet orange is reported to have arisen as a chance mutant from a seedling orange in California, USA.
5. In many species, particularly of citrus and mango, seedling (nucellar) trees breed true to type, due to polyembryony or nucellar embryony.
6. Seedlings are easier and cheaper to raise than vegetatively produced plants.

7. Seedling trees are generally free from virus diseases.

There are also several disadvantages of sexual propagation in fruit crops:

1. Seedling trees are not as uniform in their growth rates, yielding capacities and fruit quality as are, for instance, grafted trees.
2. Seedling trees take longer before bearing a maiden crop than do grafted trees.
3. Seedling trees invariably become too large for economic management, particularly in terms of greater cost of harvesting, pruning and spraying.
4. It is not possible to perpetuate the exact characters of a superior selection through seed. Therefore, to multiply superior hybrids or chance-seedlings, vegetative propagation has to be employed.
4. It is not possible to utilise the modifying influence of root-stocks on the scion as in vegetatively propagated fruit trees.

Vegetative methods

In the commercial production of most fruit crops, vegetative propagation methods are used to achieve better results. Essentially, three main types of vegetative material are used; side shoots, cuttings and graftings.

Side shoots, as used in such fruit crops as bananas, plantains, date palms and pineapples, do not usually require nursery care and can thus be planted straight into the field.

Cuttings which are planted beneath the soil (as in sugarcane) may also be planted directly in the field. Cuttings which are planted exposed to the air (for example grape vine) need special care and protection from drying out. This is often achieved by shading to reduce water loss. In stem cuttings, new shoots need to develop. Where root primordia are already present, they will grow naturally; otherwise, rooting can be promoted by the use of growth hormones. Propagation by leaf cuttings is generally used in such crops as tea.

Closely related to propagation of fruit trees by cuttings is the method often referred to as 'marcotting' or 'layering' and popularly used for such crops as mango and mandarin (Figure 5.15). To produce a marcott, a healthy side twig or branch, of about the thickness of a finger, is selected. The bark is removed for about 0.5 cm leaving the wood exposed. A twig is then bound to the exposed area in a ball of good topsoil, held in place with the aid of plastic sheeting and string. After some

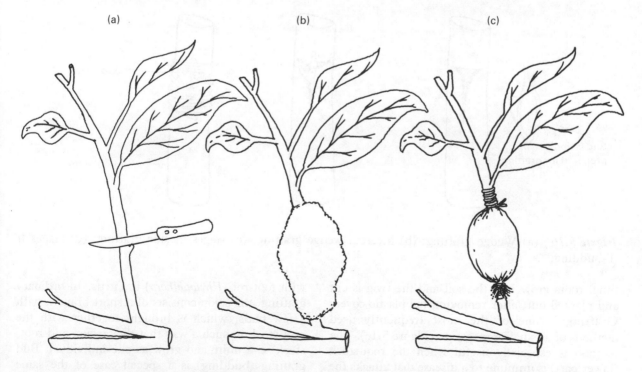

(a)   (b)   (c)

Figure 5.15 Stages in marcotting (layering).

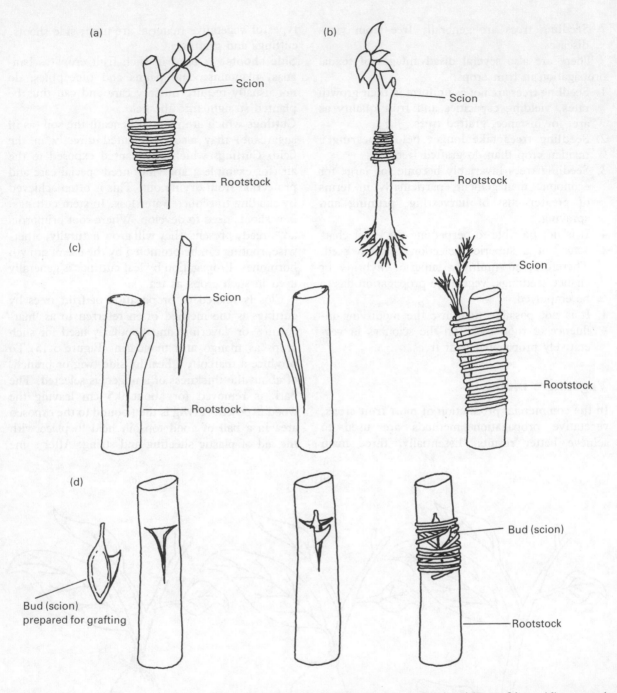

Figure 5.16 (a) Wedge grafting; (b) inverted wedge grafting; (c) stages in side grafting; (d) stages in T-budding.

time, roots grow into the soil and the twig is cut and planted out, after removing the plastic cover. **Grafting** is one of the most frequently used methods of asexual propagation (Figure 5.16). The method is especially useful when the root-stock (lower part) is immune to a disease that attacks the roots of the scion (upper part), as for example

with foot-rot (*Phytophthora*) in citrus. In ordinary grafting, the scion consists of a short branch, with a few buds, which is brought together with the root-stock in such a way that they (scion and root-stock) can unite and grow as one individual. Bud grafting (budding) is a special case of the same technique with the scion reduced to only one bud.

The method involves grafting the bud from a high-yielding and healthy tree of desirable qualities onto a normal seed-grown stock seedling. Budding is usually done at the green-wood stage when the budwood and stock seedling are about 1 cm thick. The bud is cut off, skillfully inserted under the bark of the stock seedling and taped with transparent plastic material. It is important that the bud is placed the right way up and that the inner face is firmly held in contact with the outer wood surface of the stock seedling.

It should be noted that for graft compatability, the stock and the scion must often belong to the same botanical species. However, many exceptions are known. As an example, *Citrus* spp. are not only easily grafted on each other, but are also successful on closely related genera, such as *Poncirus* and *Fortunella*, as well as such distant relatives as *Citropsis* and *Swinglea*.

Advantages of propagating fruit crops by vegetative methods are as follows:

1. Asexually propagated fruit trees are true to type and as such, are generally more uniform in growth rates, yielding capacities and fruit quality than plants sexually propagated by seed.

2. As a result of uniform fruit quality, harvesting and marketing are made easier.

3. In the case of certain crops with seedless fruits, such as in bananas, pineapple and some varieties of grape vine, the only means of propagation is by asexual methods.

4. Fruit trees propagated by vegetative means invariably come into bearing sooner than seedling trees.

5. By budding or grafting desirable, but disease-susceptible, fruit varieties onto resistant root-stocks, such varieties are able to grow without the serious effects of the disease. In citrus, for example, the sour orange variety is often a suitable root-stock where foot-rot is a problem. Similarly, the use of lemon and mandarin root-stocks is desirable in order to avoid infestation by *Tristeza* virus disease.

6. The modifying influence of root-stocks on scions can be utilised. For example, where a certain variety does not grow well in a particular type of soil or climatic regime, it is possible to adapt it to such unfavourable conditions by grafting it onto a suitable root-stock.

7. By vegetative propagation and the appropriate use of different root-stocks, it is possible to regulate the size of the tree, the fruit quality, the precocity and other characteristics.

8. Cross-pollination can be effected by grafting shoots of suitable varieties (pollinators) onto the branches of varieties incapable of self fertilisation. For example, some varieties of avocado pear, almond and plum which require cross-pollination can be made to bear if pollinating varieties are grafted on their branches.

9. The technique of grafting can be successfully used to encourage healing of wounds to trees caused by rodents, rabbits or farm implements.

10. Composite trees, each bearing several varieties or types of fruit, can be raised using grafting. For instance, on one stock it is possible to raise several varieties of sweet orange or even different citrus fruits, such as grapefruit, pummelo, mandarin and sweet orange.

11. By the use of vegetative propagation, it is possible to correct the initial mistake of using an inferior or unsuitable variety, by later re-grafting with a desirable variety, without removing or re-planting the trees.

Vegetative propagation is not without drawbacks, including the fact that the plants, particularly the budded ones, are generally not so vigorous and are shorter-lived than sexually produced plants. In addition, the transmission of viral diseases through vegetative material and the tools used, such as budding knives and tapes, can be quite a problem. The vegetative techniques of fruit crop propagation also cannot be used to create new varieties.

Crop variety and seed quality

Since the change from being a wandering gatherer of food to a settled existence as a cultivator of land, man has made use of some 3000 plant species for food, of which over 150 species are now cultivated commercially. Probably as a consequence of the search for better plant types, the tendency over the centuries has been to concentrate on relatively fewer species. In indigenous farming practices, farmers have had to select their own planting stocks from genetically variable populations and in the process have created a rich heritage of highly diverse germplasm, particularly among the more important crops which feed most of the world, including all tropical inhabitants.

As a result of the demands for increased food production and improved food quality, efforts have been continually directed towards exploring the full genetic potential of cultivated crops by employing modern scientific techniques of crop improvement. This continuous generation and release of high-yielding and adaptable crop varieties and hybrids has enabled several substantial increases in crop production relative to unimproved cultivars.

Characteristics of good cultivars

The choice of seed is one of the rare factors of production over which the farmer can exercise considerable control. For almost every major crop grown, a wide choice of varieties is available. In no case is the same variety the best under all conditions of growth.

To satisfy most farmers, a crop variety must be capable of yielding well under normal conditions and should produce a high-quality product. Additionally, a good variety should consistently produce well year after year and give a respectable yield even during years when growth conditions are adverse. This implies that a good crop variety must be resistant to, or at least tolerant of the pests and diseases which are most likely to be problematic, and must be well adapted to the climate and soil of the region. A high-yielding sorghum variety, for instance, that might be destroyed by drought or short periods of dry spells before maturity once in every four years would not be as desirable as a slightly lower-yielding and early-maturing variety that would consistently produce a mature crop. Similarly, a low-yielding and rosette-resistant groundnut variety would be preferable to a high-yielding and rosette-susceptible variety in an area where the disease is a problem.

Selecting good seeds

In choosing seed for planting it is essential to exercise care to ensure that only material of a superior and well-adapted variety is grown. Important characteristics of good seed are:
(a) It should be pure, in that it must be free of seeds of other varieties. Since it is generally difficult to identify different varieties or species by seed characteristics, the best guarantee of varietal purity is the integrity of the seed grower and seed dealer.
(b) It should germinate rapidly in order to ensure strong and vigorous seedlings.
(c) It should be large (for the variety) and plump.
(d) It should contain no noxious or objectionable weed seeds.
(e) It should contain no insects, insect eggs or disease spores in or on the seed.
(f) It should be uniform in size and shape.
(g) It should be free of stones, chaff or other forms of foreign matter.

The best way to obtain seed of good quality is to purchase only certified seed where the service is available. Certified seed is produced under conditions designed to ensure the grower the very best quality. The seed is normally inspected before it is sown. During growth, pollination is carefully controlled to avoid contamination of seed. The harvested seed is tested for germination as well as for the presence of disease organisms, pests and weed seeds.

In the absence of an agency for production and certification of seeds, a farmer may grow his own seed by following the principles outlined above.

Planting and sowing

Irrespective of the quality of seed or planting material, if a crop is sown or planted too early or too late, or if the plant stand is too thick or too thin, it is unlikely that a good crop would be obtained in terms of both high yields and quality. In order to maximise production, it is essential that the principles governing the choice of planting date and seeding rate be observed.

Seedbed

The seedbed is the place where the seeds germinate and the medium from which the plants, through their roots, secure moisture and mineral nutrients. It is therefore desirable that the seedbed should be in such a condition as to provide an abundance of moisture, nutrients and air. It must also allow full penetration of plant roots.

The best time to prepare the seedbed is before the crop is planted. In the humid tropics, a desirable seedbed generally is one that is fine, compact and without trash or growing vegetation. However, in areas that are subjected to wind erosion, clods

Figure 5.17 Ox-ridging with hilleston rotating lines, Zaria, Nigeria.

and crop residues on the surface tend to reduce the damage. In drier regions, one of the primary considerations in seedbed preparation is the conservation of moisture, a requirement which is less critical in humid areas.

Throughout the tropics, seedbeds are traditionally in the form of moulded ridges, occasionally on hills and infrequently on flat-cultivated land. Growing crops on ridges is closely associated with shifting cultivation and subsistance farming, but there is no known direct relationship between ridge-cropping and the climate, soils or topography of an area. In Nigeria, land is ridged from the wettest to the driest parts of the country, irrespective of whether soils are heavy or light and whether the topography is sloping or gentle. In traditional farming systems in the tropics ridge-making is invariably undertaken using hand-hoes, an operation which is slow and laborious. In mechanised or partially mechanised systems, appropriate tractor-mounted or animal-drawn implements are employed for ridging and other seedbed operations (Figures 5.17 and 5.18). Therefore, partly dictated by requirements of certain crops and by

socio-economic circumstances, the practice of ridge-cropping has several inherent advantages.

(a) Ridging saves manual labour, since only about half the soil is worked.

Figure 5.18 An animal drawn toolbar fabricated by a village blacksmith in Dankerzaure area, Nigeria. It can take mouldboard ridge and cultivator attachments.

81

Table 5.4. Mean yield of sorghum and maize with flat and ridge cultivation in Burkina Faso

Crop	Flat	Ridge
Sorghum	1126	1090
Maize	2791	2323

Source: Nicou, R and C. Charreau (1985). Soil tillage and water conservation in semi-arid West Africa. Appropriate Technologies for Farmers in Semi-Arid West Africa. A.W. Ohm and J.G. Nagy (Editors), 9–32.

(b) The relatively rich topsoil, with its ash and plant residues, is deposited close to the active root zone.
(c) The ridge provides a more favourable environment, particularly for crops which form their economic yield underground.
(d) Ridges facilitate free drainage and conservation of soil moisture, particularly if cross-tied.

Generally, there is no significant difference between the yields produced in flat cultivation and in ridging (Table 5.4), although in certain instances ridging is superior (Table 5.5). However, regardless of the configuration of the seedbed or its condition on the surface, it is essential that it be firm and reasonably compact beneath. A seedbed that has been properly prepared allows free movement of water and air so that plants may grow well and produce good yields.

Time of planting

Each crop species and crop variety should be planted at a time when the different stages of its development are best correlated with the expected weather pattern. In general, temperature and moisture are of special importance, as is light for many crop species. The presence of weeds, plant diseases and insects should also be considered when selecting the time of planting.

Unlike temperate regions where crop growing periods are mostly limited to the summer months, in the tropics temperature regimes are generally favourable for production for most of the year, especially if moisture conditions are favourable. Even at locations where temperatures fall below optimum for normal crop growth for two to three months in the year (as in the savanna), such cold-loving (C3) crops as wheat and barley (Table 3.3) can be grown successfully.

In general, the earlier a crop is planted at the onset of established rains, the better it is likely to perform agronomically. Since the best yields are dependent upon good vegetative development of crop plants prior to the beginning of fruit and seed development and maturation, it follows that the longer the period of vegetative growth the greater the chance of attaining an increased yield. Of course, this principle of early planting needs to be applied on the basis of sound judgement. For example, the early planting of a short-duration crop variety may cause it to mature during the rains with a consequent reduction in yield and quality, particularly in the absence of drying and other post-harvest handling facilities. In such a situation, the best combination of high grain yield and quality is achieved when planting is timed such that the produce matures and is ready for harvest at the end of the rains when problems of post-harvest handling are reduced.

Irrespective of the crop variety, the seedling

Table 5.5. Effect of tie-ridging on crop yields, Tanzania

Station	Soil	Mean Rainfall (cm/year)	Period of experiment (years)	Crop	Mean yield (kg/ha)	
					Flat	Tie ridges
Lubaga	Heavy, drainage impeded	79	12	Cotton	590	817
Lubaga	Light, free draining	79	9	Sorghum	823	1173
Ukiriguru	Light, free draining	84	6	Millet	693	745
Ukiriguru	Light, free draining	84	7	Cotton	690	817

Source: Empire Cotton Growing Corporation, 1951–56.

stage is the most critical phase in the life of the crop plant. Seeding should therefore be carried out at a time of the year when soil moisture and temperature are ideal for seedling emergence. For successful dry season cultivation of C_4 plants (Table 3.3) under irrigation and especially in the West African region, planting must be delayed until after the cool Harmattan period, and only done when the soil is warm enough for rapid germination to take place.

The influence of day-length may be illustrated by the cultivation of cowpeas which normally flower early and produce pods only when grown at a day-length of 12 hours or less. As such, the optimum planting time in Nigeria varies from July/August at latitudes 9–13°N to September at 7–9°N. It remains generally true also that any delay in the planting of crops beyond the optimum period invariably results in drastic reductions in yield, as is evident in Table 5.6.

Table 5.6. Effect of delayed sowing on the yield (kg/ha) of various crops grown under experimental conditions at Samaru, Nigeria

Crop	Sown on time (kg/ha)	Sown late (kg/ha)	Yield decrease (%)
Millet	2460 (late May)	278 (late June)	89
Sorghum	4152 (mid-May)	1883 (late June)	55
Maize	5862 (mid-June)	2616 (late July)	53
Cotton	2790 (mid-June)	1302 (mid-July)	53
Groundnut	2475 (early June)	1913 (late June)	31
Cowpea	1584 (mid-July)	862 (late July)	46
Soyabean	1758 (mid-June)	1063 (mid-July)	40
Wheat	4641 (mid-Nov.)	1888 (mid-Dec.)	59

Depth of seeding

The depth of seeding and the time of planting are interrelated. For a given crop, plantings done later than the set date must necessarily be deeper because soils, especially in the surface layers, tend to dry out more rapidly as the season progresses. In areas where weather conditions are especially difficult to predict, seeds of some crops are planted at varying depths. For example, if the weather remains relatively dry, seeds that are sown deeply, in contact with moist soil, generally germinate whereas in shallow plantings the partially germinated seedlings often dry up and die. On the other hand, with

adequate rainfall the shallow-planted seeds germinate whereas the deep-planted seeds rot.

The size of seed is generally related to the depth of planting. Sown at the same depth, seedlings from large seeds would normally emerge earlier than those from small seeds. Similarly they emerge from greater depths in sandy than in clayey soils, and sooner in warm than in cold soils. Because there is a limit to the depth at which seeds may be placed in order to enable the growing shoots to reach the soil surface, it is generally advisable not to bury seeds more than three or four times their diameter. The relationship between the depth of planting and the rate of seedling emergence in maize is shown in Figure 5.19.

Seed rate

For a given area of land, the seed rate of a particular crop species or crop variety is governed by

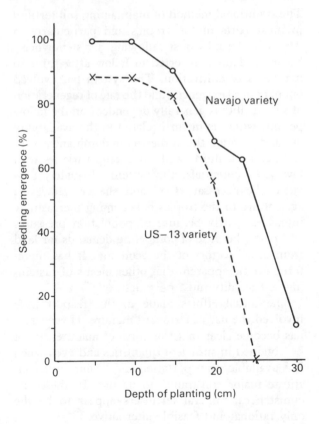

Figure 5.19 The relationship between the depth of seeding and the rate of seedling emergence in Najavo Indian maize and US-13 yellow dent maize at Urbana, Illinois (Dungan *et al.*, 1958).

83

the ultimate plant population density required to maximise yield. Seeding rate is especially critical for crops grown in rows, whether on ridges, hills or on the flat, in contrast to crops which are sown broadcast.

Where moisture is limiting, it is advisable to seed at low rates, whether dealing with broadcast or row crops. For crops which do not tiller, fertile soils and soils with a high water retention capacity are invariably able to support higher plant densities than relatively infertile and free-draining soils. For crops which tiller profusely, the seeding rate should be higher on infertile than on fertile soils with a high water retention capacity. In general, tillering is greater when the seeding rate is low and hence tends to compensate for the lowered rate.

Fertiliser use and management

The traditional method of maintaining soil fertility in most parts of the tropics and particularly in Africa has been by bush fallowing, a system whereby arable land is reverted to fallow after three to five years of cultivation. The system has evolved out of natural exigencies and the rate of regeneration of soil fertility is generally dependent on the fallow period, which is in turn related to the availability of land. Under the hitherto predominantly subsistence agriculture, and using crop varieties with low yield potentials, the system adequately provided the necessary food and shelter. However, agriculture in the tropics is becoming increasingly intensive, partly because of population pressures and partly because of competing demands for land from other sectors of the economy. It has therefore become apparent that other means of sustaining soil fertility must be sought.

The initial efforts made in this respect have involved the use of farmyard manure. However, it has become clear that this form of manure cannot be obtained in sufficient quantities and even when it is available, transportation and labour costs constitute major constraints to its use. In these circumstances, mineral fertilisers appear to be the only rational and feasible alternative.

Although, the benefits of fertilisers are well appreciated by many farmers in the tropics, high costs and inadequate and inefficient transportation and distribution systems have limited their use. Nevertheless, there has been a steady increase in

Table 5.7. Fertiliser consumption in Nigeria, 1960–1985 (every 5 years)

Year	N (tonnes)	P_2O_5 (tonnes)	K_2O (tonnes)
1960	921	716	1700
1965	1700	1400	800
1970	4482	5558	642
1975	22037	19987	8093
1980	92537	56914	23980
1985	196800	127460	65280

Source: FAO, 1960–86, Annual Fertilizer Reviews, FAO, Rome.

consumption over the years. Table 5.7 gives the mean annual consumption of fertilisers in Nigeria for the period 1953–85.

Many tropical soils are deficient in nitrogen and phosphates and so most of the fertiliser consumption has centred around these two nutrients. Potash fertilisers have been used on a much lower scale, mostly on the humid tropical oxisols and ultisols which have been heavily leached of their nutrients.

Factors affecting fertilizer use

A number of factors are known to influence the usage of fertilisers by farmers, including the following:

Crop factors

Fertilisers cannot be beneficial unless the crops on which they are used are responsive to them. Different crops remove different amounts of the various nutrients. Legumes, for instance, require large amounts of phosphorus and sulphur whereas grain crops generally require proportionately more nitrogen and potassium. The removal of nutrients by a crop is influenced not only by the cultivar, but also by the climate, the soil nutrient status and cultural practices. Genetically improved cultivars, particularly when they are high-yielding, normally respond more to increased fertiliser doses relative to unimproved, local cultivars.

Approximate amounts of nutrients removed per tonne of produce of some important crops in the West African savanna are shown in Table 5.8. While little can be done about the removal of nutrients in the economic parts of a crop, the overall loss of nutrients from the soil can be

Table 5.8. Amounts of various nutrients removed from the soil, per tonne of produce of selected tropical crops

Crop and produce	N	P	K (kg/tonne)	Ca	Mg	S
Rice						
Grain	10–15	2.6–3.5	2.0–2.6	0.4–0.6	0.2–1.5	na
Stover	5–12	0.9–1.3	5.0–29.0	1.1–2.5	1.5–3.4	1.2–2.5
Maize						
Grain	16–19	1.7–3.5	2.5–3.9	0.1–0.4	0.6–1.2	1.2–2.2
Stover	6–10	0.4–1.1	5.4–21.0	1.8–4.2	1.5–3.3	1.0–1.5
Sorghum						
Grain	14–20	1.7–3.1	2.5–3.3	0.1–0.6	1.0–1.9	1.6
Stover	3–6	0.4–1.3	5.8–24.0	1.2–3.2	0.6–2.1	1.0
Millet						
Grain	18–25	1.5–3.1	3.5–5.0	0.2–0.6	0.8–1.1	1.0–2.5
Stem	5–9	0.2–0.4	8.3–27.0	1.1–2.1	0.8–2.7	0.6–3.0
Leaf	7–10	0.4–0.9	8.3–25.0	2.9–7.9	3.0–6.0	0.8–4.0
Groundnut						
Kernel	20–50	3.3–4.8	6.7–8.3	0.3–0.9	1.5–2.1	2.2–2.7
Shell	5–11	0.3–1.1	5.0–15.0	0.5–2.9	0.5–1.2	0.7–1.7
Haulm	8–20	1.1–3.1	10.0–25.0	5.0–9.3	3.6–6.6	1.7–2.0
Yam						
Tuber	12–20	1.0–2.0	15.0–21.0	3.9–6.1	0.8–1.2	na
Tops	7–14	0.4–1.0	8.3–25.0	8.9–13.2	1.7–3.6	na

na = Data not available.

minimised if the nutrients contained in the non-economic parts are returned to the soil. The removal or burning of crop residues normally results in the loss of nutrients, particularly nitrogen and sulphur. The amount of potassium contained in crop residues is variable and reflects the fact that there is always a risk of 'luxury' uptake of this nutrient. The return of crop residues to the soil should keep the loss of potassium to a minimum.

Soil factors

The ability of soils to supply the necessary plant nutrients differs greatly from one place to another and at different times. In general, the fertility of a cultivated soil declines with time. Such soil physical properties as depth, texture and structure all contribute to its productivity. Soils inherently differ in their productive potential. Large applications of fertilisers can be profitable on soils that have a high productive potential but which are low in fertility.

The fertiliser requirement and response level

also depend on the yield potential of the cultivar. The higher the yield potential of the cultivar, the higher the fertilizer requirement. Figure 5.20 is a schematic presentation of the response levels of four cultivars, with 'A' having the highest yield potential and 'D' the lowest.

An increase in the amount of a nutrient applied to the soil may or may not increase crop yield. In general, if the amount of the nutrient in the soil is inadequate and crop growth and yield are limited by it, the fertiliser application will lead to a proportionate increase in yield in the zone of poverty adjustment. However as the supply of the nutrient is increased, a point is finally reached where maximum yield is attained and further increases in yield are uneconomical. Plants may, however, continue to take up the nutrient from the soil. This is the range of luxury consumption (Figure 5.21). In some cases, further applications of the nutrient may even lead to yield decline. For instance, a large or excessive application of potash may lead to magnesium deficiency and ultimately to a decrease in yield. However, this is not a

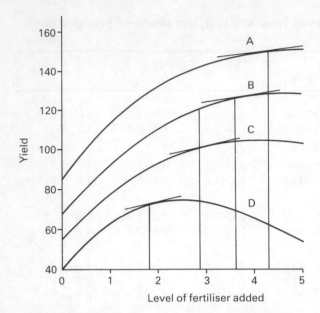

Figure 5.20 Diagram to show the crop response pattern to fertiliser application in relation to the yield potential of the cultivar (Tisdale, 1966).

common occurrence, except where the magnesium level in the soil is low or close to deficiency level.

Fertiliser recommendations may either assure the best production practices for a given region (in which case slight adjustments may be made as necessary for any specific locality within the region) or they may be based on the average production practice and management capacity of the farmers.

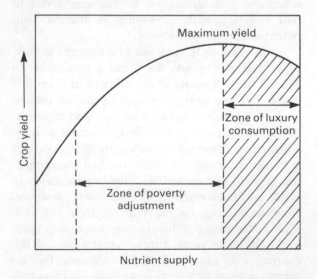

Figure 5.21 Diagram to show the effect of nutrient supply on crop yield.

The first approach is not economically safe for small-scale farmers in the tropics, most of whom are handicapped by a lack of capital and labour, often resulting in untimely farming operations. The second approach is fraught with the risk of discouraging progressive farmers whose goal is to maintain the plant nutrients at levels that will give them maximum returns. A compromise would be to assume the average level of management and make an allowance for the progressive farmers.

Climatic factors

Soils in areas of low rainfall generally do not lose nutrients by way of leaching and so their inherent fertility is relatively high. This generalisation is not necessarily true for sandy soils and where the intensity of rainfall is high. The limited amount of water available in such areas does not justify the application of high fertiliser rates. Soils in the humid tropics, on the other hand, lose their nutrients through weathering and leaching. Water supply is usually adequate for high crop production, which is limited by nutrient supply, therefore necessitating the use of fertilisers.

Economic factors

The use of fertilisers is normally enhanced when their price is low and is decreased when their price is high. Crop prices have the opposite effect, in the sense that a high price for the crop gives a profitable return from increased fertiliser applications. Increases in crop yield from fertiliser applications follow a curve of diminishing returns (Figure 5.21). Small applications of the necessary fertilisers result in the greatest return per unit of nutrient applied. Additional amounts of fertiliser give progressively smaller increases in yield. Eventually, a point is reached where the last increase in fertiliser barely increases the yield enough to pay for its cost. It is good management practice to refrain from adding more fertiliser than is necessary to reach the break-even point.

Management factors

Farm managers normally choose the input−output level at which to operate. Increased crop output usually requires increased fertiliser input. Top yields depend on many factors and the good manager learns how to either control or adjust as many of these factors as possible. Among the important factors are soil type, climate, the type of crop and the crop variety. Also of importance for the efficient use of fertilisers is the cropping history and the previous crop management practices. In addition, irrigation, drainage and erosion control are important considerations.

Methods of applying fertilisers

Plants can be expected to gain full benefit from applied fertilisers only if the nutrients supplied are within the feeding range of their roots. The soluble constituents of fertilisers diffuse through the soil, mostly up and down and only slightly in the horizontal plane. The method of application must therefore ensure an even distribution of the fertiliser in a moist soil in which the plants are growing. The various methods of fertiliser application include the following:

Broadcasting

In broadcasting, the fertiliser is spread uniformly over the field during land preparation and before planting. Of all the fertilisers, phosphates are best applied by broadcasting, usually prior to final seedbed preparation. This method gives fairly good results for a crop such as millet which is planted in narrow rows. The method is also suitable for crops whose seeds are broadcast. Although fertiliser broadcasting enables seedlings to get a good start, the method has its drawbacks in that weed growth is also stimulated, allowing weeds to grow rather rapidly.

Placement

By this method the fertiliser is placed in bands or in localised areas along the rows at a safe distance from each plant, but at a distance where the nutrients are within the reach of the root system. As a general rule, fertilisers should be placed further away from the plants or row of plants in clay soils than in sandy soils. Fertiliser placement is particularly suited to crops such as cotton, tomato and potato which tend to be sensitive to direct contact with fertilisers. The numerous advantages of fertiliser placement include the following points:
1. Because the fertiliser comes into contact with the minimum number of soil particles, losses

such as the fixation of phosphorus are minimised.
2. Fertilisers are normally within reach of the root system and the availability of nutrients is thus enhanced.
3. Fertilisers placed in bands near crop plants contribute few nutrients, if any, to weed growth near the surface.

For crops which require nitrogen, the full amount is normally split-applied. The second dose is usually applied by placement on the soil surface close to the plants when the crop is about three to four weeks old. This is often referred to as top-dressing.

A modified method of fertiliser placement is by the use of drills, at the same time as the seed is (machine) drilled (Figure 5.22). The method gives good results with wheat, barley, maize and other cereal crops which tend to withstand contact with fertilisers. Rates of nitrogenous fertilisers higher than 25 kg/ha may not be applied in this way because of the risk of burning the seed.

Split application

This refers to the application of fertilisers, particularly nitrogen, in two or more doses. The basic advantage of this method is related to the physical properties of nitrogenous fertilisers, particularly in being relatively more soluble than other fertilisers.

Figure 5.22 An animal-drawn planter and fertiliser applicator which can be pulled by a pair of oxen or donkeys. It can apply seed and fertiliser simultaneously.

If the entire nitrogen requirement of a crop for the growing season is applied in one dose, most of it would be lost through leaching. In addition, heavy applications of nitrogenous fertilizers lead to excessive vegetative growth at the expense of the economic yield. However, when split-applied, the crop is initially provided with suffent nitrogen to enable good establishment and growth, while the second application is required for later growth and development. In Nigeria, for instance, the practice is to apply fertilisers in two splits for most crops: the first dose at two to three weeks and the second at six to eight weeks after planting, depending on the crop.

Basal and maintenance fertilisation

Basal application of fertilisers is often aimed at substantially raising the fertility of the soil with one large application, the amount applied being dependent on the extent of the deficiency. In contrast, maintenance fertilisation aims at balancing only the nutrients removed by the crop and the amounts lost by leaching and erosion. Maintenance fertilisation endeavors to maintain the soil fertility level established by basal fertilisation.

Optimal fertiliser efficiency

To ensure that a high proportion of a fertiliser is used by crop plants, it is important to observe the following points:

(a) Avoid single large fertiliser applications, particularly of nitrogen or on sandy soils where rainfall could cause leaching. The latter problem is particularly important in the humid tropics, with a mean annual rainfall of more than 1200 mm. In such situations split applications of nitrogen are advocated.

(b) Avoid top-dressing (surface placement) of urea fertiliser to reduce losses through volatilisation of ammonia. If broadcast or place-banded, the urea should immediately be covered with soil.

(c) When a fertiliser is applied at planting, care should be taken to avoid direct contact with the seeds.

(d) It is best to broadcast phosphate fertiliser and incorporate it into the soil before planting. However, when the phosphorus-fixing capacity of the soil is known to be high, application by hand-placement is more effective.

(e) Where the soils are poorly drained or water-logged, it is advisable to use the ammonium forms of nitrogen fertilisers, since nitrates are readily denitrified in such conditions, owing to the enhanced activities of anaerobic microbes.

Weed management

Well over 250 weed species have been reported to be of importance world-wide. As a consequence of the large-scale problem posed by weeds, more than half of the cultural practices required for crop production in the tropics are devoted to the control of weeds. On average, yield losses caused by unchecked weeds exceed those of other agricultural problems. In Nigeria, between 10 and 87% of the losses of various crops are attributable to weeds. Indeed, complete loss has occasionally been observed in sorghum due to infestation with the parasitic weed *Striga hermontheca* (witchweed).

While part of the losses caused by weeds are the direct result of competition for limited resources, such as water, nutrients, light and space, further losses are caused indirectly, such as losses due to the presence of insect pests and disease organisms in the weeds, interference with certain farm operations and the reduction in quality of the final produce. The magnitude of crop losses from weeds depends on a number of factors, including the crop species, the crop variety, the ecology of the area of production, the level and type of production technology, the cropping system, and such socio-economic considerations as the resource base of the producer.

Approaches to weed control

Being an integral part of the crop production process, the management of weeds cannot be treated in isolation from other production practices. Indeed, for successful weed control technology it is essential to maximise other production practices, including the choice of cultivars, land preparation, sowing dates, plant population densities and the efficient use of fertilisers. Depending on the nature of the weed problem and the system of crop production in practice, approaches to weed control vary from eradication to prevention or containment.

Eradication

Eradication of weeds involves a programme to completely eliminate a problem weed from a particular area. Eradication is common practice in non-agricultural situations such as on industrial land, in sewage systems and canals, and on power line tracks, railway lines and motorways. Usually chemicals are applied at high rates to prevent subsequent regrowth of weeds.

In tropical agriculture, control of weeds by eradication is used for such noxious species as *Imperata cylindrica*, *Cynodon dactylon*, *Cyperus rotundus*, *Cyperus tuberosus* and *Digitaria ciliaris*, all of which can be perpetuated by even tiny portions of living tubers, rhizomes or stolons. Some annual weeds, including *Acanthospermum hispidum* and *Vernonia danociflora* and parasitic species such as *Striga hermontheca*, *Striga genaroides* and *Alectra volgelli*, which reduce the efficiency of production practices and lower crop yields even at low population densities, are included as species for eradication.

Prevention

Prevention of weed growth is the most commonly adopted approach to weed control in tropical agriculture. In this approach, weeds are either removed or prevented from emerging during critical periods of crop growth so that competition is avoided between weeds and the crop. The prevention of weed infestation often involves the integration of various control methods, including the use of chemicals, cultivation and other cultural practices.

Containment

When containing the effect of weeds, they are allowed to grow with the crop, but at rates that would not cause significant economic damage. The approach usually involves the manipulation of the production factors that favour good crop establishment, including the choice of cultivar, land preparation, time of sowing, crop density and the use of fertilisers.

Weed control strategy

The serious crop losses suffered annually from weed infestation necessitate the development of effective control strategies. In evolving a weed control strategy the primary objective is the manipulation of weeds such that their populations are kept at sub-threshold levels during critical stages of crop growth.

As described in the first chapter, the tropical environment has various features which make the weed control strategies used in temperate regions unsuitable. Firstly, because of the distinct division of the year into wet and dry seasons, the rapid and successive emergence and growth of weeds is greatly favoured. Weed seeds normally remain dormant during the dry season, but when the rains begin the dormancy is broken at different stages throughout the season. In this way a situation of continuous weed infestation results, in which emerged weed species grow rapidly and compete with crop plants. Secondly, the high intensity of rainfall over a relatively short period makes the timing of land cultivation difficult. In this situation, some of the applied soil residual herbicides are leached below the weed root zone, thereby reducing their selectivity and efficiency.

A characteristic of tropical soils which influences the strategy of weed control relates to their low content of organic carbon. In general, low levels of organic matter do not favour intensive cultivation, as the practice is liable to rapidly deplete the scarce soil organic matter and render the soil open to erosion. Except in soils with a high clay content, low organic matter does not normally favour the use of herbicides at high rates without reducing their selectivity in crops, primarily because of fewer absorption sites in the soil. Finally, the performance of herbicides is further reduced in tropical soils by the high temperatures which favour rapid microbial degradation.

Important ingredients of a successful weed control strategy in the tropics must include the following considerations:

(a) The control methods used should relate to the cropping systems of the area, including the rotation practices, cropping pattern and cropping sequences.

(b) The eco-edaphic situation of the area is of particular importance when determining its floristic composition as well as the degree and level of development of weeds and their subsequent potential for infestation.

(c) Any weed control strategy adopted should take note of, but be an improvement on existing practices, and should be adaptable to all categories of producers in accordance with their socio-economic backgrounds.

(d) Measures taken to contain weeds should be ones that persist long enough to avert weed competition with crops as well as prevent weed interference with farm operations.

(e) An effective weed control strategy should also ensure that weeds previously considered minor do not assume increased importance as a result of their being resistant to or tolerant of the various methods employed. In other words, the method should be self-perpetuating, self-defining and self-adjusting such that appropriate measures can be combined to control changing weed problems.

Chemical weed control

Chemicals used to kill weeds or other plants are referred to as herbicides. Herbicides hold great potential for solving the problem of weeds which is presently the greatest problem in crop production in the tropics.

Herbicides are broadly grouped into selective and non-selective types. In contrast to non-selective herbicides which invariably kill all vegetation, including weeds and crops, the selective types are generally effective against one group of plants only, while remaining largely unharmful to other species. Selectivity of herbicides is a relative phenomenon which depends, at least in part, upon the method and timing of the treatment, the dosage used, the properties of the chemical and the physiological condition of the plant.

Mode of application

Herbicides are usually applied using one or a combination of the following approaches:

1. Pre-planting incorporation of the chemical, about two days before sowing. In general, application of the herbicide is immediately followed by shallow cultivation to ensure that it does not loose its activity due to volatilisation. Suitable mechanical equipment for incorporating herbicides before planting include rolling cultivators and the spike toothed harrow.

2. Pre-emergence application involves applying the herbicide before the emergence of both crops and weeds. Most soil residual herbicides are applied pre-emergence, often one to three days after sowing.

3. Post-emergence application occurs after the crop or weed has germinated. This mode of application is usually done to increase the selectivity of the herbicide.

Some foliar translocated herbicides are applied post-emergence to the weed, since translocation is usually with photosynthetic products, through the phloem to the underground organs. It is important to note that whatever mode of application is employed, the effectiveness of the herbicide is largely dependent on using the correct quantity of a properly mixed spray. About the only way to ensure this is by proper calibration of equipment, particularly to indicate the amount of herbicide released per unit area of land.

Although some herbicides are applied broadcast in the dry form, liquid formulations, using sprayers, are more common. Various types of equipment are used for liquid applications, depending on the volume to be applied (Figures 5.23 to 5.25). For example, high volume applications involving 150–5000 litres/ha employ the use of Knapsack sprayers or tractor-mounted Boom sprayers, especially on large farms. Low volume applications usually involve 50–156 litres/ha. Very low volume applicators, supplying 10–50 litres/ha, invariably have spinning discs which break the liquid into fine droplets of a uniform size.

Pest and disease management

The term 'pest' is used to include any organism which is detrimental to a crop. The tropical climate provides more favourable conditions for the development of pests and diseases than are found in temperate countries. Damage caused to crops by pests and diseases is therefore generally greater, as are the challenges for their effective management.

Being inherent components of the agro-ecosystem, pests must be managed on a continuous and well informed basis if the damage they cause is to be maintained below an economic threshold.

Pest management criteria

Before any feasible pest management decisions are made it is essential to satisfy the following basic criteria:

Figure 5.23 A knapsack compression sprayer. This is a high volume pesticide sprayer which is characterised by the production of a wide range of droplet sizes.

Figure 5.24 A tractor-mounted boom sprayer is a low volume multi-purpose sprayer for liquid insecticides, fungicides and herbicides. It has potential for use on flat surfaces and by small-scale farmers.

Figure 5.25 A battery operated, hand held, spinning disc, ultra low volume sprayer. It has a plastic spray head, a small electric motor which drives a rotating disc, a liquid reservoir and a long handle. It is used mainly where there is insufficient water for high volume sprayers.

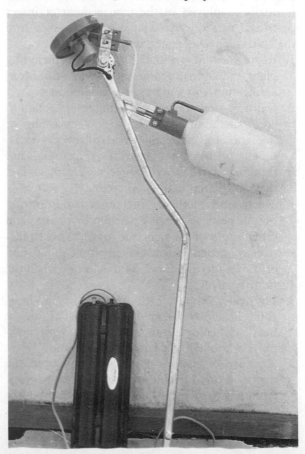

(a) It is important to correctly identify the pest and to provide the ecological information necessary for the development of control measures.

(b) It is important when determining the mode of control to have a proper definition and clear understanding of the agro-ecosystem, particularly as it relates to the capacity of the pest to move from place to place.

(c) It is necessary to have a determination of the socio-economic benefits to be derived from pest control particularly when viewed against the short and long-term risks of not effecting any control measures. Important as this criterion is, it is difficult to achieve, since it involves determining the economic threshold of individual pests as well as their combined effects.

(d) The use of monitoring techniques normally allows proper and effective timing of control measures. To be able to predict the occurrence and severity of pest attacks, the collection and integration of several sets of data, each relating to a dynamic biological, physical, meteorological and socio-economic sub-system, is a necessary management practice.

(e) The ability of the farmer to undertake a particular pest management strategy, in terms of both the means and the understanding, is an important consideration.

Management objectives and methods

Irrespective of ecological conditions or methods of control, each pest management practice aims to achieve one or more of the following objectives:

(a) Maintaining the exclusion of the pest from the agro-ecosystem if it is not already present.
(b) The eradication of the pest from the agro-ecosystem if and when it is present.
(c) The protection of crops from the pest.
(d) Therapeutic actions aimed at 'healing' those crops and crop components to which damage has already occurred.

Although the control of crop pests is an integral part of the crop production process, the methods normally employed in the management of insect pests and disease organisms can be classified into the following seven categories:

1. Regulatory or legislative methods, including plant and animal quarantine, seed certification, eradication and suppression programmes.
2. Cultural operations, including crop rotation, crop sanitation, routine use of resistant or tolerant crop varieties, good tillage practices, timely planting and harvesting, pruning and thinning, fertilisation, water management and the use of trap and catch crops.
3. Mechanical methods, including hand-destruction of pests and their exclusion by screens and other types of barriers and traps. These practices are generally feasible only if the population of the (primarily insect) pest is low or when the farm is small.
4. Physical methods, including the use of heat, cold, humidity, sound and light. The use of heat and cold are more common with post-harvest produce than in field situations.
5. The use of biological methods has been the major means of natural control of crop pests. Although biological control of pests produces

relatively stable results, it is slow-acting and not very effective for above-ground pest control in annual crops. Perennial crops, such as those grown in plantations, benefit more from this method of pest control. Essentially, the method employs natural enemies to reduce the population of the target pest. The predators or parasites utilised are either locally present or are deliberately introduced.

In one example, several species of wasp of the genus *Aphytis* parasitise citrus scale insects. The relationship is very specific as each wasp species only controls one species of the citrus scale insect. There are many other examples of biological control. In the humid regions of the tropics, certain fungi, including *Aschersonia* sp. and *Fusarium* sp., also attack citrus scale insects. Adults and larvae of lady beetles, as well as larvae of hover flies (*Syrphidae*) are reported to control a large number of aphid species. Certain types of birds help to reduce infestations of caterpillars. Rattle pod (*Crotolaria*) and marigold (*Tagetes*) are often used as trap crops, to trap and kill nematodes. Such trap crops are useful in crop rotations. A recent example of the use of biological control has been the successful control of cassava mealybugs, *Phenacoccus manihoti*, in infested areas of West, East and Central Africa using natural insect predators from South America.

6. Chemical methods include the use of insect attractants, repellents, pesticides, sterilisers and growth inhibitors. By far the most commonly employed chemical practice is the use of pesticides, primarily because of their fast action and the subsequent dramatic results in terms of increased crop yields and quality.

Pesticides are marketed in various formulations both as solids and liquids. A range of pesticide applicators currently available are shown in Figures 5.23 to 5.25. The difficulties often associated with conventional high-volume spraying techniques, especially as they affect small-scale farmers in the tropics faced with the problem of finding sufficient water in semi-arid and arid conditions, have recently been overcome with the development of very and ultra low volume pesticide applicators, some of which use specially prepared formulations (Figure 5.25). Most of these sprayers are hand-held and battery operated and are adaptable for use in large and small-scale farm enterprises.

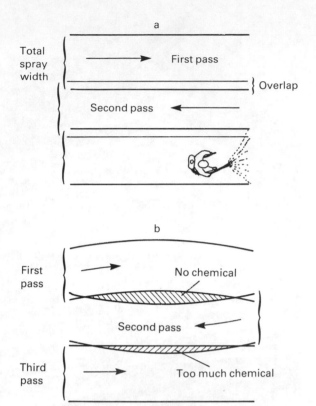

Total spray width — First pass

Overlap

Second pass

a

First pass — No chemical

Second pass

Third pass — Too much chemical

b

Figure 5.26 Diagram to illustrate (a) even and (b) uneven spraying of chemicals.

The efficient application of chemicals by spraying is illustrated in Figure 5.26.

7. Genetic methods, in which pests (particularly insects) are genetically manipulated for population control. An example of this approach is the mass release of sterile or genetically incompatible specimens of a pest species.

Integrated control

The realisation that the exclusive use of any one of the approaches referred to above rarely maximises its effect has led to the concept of managing crop pests by integrating various suitable techniques and methods into a workable, economic and environmentally desirable unit. As an example, the spray-banding of certain tree crops with dieldrin to keep out ants also serves to protect aphids for their effective elimination by natural enemies. Similarly, the Mediterranean fruit fly is effectively controlled by a combination of three methods: fly parasites on wild hosts, bait sprays on citrus and a sex pheromone which lures males to the poisoned bait.

Harvesting

In such technologically advanced regions as Europe and North America, in no area has the impact of the industrial revolution on agriculture been more dramatically shown than in the harvesting of crops. The development of labour-saving and suitable equipment has allowed increasing hectarages to be utilised and has ensured improved efficiency, particularly in crop harvesting. The present scarcity of harvesting and other technologies appropriate to farming and farm conditions in the tropics constitutes a serious limitation to agricultural development.

Although each crop has special harvesting problems associated with it, there are some broad principles which are applicable to all crops irrespective of the mode of harvesting. Among the most important factors are the following points.

Maturity of the crop

It is vital that each crop should be harvested as near to maturity as possible. Since the maximum ability of seed to germinate vigorously is obtained at full maturity in the field, it could be especially expensive for a crop grown for seed to become over-mature. For food crops, maturity refers to the stage of development at which a crop is most desirable to the consumer. In certain crops, for example sorghum and wheat, market maturity is attained when they are also botanically ripe. But in such crops as cucumbers, they are green at maturity in the market sense, but botanically immature. The general case for seed crops is that market and botanical maturity are identical.

In practice it is often necessary to harvest a crop before it reaches the stage of development desired by consumers. The time to harvest is therefore a decision each farmer has to take. Reasons for harvesting a crop before it reaches full maturity are as follows:

(a) To avoid shattering or other damage when mature.

(b) The entire crop may not mature at the same time, so harvesting is carried out while part of it is immature to prevent the remainder becoming overmature.

(c) The need to transport the crop to distant places while maturation may continue in transit.
(d) Certain crops yield better quality products if they mature in storage.
(e) By marketing early, the farmer may take advantage of sufficiently higher prices to compensate for the lower yield or lower quality associated with immature harvests.

Timeliness in crop harvesting

It is important to harvest a crop as early and as rapidly as possible to avoid losses associated with over-maturity. Grain crops generally shatter, pastures grow unpalatable, fruits become soft and easily bruised, and other crops suffer similar losses in quantity and quality if allowed to become overmature. The problem is complicated by the fact that in general it is the highest-yielding varieties that mature latest.

Moisture content

As soon as a crop is harvested, it becomes susceptible to insect and fungus damage if it is allowed to remain at relatively high moisture levels. In arid tropical conditions this problem is less serious than in humid climates. In either case, it is important to take great care to bring the moisture content of crops down to safe levels before being stored. In general, grains and seeds can have a moisture content for storage of between 8–12%. The lower end of this range is for oilseeds, while the upper end is for cereals and pulses.

Harvesting expenses

Partly because of the need to harvest the entire crop over a relatively short period, harvesting operations generally tend to be expensive in relation to other farm operations. The use of special farm machinery and techniques may considerably reduce harvesting expenses.

Crop storage

After a crop is harvested, it is of relatively little value until it is processed for consumption or

Figure 5.27 A maize store on stilts, Lukimba, Ravuma Region, Tanzania.

marketed. In marketing the produce, there are many decisions the farmer must make, among which is whether immediate marketing at harvest time will be most advantageous or whether he should hold back the crop and market it later when prices are higher. When holding the produce, facilities must be provided to store it effectively. In general, agricultural products are stored for various reasons, including the following:

1. To balance periods of plenty against periods of scarcity.
2. To make the products available the whole year round.
3. In some cases, to improve on quality.

Some products are more easily stored than others. For example, grain crops and cotton are easily stored, with relatively high proportions held over from one year to the next (Figure 5.27). The safe moisture content for the effective storage of grains is 8–12%. In contrast, fruits and vegetables are stored with great difficulty in their natural forms, seldom keeping for more than a few months. Nevertheless, a considerable part of the annual crop of some fruits and vegetables is stored either in the dried form or as canned or frozen goods. Tobacco is an example of a crop in which the quality is enhanced by proper storage.

The most obvious requirement of storage is that the produce be provided with shelter so that rain, wind and direct sunlight are kept out. In addition, adequate storage facilities must provide crops with proper temperatures and relative humidity, protection against rodents, insects and plant disease organisms and protection against fire, climatic hazards and floods. For those crops which emit ethylene and other gases, to enhance maturation (and hence possible over-maturation), provision must be made for controlled atmospheric conditions.

Crops such as potatoes are quite exacting in their temperature requirements. If stored at too high temperatures, respiration losses result in decreased dry weight, but if stored at too low temperatures, respiration is not rapid enough to utilise the products of metabolism and the tubers accumulate sugar which results in an undesirable sweet flavour and a tendency to darken upon cooking. Fruit and vegetable crops are also exacting in their storage requirements. In contrast, grain crops are not at all exacting and are injured neither at relatively high temperatures nor at temperatures below freezing. However, grain crops are more exacting in their humidity requirements than vegetable crops and potatoes. The relative humidity for grain storage, if losses are to be avoided, should be relatively low (Figure 5.28, after Ngugi, 1978).

Marketing of agricultural crops

Even though the bulk of the agricultural output in the tropics is from a large number of small-scale producers who are engaged in farming primarily to meet their families' daily requirements, a reasonable proportion of their produce is marketed.

In many tropical countries faced with agricultural marketing problems such as inadequate transport and storage facilities, it is often the case that more people are involved in marketing than production. In other words, between the producer and the consumer, there is often a long chain of middlemen who perform such functions as assembling the produce and transporting, storing and selling it. The large number of intermediaries should be expected, considering that the output comes from farmers spread over a wide geographical area, in scattered farms and markets. Together with inadequate production levels, the overhead costs resulting from the activities of middlemen is thus an important contributing factor to the rising market prices for most food and cash crops in many developing countries of the tropics (Table 5.9).

Characteristics of agricultural products

Agricultural products have certain characteristics which necessitate special handling and marketing arrangements.

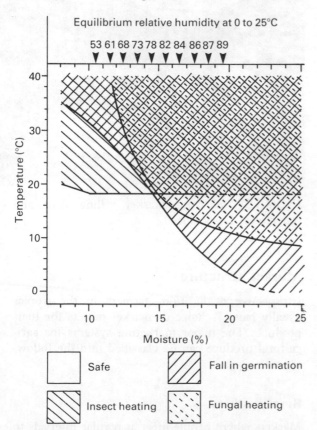

Figure 5.28 Effects of relative humidity, moisture and temperature on stored grain and grain pests.

95

Table 5.9. Annual wholesale prices for various crops in the Zaria area of Nigeria, 1973–85

Year	Millet	Sorghum	Rice	Wheat	Maize	Cowpea	Groundnut	Cassava flour (Gari)
					Cost/kg (in naira)			
1973	0.13	0.12	0.27	0.23	0.10	0.22	0.12	0.12
1974	0.13	0.13	0.35	0.24	0.13	0.56	0.20	0.12
1975	0.14	0.12	0.39	0.27	0.14	0.27	0.23	0.13
1976	0.17	0.17	0.53	0.23	0.19	0.27	0.39	0.25
1977	0.24	0.23	0.60	0.30	0.29	0.39	0.41	0.37
1978	0.36	0.30	0.58	0.40	0.30	0.66	0.44	0.48
1979	0.30	0.24	0.56	0.30	0.37	0.45	0.48	0.34
1980	0.29	0.24	0.85	0.31	0.29	0.56	0.62	0.43
1981	0.46	0.42	1.10	0.43	0.39	1.00	0.58	0.97
1982	0.45	0.36	0.96	0.42	0.28	1.01	0.65	0.83
1983	0.45	0.31	0.99	0.44	0.38	1.03	1.05	0.99
1984	0.96	0.85	3.17	0.85	0.85	1.94	1.92	1.10
1985	1.25	1.01	2.18	0.80	0.84	3.36	1.94	0.89

Perishability

Certain crops, especially fruits and vegetables, are highly perishable and need special storage facilities. Hot and humid tropical conditions normally favour the rapid decay of some crops, as the result of infestation by pests and diseases. In the absence of adequate facilities for their preservation, farmers are forced to dispose of such crops at relatively low prices.

Seasonality

In most parts of the tropics agricultural production is almost entirely dependant on rainfall. During each harvesting period and in the absence of adequate storage facilities, the supply of farm products is usually greater than the demand.

Risk and uncertainty

The production of most crops in the tropics fluctuates markedly from year to year, primarily due to weather factors. Uncertainty about the level of production in any one year creates problems, particularly with respect to accurately predicting the market supply, demand and price.

Bulkiness of produce

The bulkiness of agricultural produce often causes problems with transportation and storage.

Figure 5.29 A rural market selling fruit and vegetables, Ivory Coast.

Market structure

Irrespective of location, farmers in the tropics usually have a choice of market outlets for their produce. The major marketing systems for agricultural produce can be classified into the following five types.

Rural markets

Markets where people meet at regular intervals to sell and buy farm produce are a common feature in the rural areas of the tropics. Such markets may or may not be isolated, depending on whether or not they are located along or near good roads. Agricultural products handled in village markets are mostly food crops which have undergone little or no processing (Figure 5.29).

A large number of people, particularly women, normally carry head-loads of farm produce to the market. Other modes of transportation for farm produce include various types of animal power, bicycles and motor-cycles. While some people attend village markets to shop for family requirements, others are traders who buy produce for sale elsewhere.

Urban markets

Urban markets are often linked with a well-developed network of roads and occasionally railway lines. Because they operate daily and offer a wider choice of items than village markets, urban markets invariably attract a larger number of people. The markets also serve as venues for inter-state and inter-regional trade.

Roadside markets

With improvements in roads in many tropical countries, roadside markets are becoming increasingly important outlets for farm produce. It is now a common feature to see farmers and small traders displaying farm produce along routes to major cities for the attention of prospective buyers (Figure 5.30).

Marketing cooperatives

Agricultural cooperatives are organisations owned and run by farmers who use the service to reduce marketing costs, by collectively performing the functions previously undertaken by middle-men. In Nigeria, marketing cooperatives are by far the most numerous of all types of agricultural co-operatives.

In many tropical countries agricultural marketing cooperatives play a prominent role, particularly in the marketing of export crops, fruits and vegetables. Cooperative societies in rural areas often buy, assemble, pack, grade and store members'

Figure 5.30 A roadside fruit and vegetable market.

97

farm produce, which they transport to cooperative unions in bigger cities and towns for sale to crop commodity (marketing) boards. Cooperatives often advance loans to their members to enable them to hold their produce long after harvest to ensure better prices.

Crop commodity boards

Where they exist, commodity (marketing) boards function as public trading agencies which handle marketing, primarily of export crops. The boards usually control the internal marketing arrangements through licenced buying agents who undertake the purchase and collection of the produce from the farm to storage facilities owned by the board. Commodity boards participate in the setting of producer prices which are reviewed annually.

In some developing countries the mandate of commodity boards to market export crops has undergone review to include produce that has traditionally been classified as food crops. For example, in anticipation of surplus annual harvests resulting from Nigeria's various agricultural development programmes, a Grains Board is now in existence in Nigeria to purchase surplus grain, which is held and released for sale at times of food scarcity.

Intermediaries in distribution of farm produce

The intermediaries associated with the marketing of farm produce in most tropical countries can be classified into several categories.

Traders

Traders generally belong to two categories: local and urban. The local trader lives in the village and purchases farm produce directly from farmers who are unable to take their produce to market. He or she then sells this collection in village markets, either through village retailers or directly to consumers.

The urban trader, on the other hand, purchases farm produce in rural markets from retailers or directly from farmers, and transports the material to other markets or to urban centres for sale. This type of trader normally covers a wider area and his purchases are larger than the local trader.

Retailers

Retailers generally purchase produce from wholesalers and sell to those consumers who are only able to buy small quantities at a time. The functions performed by retailers of farm produce are as follows:
(a) Reducing produce to small units which relatively low-income consumers can more easily afford to buy.
(b) Buying and displaying produce for sale at places that are convenient to consumers.
(c) Sorting, processing and repacking produce to suit consumers' needs.

Wholesalers

Wholesalers are individuals who rent stalls in urban markets and handle large quantities of farm produce. They perform such marketing functions as buying, storing and financing the exchange produce. Marketing through wholesalers channels has important advantages. Firstly, wholesalers buy produce from farmers or local traders and sell to retailers, to other wholesalers in domestic and foreign markets and to manufactures and agricultural processors. Secondly, wholesalers often finance the movement of produce and invariably bear most of the marketing risks.

In general, farmers who are able to sell their produce directly to wholesalers at the village market level without incurring additional selling costs stand a better chance of receiving higher prices than those who dispose of their produce through local traders to wholesalers.

Transporters

The transporter is another intermediary in the chain of middle-men who perform marketing functions. The transporter conveys farm produce from place to place by means of animal power and various types of motor vehicles.

Forage crops

Forage is used in this context to include all fresh or preserved vegetable matter which is used as feed for animals. Most forage crops are perennials with suitably developed root systems which are well adapted to holding the soil in place. Because of this, forage crops are also recommended by conservationists as an aid in the fight against soil erosion. However, in the context of this book, the discussion of forage crops covers both the grass crops and legumes that are used for grazing and browsing purposes.

Natural grassland

The vast majority of tropical grasslands consist of natural and unimproved species which are invariably nomadically maintained, extensively grazed, indiscriminately burnt and generally poorly managed. By far the greater part of natural grasslands are those in the broad-leaved woodland, thorny woodland and thicket zones where climatic conditions limit productivity and pose management problems. Over most of these areas, the low and often unreliable rainfall and the subsequent dry season result in limited grass growth. Many natural pastures are dominated by coarse grasses that are palatable and nutritious only when young. The occurrence of forage legumes is not widespread. The carrying capacity of the land is therefore invariably low at all times, although it is considerably greater during the warm, wet growing season than in the dry season when growth stops rather abruptly and the grass dries out rapidly to give poor-quality hay. Where bush burning is indiscriminate and rampant during the dry season, most grasslands are affected so that even low quality hay becomes unavailable.

In general natural grasslands tend to be unstable in that they are continuously invaded by trees and shrubs that are better adapted to the climatic and edaphic conditions. This tendency is generally greater in drier regions and is encouraged by a weakening of the competitive power of the grass and over-grazing.

Sown pastures

Sown pasture, also variously referred to as temporary grassland, ley or grassland farming, involves a balanced cultivation of grasses and legumes of desirable characteristics for pasture, hay and silage. Sown pasture also serves to conserve and improve the soil for high production and profitable farming. Therefore the principle of sown pasture is aimed at achieving a balanced system of agricultural production which permits maximum output from soil resources while maintaining or improving them. The system assures the most economic production of livestock products while ensuring the maintenance of the productive capacity of the soil resources for generations to come. In many ways the principle behind sown pastures is the foundation of a permanent sustainable agriculture.

Forage mixtures

Legumes are generally of better nutritive value than grasses, having higher contents of crude protein, calcium and phosphorus, and often slightly lower crude fibre values. Forage legumes are not grown in pure stands as extensively as grasses. Apart from giving much lower dry matter yields, legumes are also generally unsuitable for making silage, unless mixed with other crops. For their part fodder grasses yield high amounts of dry matter, but have the disadvantage of a relatively low protein content. Therefore it is advantageous if the low protein content of forage grasses can be balanced by growing them in mixtures with suitable forage legumes.

The advantages of mixtures as compared with pure stands are as follows:
(a) Forage mixtures establish more rapidly and give better land use.
(b) There is better seasonal distribution of growth with mixtures. Indeed, the grazing season may be extended by the inclusion of both early and late-maturing species.
(c) The practice of growing mixtures allows for increased production with greater palatability.
(d) Leguminous plants may be associated with the grasses to the advantage of the grasses.

When selecting crop species for inclusion in forage mixtures, the following points should be considered:
1. As far as possible the species to be sown should be indigenous or adapted to the local climatic and soil conditions. For example, the length and severity of the dry season have important bearing on the selection of species.
2. It is important to decide the intended duration of the sward.
3. It is necessary to know the intended use; whether for grazing, hay, silage or any combination of these.
4. If intended for grazing, the type of stock should

be known; whether fattening or milking cattle, or sheep, pigs or poultry.

5. The season of the year for which the sward is intended to provide should be known.

Characteristics of forage species

As a general rule, forage grasses and legumes have certain desirable characteristics which lend them to use in grassland farming. In particular, forage crops should show the following points:

(a) They should be relatively easy to establish, particularly in terms of high germination rates, good seedling survival rates and high capacity to suppress weeds.

(b) They should yield high total amounts of dry matter, with the production distributed through most of the year.

(c) They should be of high quality, particularly with respect to palatability, digestibility, crude protein content and starch equivalent.

(d) They should be as drought-resistant or tolerant as possible.

(e) They should have high persistency and be able to withstand grazing.

(f) They should be able to produce good yields of viable seed which is relatively easy to harvest.

(g) They should not be difficult to eradicate when the ley is to be ended.

PART THREE

Production practices

To increase crop output per unit area of land requires efficient crop management, using seed with a high yield potential. In the context of this section, the management of crops covers all man-controlled operations in the crop environment. The effect of management and the environment on crop production are in the modification of the phenotypic expression of the crop potential.

CHAPTER 6

Cereal and sugar crops

The term cereals includes all the cultivated grains belonging to the large monocotyledonous grass family Gramineae (Poaceae). The cereals are native to the 'Old World' except perhaps maize, which originated in America.

The most valuable part of the cereal is the grain, the botanical structure and basic features of which are essentially the same in all cereals (Figure 6.1). The seed is a modified fruit called a caryopsis, in which the pericarp and the testa are inseparable. Seeds develop from small bisexual flowers borne in an inflorescence, either in a spike or in a panicle. The embryo occupies only a small portion of the seed, the bulk of it being taken up by the floury portion, the endosperm, which makes up the food reservoir. The endosperm of a wheat grain, for instance, represents about 80–85% of the caryopsis. All cereals are therefore said to have endosperm seeds. The outer layer, the bran, is usually high in protein, cellulose, hemicellulose and mineral con-stituents. The endosperm consists largely of starch granules embedded in a matrix of protein.

Cereals are the most concentrated and cheapest source of food energy known to man. They have long been grown and depended upon by man for the major portion of his food supply. Today, cereals together with legumes, still constitute the world's most important sources of food and feed-ing stuffs. Although their chemical composition varies widely, cereals are characterised by their generally low protein content and low nitrogen extract, being composed of over 90% starch. The protein content of cereals is often an important index of quality in certain types of food product. As a class, cereal proteins are not given as high a biological value as those of certain pulses and oil-seeds. Nevertheless, cereals generally provide as much as 40% each of the food calories and protein available to man from vegetable sources. The pro-minence of cereals as food plants is also related to

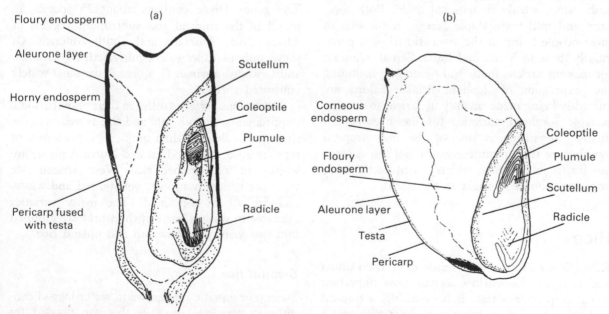

Figure 6.1 Longitudinal section through (a) a maize grain and (b) a sorghum grain.

Table 6.1. World production of major cereal crops, 1982–1984

Crop	Total production (million tonnes)			Land area cultivated (million ha)			Grain yield (tonnes/ha)		
	1982	1983	1984	1982	1983	1984	1982	1983	1984
Wheat	485	495	522	239	231	232	2.0	2.1	2.3
Rice	424	450	470	142	144	148	3.0	3.1	3.2
Maize	448	348	449	126	118	130	3.6	2.9	3.5
Barley	162	169	172	78	80	79	2.0	2.1	2.2
Sorghum	68	60	72	48	46	49	1.4	1.3	1.5
Oats	45	44	43	26	26	26	1.7	1.7	1.7
Millet	29	32	31	41	41	42	0.7	0.8	0.7

Source: FAO production yearbook, 1984.

their wide adaptability to a variety of growing conditions, their high yielding ability and ease of cultivation and management as well as their good handling and storage properties.

The most important cereal crops in the tropics include rice, maize, sorghum and millet (Table 6.1). Although essentially a temperate crop requiring a cool climate, wheat is also rapidly becoming an important crop in the tropics. Its production is presently restricted to high altitude areas and semi-arid regions where it is grown largely under irrigation during the cool dry season. While wheat, rice and maize are grain commodities in international markets, sorghum and millet have a low cash value, mostly in national trade. Both sorghum and millet are staple cereals in the diet of most people living in the semi-arid tropics, particularly those in Africa and India. Cereals are used for making various foods and beverages, including the preparation of alcoholic drinks. Efforts are currently being made, mainly in Africa, to develop suitable sorghum varieties for the brewing industry. Sugarcane is one of the only tropical members of the Gramineae grown not for its seed, but for its juicy stems, which constitute the most important source of sugar in the world.

Rice

Rice (*Oryza sativa* L.) originated in Indo-China and is today the world's second most important cereal crop ofter wheat. It is essentially a tropical and subtropical crop, but can be cultivated over a wide range of climatic conditions. Some 380–400 million tonnes of paddy rice (unhusked grain) are produced annually throughout the world, mostly from the Far-East and India. Substantial quantities are also produced in tropical North America as well as Central and West Africa. Rice is the main constituent of the diet of more than half of the world's population, and its production is extending rapidly, even to areas which are not traditional producers of the crop. In Africa, rice is used for preparing many types of dishes while in industrialised countries it is also used for the production of starch and alcohols.

Types and ecological adaptation

The genus *Oryza* contains about 25 species all found in the tropical and subtropical regions of Africa, Asia, Australia and South America. Of these, only two species are cultivated; *Oryza sativa* and *Oryza glaberrima*. *O. sativa* is the most widely cultivated species.

The various varieties differ in their photoperiodic requirements, plant height and size as well as in the shape and colour of their grain. The predominant types in Asia, the Far-East and North America are white and long-grained rice. West African rice types are generally brown, yellow, red and white, and are all short-grained. The many varieties available in various parts of the world are grouped into two main types; swamp and upland rice.

Swamp rice

Swamp or aquatic rice grows in water-logged conditions, especially on soils that are flooded for most of the growing season. Where a field cannot be flooded by rainfall, this must be achieved through supplementary irrigation. In most parts of Asia and Africa, rice is grown in mangrove and inland swamps, under irrigation, along river valleys or in areas of topographic depressions, called 'fadamas'. Two types of swamp rice are generally recognised:

Shallow swamp rice requires four to five months of flooding and generally matures in 145–160 days. Varieties are available which are early-maturing, requiring only 110–115 days to mature. The number of days to maturity is a critical factor to consider when selecting a site for non-irrigated shallow swamp rice, so as to ensure the availability of adequate surface water throughout the growing period.

Deep flooding varieties of swamp rice can withstand up to two metres depth of water. Deep flooding varieties are generally recommended for deep-flooded areas, as they require up to six or seven months of rainfall. They are, however, longer season, maturing in about 175–200 days.

Upland rice

Upland rice is grown on normal upland soils, sometimes in rotation with other annual crops. The crop is entirely rainfed but like swamp rice it requires heavy rainfall (800–1200 mm) and long sunny periods. In some parts of West Africa, notably the humid and sub-humid savanna, the furrows between sorghum or cocoyam ridges are often planted with upland rice. In general, upland rice matures in 100–125 days. The yield is dependent on the variety used, the soil, the climate, the cultivation methods and the level of management employed. On average, upland rice yields 1000–1200 kg/ha compared with 1500–2000 kg/ha for swamp rice.

Although rice grows on a wide variety of soils, the preference is for soils that are slightly acidic, especially heavy and fertile clay loams which are rich in organic matter. Sandy soils with low water retention capacity are not suitable.

Planting

Rice is generally planted on the flat. The field is first cultivated (Figure 6.2) or ploughed to about 15–20 cm deep, harrowed (if tractor cultivated) and then made as level as possible to allow even distribution of water. Rice may be sown direct on

Figure 6.2 Cultivation of rice using a mechanical cultivator.

the farm either by broadcasting or by drilling. Seedlings may also be raised in a nursery and then transplanted. Transplanting of upland rice is not a common practice, but where this is done the seed rate for the nursery ranges between 40 and 50 kg/ha. Scattering of the seeds directly on the field (broadcasting) is a less laborious and involved method of planting, although weeding may be a problem later. When drilled, the spacing for upland rice is 25–30 cm between rows and 25–30 cm between stands, depending on the variety. The seed rate for broadcasting or drilling is 50–70 kg/ha. A planting depth of 2–4 cm is usually recommended for both seeds and seedlings.

In swamp rice production, seedlings are almost always raised for transplanting. About 35–40 kg seed/ha planted in a nursery should be enough to transplant one hectare of swamp rice. The seedlings are transplanted about three to four weeks after they have germinated. At this stage the plants are 15–20 cm high. The field should be well worked and puddled. Transplanting of swamp rice should be carried out when the land is flooded with water. For upland rice, a wet puddled condition is adequate for transplanting. Seedlings are transplanted 15 cm apart in 30 cm rows. The number of seedlings planted per hole varies from two to four, depending on the variety. If the field is moist, but not flooded, at the time of transplanting, water must be allowed onto the field soon after. The field should remain flooded until the grains are formed, when the water may be drained to hasten ripening. Many rice fields in West Africa and Asia are not irrigated, but are flooded naturally during the rains and remain flooded until the grain filling stage which coincides with the end of the rainy season. In northern Nigeria, such seasonally flooded fields are referred to as 'fadamas'. They are usually fertile, dark coloured or greyish-white heavy clay loams which are rich in organic matter.

Fertiliser application

Fertiliser rates for rice cultivation normally depend on the fertility of the soil. In general, rice responds well to phosphate, but particularly to nitrogen. About 75–100 kg N/ha, applied as calcium ammonium nitrate or sulphate of ammonia, together with about 30–40 kg P/ha, applied as single superphosphate, should give good economic returns. Phosphate fertiliser should normally be applied before the seedlings are transplanted to the field. In the case of swamp rice, the water must first be drained off the field before fertilisers are applied and worked into the soil. The field is then flooded again within the same week. Nitrogen fertiliser is often split-applied in two or three doses. Potash fertiliser is not always a basic requirement, but when it is required, about 30 kg K/ha is usually adequate. Where the field is flooded naturally, the timing and scheduling of fertiliser applications must be based on practical experience, common sense and expediency.

Weed control

The control of weeds is one of the most important management practices in rice production. The use of chemical weed killers (herbicides) has been widely adopted for large-scale rice plantations in the Far-East, India and North America. However, at the level of the small-scale farmer, particularly at the subsistence level, weed control chemicals may be difficult to procure in terms of both cost and availability. This is the prevailing situation in many parts of Africa. In such cases, weed control in rice fields is effected manually by hand-pulling, a tedious and time-consuming operation.

For chemical weed control in rice, the use of propanil in combination with bentazone, fluodifen, thiobencarb and oxadiazon applied post-emergence (about 14–21 days after seeding or transplanting) is reported to be effective. Some of the common and notorious swamp weeds, including sedges, *Polygonum*, *Hydrophylla* and swamp grasses are hardly influenced by the application of herbicides and are therefore hand-pulled, sometimes several times within a growing season. In general weed control for upland rice is less demanding than for swamp rice. This is because in upland conditions hoe-weeding (as opposed to hand-pulling) is widely practised.

Pests and diseases

Birds are important pests of cereal crops, particularly in Africa. Grain-eating birds, including doves, pigeons, and 'canaries', can cause serious damage to rice and to other cereal crops. From the time the crop flowers until the grains are harvested bird-scaring is essential if good yields are to be guaranteed. Rodents, such as rats, grass cutters and squirrels, are also capable of causing considerable damage to rice. The construction of screened fences to keep off birds is known to be effective, but involves high costs. Traps can be set up for rodents where there is no risk to domestic animals.

Lepidopterous (*Maliarpha* spp.) and dipterous stem borers (*Diopsis* spp.) are the commonest and most important insect pests of rice in the field (Figure 6.3). While the use of such systemic insecticides as granular diazinon applied to the seedbed ensures a measure of control, the cheapest and most successful control method appears to be in cultivation practice. The field should be flooded for about a week to completely submerge the stubble to kill carry-over populations. All stubble and crop residues should be burnt after the grain harvest.

The commonest and most serious disease of rice is rice blast which is caused by the fungus *Piricularia oryzae*. It attacks the leaves, stems and flowers of young rice plants. The characteristic symptoms are long red spots which appear on the leaves. The leaves turn yellow and photosynthesis is impaired, resulting in the failure of the seeds to form

properly. The disease is encouraged by an over-supply of nitrogenous fertiliser. It is therefore important to guard against excessive application of nitrogenous fertilisers. When the disease has been observed, a field should not be planted to rice again for at least three years. The most effective control measure against blast is the use of resistant varieties. However, even where resistant varieties are used, infested fields must be avoided because the resistance can easily break down.

Rice smut is a fungal disease which attacks the grains and reduces them into a mass of black spores. It also attacks many other cereals, notably sorghum, maize and millet. The symptoms are similar in all cases, including the appearance of rust-coloured pustules on the leaves, after which the disease spreads to the grains which eventually turn into a mass of black spores. Rice smut can be controlled by the use of appropriate seed dressing. However, the most effective control method is to destroy the affected plants, preferably by burning, as soon as they are detected. Wherever possible, the farmer is encouraged to use varieties resistant to smut.

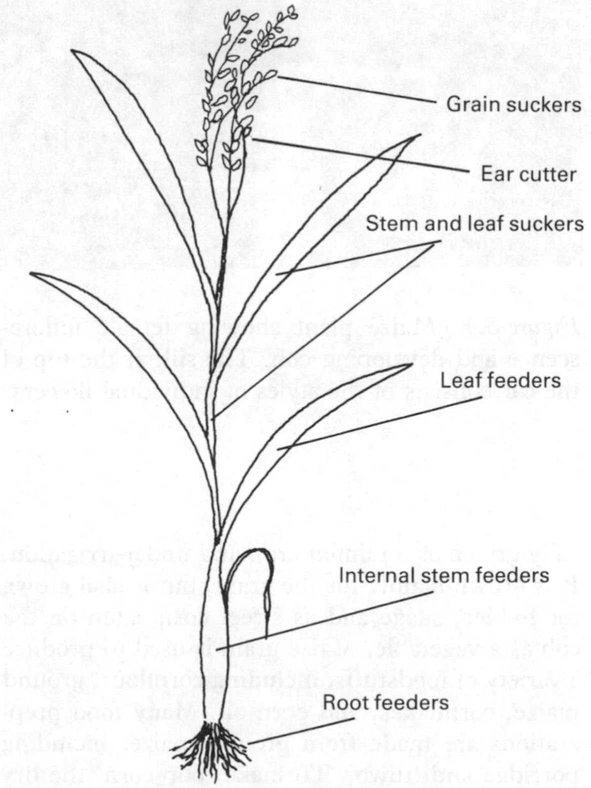

Africa	Asia
Stenocoris spp.	*Leptocorisa acuta*
Aspavia spp.	
	Mythimna spp.
Nilaparvata meander*	*Nilaparvata lugens*
Nephotettix spp.*	*Nephotettix* spp.
Sogatodes cubanus*	*Sogatella furcifera*
White flies	*Brevennia rehi*
Mites	Thrips
Epilachna similis	*Dicladispa armigera*
Nymphula stagnalis	*Cnaphalocrosis medinalis*
Hispids	*Hydrellia* spp.
Hydrellia spp.	*Nymphula depunctalis*
Spodoptera sp.	*Spodoptera mauritia*
Chilo spp.	*Tryporyza inceretulas*
Maliarpha separatella	*Chilo* spp.
Sesamia spp.	*Sesamia inferens*
Diopsis spp.	*Orseolia oryzae*
Orseolia oryzivora	
Termites	Mole cricket
Mole cricket	Termites
Root aphid	

* Potential pests

Labels on figure: Grain suckers; Ear cutter; Stem and leaf suckers; Leaf feeders; Internal stem feeders; Root feeders

Figure 6.3 Insect pests of major importance to rice in Africa and Asia.

Harvesting and storage

Rice is harvested when the heads are mature and nearly dry. From sowing to harvesting takes an average of five to seven months, depending on the variety. On large-scale farms, machines such as combine harvesters may be used. But at the level of the small-scale farmer, harvesting is usually done manually by cutting the stalked inflorescence with a sickle, knife or scythe. The paddy (hulled seed) is then allowed to dry before threshing. The paddy may then be stored in this form or subjected to further processing.

Rice processing

1. To reduce cracking of the seeds during milling (dehusking) the rice is parboiled (heated until partly cooked) by steeping in hot water or by steaming. The process of parboiling helps to loosen the husk. Soaking the paddy in cold water for 24–36 hours before parboiling is claimed to toughen the grains and minimise cracking. Once parboiled, the rice is thoroughly dried and then hulled, either using appropriate implements or by manual pounding in a mortar. The loosened husks are winnowed, where threshing is manual, and the rice grain is stored.
2. If desired, the rice grain may be subjected to a polishing process to give it a shiny white appearance. Polishing removes the outer (aleurone) layer of the grain which is rich in protein and vitamin B. Polished rice is therefore of lower food value than the unpolished product.

   When bagged in insect-proof and air-tight containers, dry rice keeps well in cool storage. Provided it is well dried, it is normally not attacked by diseases in storage. The likely storage pests are rats, mice and weevils, although parboiled rice is usually less prone to weevil attack.

Figure 6.4   Maize plant showing female inflorescence and developing cob. The silk at the top of the ear consists of the styles of individual flowers.

Maize

Maize, *Zea mays*, originated in tropical America, but is now one of the world's most widely cultivated food crops (Figure 6.4). It has a remarkably adaptable physiology and is rightly described as both a tropical and temperate crop. However, being a crop of tropical origin (Table 3.3), it thrives best in warm to hot sunny climates. It can be successfully grown as a rainfed crop and under irrigation. It is grown mainly for the grain, but is also grown for fodder, silage and as sweet corn eaten on the cob as a vegetable. Maize grain is used to produce a variety of foodstuffs, including cornflour, ground maize, cornflakes, and corn oil. Many food preparations are made from ground maize, including porridge and 'tuwo'. To make 'pop-corn' the dry grains are burst by roasting or frying. Other products manufactured from maize grains include glucose, starch, alcohol and acetone.

Types and ecological adaptation

Maize is normally classified into grain types, mainly for flour, and sweet corn types for consumption directly off the cob when boiled or roasted. The grains of both types may have a white, yellow or purplish-red endosperm. The white grains are usually preferred by millers and for livestock feed. Yellow grain maize contains more carotene, a precursor of vitamin A, and is therefore nutritionally more valuable. The maize for flour may be either flint type (having hard wrinkled grains) or dent type (with depressions on the outer ends of the grains). The dent is usually caused by shrinkage of the starchy endosperm. In most tropical countries, the traditional low-yielding local varieties of maize are being replaced by high-yielding composites and hybrids.

Maize grows well on a wide range of soils, but the preferred types are deep, easily-worked, fertile and well drained loams. As far as possible, shallow, sandy or very clayey soils should be avoided. The crop is tolerant of a wide range of values for soil pH, from 4.5 to 6.8. Plantings in freshly cleared fallows or in forest soils in the humid tropics generally require low applications of nitrogenous and phosphorus fertilisers. Maize is a suitable crop for rotation with legumes. The advantages of this rotation were discussed in chapter four.

Planting

Maize may be planted either on ridges or on the flat. In either case, the soil must be loosened sufficiently to permit easy root penetration, particularly during the seedling stage. For unmechanised cultivation or when ox-drawn ploughs are used, the ridges are usually placed about one metre apart. Two to three seeds per hole are planted about 3−4 cm deep. The seedlings are later thinned to two plants per stand in the case of wider (90 × 60 cm) spacings, or one plant per stand for narrower (90 × 30 cm) spacings. In a mechanised system, much closer spacings (75 × 25 cm) may be used, especially where the crop is grown in monoculture. In practice, a population density of 60−90 thousand plants (for grain maize) and 95−100 thousand plants (for silage) per hectare is considered suitable, depending on the variety. Where germination fails, replanting should be effected within seven to ten days of the first planting.

A well managed maize crop should mature within 10−12 weeks of planting. It is therefore possible to have two rainfed crops in one year where the rainy season lasts for seven months. Where this type of condition exists, cowpeas can be intersown with the maize crop when it begins to flower. As the maize matures and dries, the cowpeas flourish and provide not only a good grain yield, but also an enriched soil. In subsistence farming in the tropics it is common practice to cultivate maize in a mixture with grain legumes. Drying the cobs may be a problem if the maize crop is harvested at the peak of the rains. In some parts of the tropics with bimodal rainfall, notably the West African region, the problem of drying is minimised since harvesting of the early (first) maize crop coincides with a 'dry spell' which occurs in August.

The planting date for maize is normally dependent on the rainfall pattern and the length of the rainy season. Under sub-humid savanna conditions where the rainfall lasts for about five months, the crop should be planted as soon as the rains are established. In the more humid tropical zones where the rainy season lasts for up to nine months, plantings should be such that the crop is harvested before or preferably after the peak of the rains in order to minimise problems with drying of the cobs. In general, the choice of planting date is often a compromise between socio-economic demands and other forces faced by the farmer. For instance, in many parts of tropical Africa where maize is grown essentially for food, it is the first crop to be planted because, as well as being an easy crop to manage relative to other food and cash crops, it matures early enough to provide the farmer with needed grain or green cobs.

Fertiliser application

Optimal fertiliser requirements for maize depend on the productive potential of the cultivar, the previous cropping history and the general fertility of the field used. In general, the crop almost always responds to nitrogen and phosphate fertilisers. Of the major nutrient cations, a response to potash is the most common, followed by magnesium. For any given geoclimatic zone, it is desirable to ascertain, by soil analysis, the need for these two nutrient elements before applying them. In general, the fertiliser requirements of maize in tropical conditions are about 100−120 kg N, 40 kg P and

50 kg K per hectare. With a higher level of technology involving high-yielding composites or hybrids and effective weed, pest and disease control, as well as other improved management practices, considerably higher rates of fertiliser may be necessary, especially where moisture and solar radiation are not limiting.

Single superphosphate is the commonest phosphate fertiliser applied to maize. It is usually applied (broadcast) to the seedbed and worked into the soil, preferably one to two weeks before ridging and planting. Nitrogen fertiliser for maize is best supplied in two doses. The first, comprising about a third of the total requirement, should be applied at planting, while the remaining two thirds should be top-dressed at the commencement of flowering. Potash and magnesium fertilisers, if required, should be mixed and applied with the first dose of nitrogen fertiliser.

Weed control

Maize is sensitive to weed infestation and as such effective weed control is essential. Post-planting cultivation techniques, such as earthing-up of the ridges, often provide satisfactory weed control. The use of suitable herbicides significantly reduces labour costs. Atrazine applied pre-emergence, either alone or in combination with other herbicides such as metolachlor, pendimethalin and butylate, provides satisfactory control if the right dosage is used (Table 6.2). It may, however, be

Table 6.2. Relative effects of various weed control treatments on the yields of maize at Kadawa in the savanna zone of Nigeria

Treatment	Active ingredient (kg/ha)	Grain yield (kg/ha)
Atrazine	1.0	3857
Atrazine	2.0	3757
Cyanazine	1.0	3221
Cyanazine	2.0	3544
Atrazine + simazine	0.5 + 0.5	3226
Atrazine + simazine	1.0 + 1.0	3982
Atrazine + metolachlor	0.33 + 0.67	3120
Atrazine + metolachlor	0.67 + 1.33	4004
Hoe-weeded at 3 weeks	–	2303
Hoe-weeded at 4 weeks	–	1896
Unweeded (control)		654
LSD (at 5%)	–	1539

necessary to follow this with one supplementary weeding during the season.

Pests and diseases

Maize is prone to pest attack both in the field and in storage. The common pests in the field include stem borers, such as *Busseola fusca*, which eat the juicy centres of the stems causing considerable damage, particularly when the crop is planted late. For effective control of stem borers, several insecticides can be used. Effective cultural control measures should also be encouraged, including early planting, the use of resistant varieties and the burning of stalks after harvest.

In storage, the most serious pests of maize and most other cereals are weevils, *Sitophillus* spp. The adults lay their eggs in the grain while still in the field, but the larvae only emerge in storage. Besides the use of suitable insecticides, such as carbon disulphide and malathion, storage in air-tight polyethelene is also effective in killing off the emerging larvae.

The common diseases of maize include smuts, rust, bacterial blight and streak. Maize smut is similar to the smuts of other cereals, which are controlled by the use of chemicals as well as by seed selection. Bacterial blight is easily recognised by the appearance of isolated necrotic areas on the leaf which spread until the entire leaf looks burnt. The photosynthetic capacity of the plant is therefore impaired, with a consequent drop in grain yield. The most effective methods of control are crop rotation and the use of resistant cultivars. Streak is a viral disease which appears as a light yellow streak on the leaves. Viruses are ultra-microscopic organisms which cannot be seen with even the most powerful light microscopes. Their presence in plant (or animal) systems is evidenced by the disease they cause. Virus diseases are extremely difficult to control because the virus cannot be killed without killing the host plant. Crop rotation, elimination of affected plants and the use of resistant cultivars provide the only practical and effective control measures.

Rust is a fungal disease, caused by *Puccinia* spp., which attacks many cereal crops, including maize. Symptoms of rust include the appearance of rusty or red coloured spots or pustules on the leaves. Early planting and the use of resistant varieties are effective control measures. Some rust species are known to have alternate hosts, which may be shrubs or weeds. In such cases, the most effective control measure is to remove the alternate host from the vicinity of the crop.

Harvesting and storage

The time of harvesting is obviously dictated by the planting time, but in general maize requires up to 120 days to reach maturity. Short season or early-maturing varieties take 75–80 days. Maize should be harvested as soon as the grain is dry, usually at 15–20% moisture content. For fresh corn, it is best to harvest just before the stigmas dry out or, even better, as soon as they turn brown. The ears are harvested whole and the husks or sheathing leaves are peeled off. The cobs are then tied and sun-dried for storage (Figure 6.5). The moisture content of the grain should not exceed 14–15% at the beginning of storage. In most of tropical Africa, harvesting is done by manual labour. The cobs are usually removed from the stems, which are cleared off the field later. However, some

Figure 6.6 A maize 'corn crib', Uganda.

farmers perform the two operations at the same time in order to prepare the land for another crop. When maize is grown as an intercrop, the stalks are usually left on the field after the cobs have been harvested. Where maize has been attacked by stem borers, rust or streak, it is advisable to burn the stalks at the end of the harvest. Average yields of 3000–5000 kg/ha of grain have been obtained by small-scale farmers in Nigeria, but yields vary depending on varieties and location. Maize is stored in brick silos in many tropical countries (Figure 6.6). With subsistence farming, it is not uncommon to see cobs tied in bundles and hung to dry.

Separating the maize grain from the cob is done either manually or with mechanical devices (Figures 6.7 and 6.8). While simple hand shellers may adequately meet the needs of farmers with small holdings, more sophisticated machinery is necessary for large-scale businesses. Storage should be in air-tight plastic bags and it is advisable to treat

Figure 6.5 Maize stacked to dry before the removal of husks.

111

Figure 6.7 Manual threshing of maize with sticks, Songea Region, Tanzania.

maize grain in store with appropriate insecticides to control weevils.

Sorghum

Sorghum, *Sorghum bicolor*, also referred to as guinea corn, is an indigenous crop of tropical Africa, originating in Ethiopia. It is now produced in all the continents of the world under varying cultural conditions and for different uses. In terms of both world production and as a grain food, sorghum ranks fourth in importance after wheat, rice and maize. Nearly 70 million tonnes of sorghum are produced annually world-wide from about 51 million hectares of cultivated land.

Sorghum requires a dry and sunny season in which to mature and ripen, and is grown mainly in Asia, particularly India, Africa, the southern parts of the United States of America and Latin America, particularly Mexico and Argentina. About 45% of world production comes from the semi-arid tropics. Sorghum has a low cash value and except perhaps for the United States of

Figure 6.8 A maize chopper.

America, where the grain is produced almost exclusively for livestock, its cultivation is for human subsistence. Presently trade in sorghum is restricted to within each country and area of production. Grain yields in the semi-arid tropics range from 800–1000 kg/ha. In the sub-humid tropical areas of Africa mean yields of nearly 1800 kg/ha have been obtained while over 4000 kg/ha have been reported for some of the hybrids grown in the United States of America.

Sorghum is a staple food for millions of people around the world. It is used for making thin gruels such as 'ogi' and 'kunu' and thick porridge such as 'tuwo' and 'kafa' in West Africa, 'ugi' and 'bogobe' in East Africa and 'injera' in Ethiopia. Nutritionally, sorghum is a rich food containing most of the requirements for a balanced diet. It is also used for making various fermented and non-fermented beverages, particularly in Africa. The stalks are often used as building and fencing materials and as fuel. Other industrial uses of sorghum include the production of sugar, syrup and alcohol.

Types and ecological adaptation

The various species of sorghum may be broadly grouped into the following two types:
1. **Forage sorghums**, for example *S. vulgare*, are grown principally for animal grazing. They have a distinctive ratooning characteristic and the more they are grazed the more they grow and ratoon. The occurrence of forage sorghums is greater in the United States of America than in Africa and Asia. The sweet sorghums, for example *S. mellitum* and *S. nigricans*, have sweet juicy stems and are sometimes chewed like sugarcane, in addition to being used for silage and forage. They are also crushed in the manufacture of molasses and syrup.
2. **Grain sorghums**, such as *S. bicolor*, include most of the cultivated types. In this group, the grain is the major economic product. The commonly cultivated grain sorghums include: Durra (Egyptian corn); Caudatum; Guinea sorghums, such as Kaura; Guinea-Caudatum, such as Farafara; Guinea-Durra; the Zera of southern Sudan; Kaffir corn of eastern and southern Africa, and the converted dwarf sorghums of the United States of America and India.

Grain sorghums differ in plant height and the shape of their fruiting head as well as the size, shape and colour of their grain. It is mainly on the basis of these parameters that the varieties are classified. For example, dwarf varieties are mostly short-season and are 130–160 cm in height. These varieties generally head in 60–66 days and mature in 100–110 days. They have compact, medium-sized heads and cream-coloured grain. The short-Kaura sorghum, widely grown in the West African savanna, is a typical member of this group. Egyptian corn is essentially a dry-season type found mainly in the arid and semi-arid zones of West Africa, particularly the Lake Chad area, and India (Figure 6.9).

The medium-long season varieties are also medium-tall, growing 180–200 cm in height. They generally head in 95–100 days and mature in 140–145 days. The heads are usually medium-sized and compact. Varieties that are long-season are invariably tall. They average 360–400 cm in height. They head in about 150 days and attain maturity after 180 days. Seeds are either white, creamy-white or yellow. 'Farafara', a white-seeded variety popularly cultivated in West Africa, is a typical example of a tall, long-season sorghum (Figure 6.10).

When selecting a suitable variety of sorghum for cultivation, the important factors to consider include the yield potential, resistance to disease and insects and the quality of the grain, particularly in terms of colour and nutrient value. Most often, a high-yielding variety without acceptable grain qualities is not favoured for production. However, the factor of colour acceptability varies according to local values. For example, white-seeded grains may be preferred to creamy sorghums in one locality, while the reverse may be the case elsewhere.

Other factors which contribute to cultivar selection include the length of the growing season, the ability of the cultivar to grow with other crops in mixtures and the usefulness of such by-products as the stalk. In many areas of Africa tall sorghums are preferred because of the use made of their stalks for fencing, house construction and as fuel wood.

Sorghum is generally highly versatile, hardy and dependable in yield under adverse soil and climatic conditions. For most varieties, a rainfall of 500–1000 mm is adequate. Under optimal rainfall conditions, the plant has an extensive fibrous root system which allows extraction of water from a considerable soil depth. Moisture is essential at

the seedling stage, although it is considered that moisture stress during the formation of flower primordia and flowering is equally critical. A yield loss of 60−70% can result in either case. Well drained, light to medium-heavy, slightly acid loams (pH 6.5−6.9) are the best soils for sorghum cultivation. In general, however, sorghum grows well on a wide variety of soils when good crop husbandry practices are enforced.

Planting

Sorghum is often grown in rotation with a legume, to fully exploit the residual benefits of the nitrogen fixed by the legume crop. Sorghum should be planted 30 cm apart on the flat, or on ridges made 75 cm apart. Where moisture conditions are less favourable, the spacings can be wider (60 × 90 cm).

Seeds should be dressed with the appropriate fungicide before planting to ensure control of smut disease and rust. Four to five seeds are planted per hole. The seedlings are thinned to two plants per stand at 2−3 weeks after sowing. Where seeds have failed to germinate, the gaps may be filled with plants removed during thinning.

Fertiliser application

Sorghum responds well to nitrogen and phosphate fertilisers. About 50−60 kg N/ha and 25−40 kg P/ha should give a good return. Farmyard manure applied at 5−7 tonnes/ha has also been shown to give a substantial increase in yield. The fertiliser rate depends on the variety used, the soil fertility and the standard of management. The short season American hybrids are generally more nutrient-demanding than the local varieties which have been bred and developed under conditions of low fertility.

Superphosphate should be applied to the seedbed so that it becomes available to the young seedlings as early as possible to ensure good root formation. If old ridges are used, the fertiliser should be placed in the old furrows before splitting the ridges to form new ones. Where old ridges do not exist, phosphate fertilisers should be broadcast and worked into the soil before planting or before the ridges are formed.

In sandy soils or where conditions could cause the loss of applied nitrogen by leaching, calcium ammonium nitrate should be split-applied in two

Figure 6.9 Two varieties of sorghum (a) tall and (b) dwarf, both fully mature.

Figure 6.10 A tall white-seeded variety of sorghum at full maturity.

equal doses, at two and six weeks after planting (Table 6.3). Where urea is used, it should be applied in one dose, three to four weeks after germination. Where a compound fertiliser is used, the rate must be calculated to supply all the

Table 6.3. Effect of time and rate of nitrogen application on the grain yields of two varieties of sorghum in Nigeria

Treatment	Yields (kg/ha)	
	(Guinea-caudatum)	(Durra-caudatum)
Half rate, at planting *	1995	1602
Full rate, all at planting	3042	2630
Half at planting + half 4 weeks	3692	3046
Half at planting + half 8 weeks	3531	3352
Half at planting + half 12 weeks	2933	2465

* Recommended rate = 32.5 kg N/ha.

phosphate needed, with the balance of the nitrogen required being supplied by a suitable source.

Weed control

For maximum yield, it is essential to keep a sorghum field free of weeds. The most important weed of sorghum is striga, *Striga senegalensis* and *Striga hermontheca*. This is a parasitic weed which feeds from the roots of sorghum and can constitute a major source of yield reduction (Figure 6.11). *Striga* also attacks other cereal crops such as maize and millet.

It has pink-purple coloured flowers and is the commonest weed found in sorghum fields in West Africa. Application of pre-emergence herbicides, such as atrazine and terbuthylazine, followed by one supplementary weeding where necessary, should give satisfactory control. The best cultural method of control is the use of resistant varieties of sorghum and the practice of crop rotation. The effect of *Striga* in any one season can be reduced by using high rates of nitrogenous fertiliser which encourage the early and mass emergence of the weed seedlings. These can then be removed before they are able to infect the sorghum plants.

Pests and diseases

There are several important pests of sorghum.
1. The migratory locust, *Locusta migratoria*, the

Figure 6.11 A *Striga* infested sorghum−cowpea−pepper mixed crop.

desert locust *Schistocerca gregaria*, and the red locust *Nomadacris* sp. can constitute serious problems to grain sorghum in the field. These grasshoppers attack the crop by biting off seedlings and feeding on the leaves. A swarm of locusts is capable of destroying a whole crop in a few hours. In the event of a locust invasion, aerial insecticide spraying is recommended using fenitrothion and deltamethrin.

2. Of all the stem borers which feed on guinea corn, *Busseola fusca* is the most important. However, with good cultivation practices, the crop is usually able to counter this pest because the stems killed by the borers can be replaced by tillering. When a crop has been affected, the stems should be removed from the field after harvest or should be burnt on the field to kill the diapausing larvae, thereby reducing the carry-over population. An equally effective control measure is to rotate sorghum with other crops, preferably legumes. The use of insecticides to control stem borers is rarely economical, but where such a measure becomes inevitable, carbaryl (wettable powder) or granular formulations of endosulfan (Thiodan) are recommended.

3. Sorghum shootfly (*Atherigona socatta*) and sorghum grain midge (*Contarina sorghicola*) are both important pests of sorghum, although not as serious as stem borers. Both the shootfly and the midge can be successfully checked by planting early to avoid peak population periods. The use of a variety that matures uniformly also helps to prevent infestation with both of these pests.

4. Termites may be a problem where sorghum is left unharvested in the field for long periods after maturity or where the harvested crop is left on the field for long periods. These are common practices in West Africa, particularly among small-scale farmers.

5. The major storage pests of sorghum, and indeed most cereal grains, are weevils, *Sitophilus* spp. Infestation may take place in the field, from an already infested granary or by cross-infestation between granaries during storage. The larvae bore into, pupate in and feed on the seed, reducing the grain to a powder. Weevil infestation can cause as much as 5−7% yield loss in store. Weevil attack can be effectively controlled by fumigating the stored grains with carbon disulphide. In addition, the granary should be thoroughly cleaned and fumigated before the produce is stored.

The occurrence and severity of any of the diseases affecting sorghum varies from one location to another, from year to year and between seasons. The severity of a disease attack is determined largely by the availability of inoculum, weather conditions, varietal susceptibility and standards of crop husbandary and management. The important diseases of sorghum are as follows:

1. Grain or covered smut is a fungal disease caused by *Sphaecelotheca sorghi*. It occurs mostly in West Africa and attacks individual grains, filling them with masses of spores. Usually the affected plant retains its normal appearance, but at threshing the infected grains break and release the black spores which impart a dark, dirty colour to the remaining grain. Control is effected by dressing the seed before planting using an appropriate fungicide such as copper carbonate, Fernasan-D or Aldrex-T.

2. Head smut is a fungal disease caused by *Ustilago reliana*. It causes the whole head to become a mass of fungal spores. As with covered smut, the mycelia infect the seedlings from the soil and grow within the cereal tissue until they eventually reach the head and attack the grain. As a control measure, all infected heads should be cut off and burnt before the membrane breaks to shed the fungal spores. Crop rotation also helps to reduce head smut since it is soil-borne.

3. Loose smut is caused by *Sphaecelotheca cruenta*. It occurs as frequently as covered smut. The thin membranes covering the smut sacs break easily, giving a mass of loose dark-brown spores which are easily blown away by wind. Seed dressing, crop rotation and the use of resistant varieties are the recommended control measures.

4. Downy mildew is a fungal disease which causes yellow streaks on the leaves. These eventually turn brown, reducing the photosynthetic capacity of the crop, and ultimately giving a reduction in yield. The burning of infected plants is an effective control measure.

Harvesting and storage

In general sorghum matures in five to seven months after planting. The crop requires about eight weeks of dry weather at the end of the rains prior to harvesting. The cut sorghum heads are allowed to dry before being stored. In many tropical areas and especially by small-scale farmers in Africa, the grain is left on the heads for storage. Otherwise, the crop is threshed and the grain stored. A layer of woodash spread on the floor of the bin is reported to check insect pest infestation in the stored sorghum heads. In general, the yield range of grain from sorghum crops is 900–1500 kg/ha.

Millet

Millet, *Pennisetum typhoides* and *P. americanum*, probably originated in the West African savanna region. It is grown on more than 43 million hectares world-wide and ranks sixth in the important cereal crops in the world. Its cultivation is principally in the semi-arid tropics where it is second only to sorghum as a major staple food crop. Millet is a very hardy crop which thrives in conditions which most other crops would not survive. Africa and Asia together account for 98% of world output. Areas of major production are the Sahelian-Sudanian zones of West Africa (Nigeria, Chad, Mali, Sudan, Niger and Senegal) and the semi-arid regions east and south-east of the Thar desert in India. The areas of cultivation of millet are defined by the mean annual rainfall isohyets of 200–600 mm in both continents. A considerable amount of millet is also grown in mainland China, Ethiopia and Tanzania. These zones are generally characterised by short growing seasons of 3–4 months duration. As for sorghum, millet has a low cash value in that it is grown mainly as a subsistence crop for family use and internal trade (Figure 6.12).

Millet is used primarily as a grain crop in Africa and India and as a forage crop in south-eastern United States of America. The grain of millet is more nutritious than the sorghum grain but the fodder is inferior in quality. The grain is used for making various types of food, similar to those made from sorghum.

Types and ecological adaptation

The name 'millet' embraces several species, mostly in the genus *Pennisetum*. The major cultivated species are *P. americanum* (pearl or bulrush millet) and *P. typhoides*. Other cultivated, but less

Figure 6.12 Millet growing at the Gusan Demonstration Farm, Nigeria.

important millets include *Setaria italica* (foxtail millet), *P. inloceum* (proso millet) and *Eleusine coracana* (finger millet).

The millet types cultivated in the tropics include 'gero' millet and 'maiwa' millet in West Africa, and 'Bajra', 'Bajri' and 'Sajje Cumbu' millets in Indo-China. The gero millet of West Africa is relatively short-season (75–100 days maturity) and non-photosensitive. Maiwa millet, however, is photosensitive and long-season, maturing in 150–180 days. Gero millet is more adaptable than maiwa millet and is therefore more widely cultivated throughout West Africa. A third less important millet is 'dauro' millet. Dauro millet is basically similar to maiwa millet, except that the former is a transplanted crop while the maiwa millet crop is grown directly from seed.

117

Millet does well on a wide range of soils, but well-drained, light-sandy soils are best. The crop has the unique ability to thrive in areas of low rainfall and on soils too low in fertility for the successful cultivation of other cereal crops. However it is sensitive to waterlogged conditions.

Although millet does reasonably well on relatively poor soils, it does respond to good crop husbandry, good soil management and high soil fertility. Research has shown that millet grain yields are enhanced by applications of phosphate and particularly nitrogenous fertilisers. With good husbandry practices, including well timed planting, 30−40 kg/ha each of N and P are considered adequate. Millet benefits most from fertilisers applied just prior to or at planting.

Planting

Millet can be successfully grown both on the flat and on ridges. For sole crop millet, the suggested spacing is 60−75 cm apart on 90 cm rows or ridges. About 6−10 seeds may be sown per hole, but seedlings should be thinned to two plants per stand at about two weeks after germination. In West Africa, the best yields of gero millet are obtained when the crop is planted as soon as the rains are established. In the Sudan and Sahel zones of West Africa, the rains come rather late and plantings are invariably done in June (Table 5.6). It is clear from these results that late planting has a considerable negative effect on the grain yield of millet.

In many parts of the semi-arid tropics, millet is an early crop, intended to provide grain to sustain the subsistence farmer between the end of the dry season and the first few months of the rainy season. For this reason, millet is usually the first crop to be planted, most often with the first rains, whether or not they are established. There is some element of risk in this practice, especially when there is a long interval of rainlessness when many of the seeds planted may fail to germinate. Also many of the seedlings that manage to emerge become stunted and, by the time the rains become established, a large porportion of the planted crop may be lost. The farmer then has to replant the field.

Most subsistence farmers in the West African region intercrop millet with other crops, such as sorghum and cotton. Typically, the millet is planted with the first or second rains and the second crop (sorghum or cotton) is interplanted about a month later. In Nigeria, the millet is sown about two metres apart in old furrow bottoms, spaced about one metre apart. The seeds are sown without any land preparation. Later, however, when the seedlings are well established, the old ridges are split open to form new ones along the row of millet plants. At the appropriate time the sorghum or cotton is planted in between the millet stands. Both maiwa (long-season) and dauro (transplanted) millets are usually grown as sole crops. Millet has a remarkable ability to compensate for missing stands by tillering.

The millet seed contains only a small food reserve and has a short coleoptile, so must not be sown too deep. In general, sowing depths of 15−20 mm, 20−25 mm and 30−40 mm are suggested for heavy, medium-heavy and light-sandy soils, respectively. Where it is necessary to sow deeper than 40 mm for fear of picking by birds, more seeds must be sown to compensate for the possible reduction in germination. Where a farmer has to plant during dry conditions, as a result of rainlessness, a heavier seeding rate (20−30 seeds per hole) may be necessary.

Fertiliser application

Most millet varieties presently cultivated are not very responsive to fertiliser applications. However, improved cultivars respond well to fertilisers. The data in Table 6.4 show the response of millet to different levels of nitrogen, phosphorus and potassium.

Nitrogen fertiliser is best applied at planting time, about 20−25 cm away from the planting hole. It may also be top-dressed about three weeks after planting. Most farmers growing millet mixed with sorghum, prefer to delay the application of fertilisers until the sorghum crop is sown. This procedure has evolved out of expediency and is designed to minimise labour costs and to make maximum use of the limited amount of fertiliser available. Unfortunately, the approach tends to be counter-productive because it means that the best time for applying fertilisers to millet is often missed, with the consequence of reduced grain yields.

Where millet is grown in a mixture with sorghum, nitrogenous fertiliser should be applied

Table 6.4. Response of millet to different levels of nitrogen, phosphorus and potassium at Kano, Nigeria

Fertiliser rate kg/ha	1973	1974	1975	Mean
		Yield of millet (kg/ha)		
kg N/ha				
0	1282	1308	1606	1398
50	1795	2154	2501	2150
75	1758	2116	2389	2087
100	1706	1972	2142	1940
kg P/ha				
0	1045	1163	1648	1285
29	1789	2129	2459	2125
44	1932	1968	2042	1880
58	1643	1891	1842	1792
kg K/ha				
0	1662	2040	2218	1873
27	1693	2132	2001	1942
54	1643	1956	2203	1934

about four weeks after planting. When sowing dates for the crops in the mixture differ, fertilisers should be applied to each crop. If sown together, a second application of nitrogen will benefit the sorghum crop. Research on the effective fertilisation of mixed crops is presently in its infancy in most countries of the world. In the absence of any concrete information or recommendations, the amount applied should be proportional to that used for sole crops.

Weed control

The data in Table 6.5 illustrate the importance of weed control in millet production. Ox-drawn equipment, such as ploughs, may be used to earth-up ridges and thereby ensure effective weed control. The most convenient and economic way of controlling weeds, however, is by the use of chemical weed killers. Typical herbicides often used for weed control in millet under tropical conditions include atrazine and terbuthylazine used together or with other herbicides such as linuron and terbutryne. The correct formulation and accurately timed applications are critical.

Pests and diseases

Contrary to the belief that pests and diseases are not important in millet cultivation, many farmers, especially in the West African region, find that this is one of the worst problems they encounter. Stem borers, mainly *Coniesta* spp. and *Acigona ignefusalis*, can cause severe damage, particularly to long-season or late-planted millet such as the maiwa millet of West Africa. Early planting reduces the likelihood of borer attack. Millet head caterpillars, *Raghuva* spp. and *Lepidoptera noctuidae*, are also serious pests, particularly in seasons when the rains finish early. A number of contact insecticides, especially synthetic pyrethroids, are effective in controlling these pests. The destruction of the crop residues by burning is one effective control measure while a systemic insecticide applied to the soil in the right dosage also gives adequate control.

The most important diseases of millet are caused by fungal pathogens. Amongst these are millet downy mildew, smuts (both head and grain types), rust and blast. The use or resistant varieties, seed dressing, effective crop husbandry and the burning of stubble after harvest are the main methods of controlling millet diseases.

Harvesting and storage

Millet stalks are cut with the heads intact and the plants are stacked in bundles with the heads up. The harvested crop is then allowed to dry for

Table 6.5. Effects of weed control on the grain yields of two millet varieties in the Sudan savanna zone of Nigeria

Treatment		Ex-Bornu		Ex-Ghana	
	1975	1976	1975	1976	
			(kg/ha)		
Unweeded (control)	92	254	59	109
Hand-weeded once + one hoe weeding	1283	1806	928	1368
Hand-weeded once + two hoe weedings	1503	2042	1269	1790
Weed-free throughout	1588	2248	1404	1809

Figure 6.13 Harvested millet heads tied in bundles for storage.

about two weeks before the heads are cut and tied in bundles (Figure 6.13). Millet is usually stored in this form until needed for use, when it is threshed to separate the grains. In storage, millet should be treated with Gammalin 'A' dust to protect it against storage pests.

Figure 6.14 Wheat, the most important grain crop in the world.

Wheat

Wheat, *Triticum aestivum*, is the most important grain crop in the world, providing nearly 20% of total food requirements. The crop utilises about 30% of the land area under cereal cultivation and accounts for about 27% of world cereal production. In temperate areas, wheat has largely become the staple food grain. It is used to a limited extent as a concentrate in livestock feeds, mostly in the United States of America. In the tropics as elsewhere its principal use is in the making of bread and confectionaries. It is also used for preparing various porridges and gruels in some African communities (Figure 6.14).

Types and ecological adaptation

The genus *Triticum* comprises several wild and cultivated species, including *T. aestivum*, *T. spelta* and *T. durum*. *Triticum aestivum*, the most widely cultivated wheat in the world, is used primarily for bread making. It is a hexaploid species from which most modern cultivars have been derived. Another cultivated species, *T. spelta*, is also hexaploid and is often found in areas of climatic extremes. It is usually very resistant to fungal diseases. *T. durum* is the most important of the tetraploid wheats. The grains of this species are usually red or white and are very hard.

Wheat is basically a temperate crop, requiring a cool temperature of about 25–29°C during its growing period. It does not do well in warm humid tropical and subtropical environments primarily because of unfavourable temperatures during the growing season. Production is mainly concentrated in the northern hemisphere, important producing countries being the USSR, China, the USA, India, Canada, European countries, Australia and Argentina. The African share of world wheat production is 1–2%, mostly from Egypt, Morocco, Libya and Algeria. In most parts of the tropics, other than high altitude areas such as are found in East Africa, the only period when temperatures are suitable for wheat production is during the dry season. In Nigeria, for instance, production is limited to the semi-arid zone north of latitude 12°N and is only carried out using irrigation during the cool Harmattan period between November and February. In West and Central Africa the crop has been cultivated for several centuries on small irrigated plots during the dry season in the Sudan and Sahel zones. The level of production and economic importance of wheat has increased considerably in recent years following the development of large irrigation schemes. Wheat is also grown in many high altitude tropical countries where the temperatures are generally lower for most of the year.

The wheat crop requires soils that are well drained, well pulverised, slightly acidic to mildly alkaline (pH 6.5–7.5), heavy loams or clayey loams. The popular cultivars of wheat grown in the semi-arid tropics are the semi-dwarf and short-strawed types which are quick-maturing. The tall varieties are invariably long-season and often do not ripen sufficiently early.

Planting

The land should first be cultivated and then disc-harrowed twice to produce a well pulverised and compact surface. As far as possible all undulations should be flattened. Land levelling is the most critical and tedious part of land preparation. Pre-sowing irrigation can be employed as a finishing touch in land levelling. Border checks can be constructed only on well-levelled fields of uniform slope; otherwise basins of 10 × 10 m, 10 × 5 m or 5 × 5 m should be used for wheat planting. Levelling and maintenance of slopes are easier with basin than with border checks.

Wheat is usually planted on the flat, in rows 20–25 cm apart. A seeding rate of 120–150 kg seed/ha is suggested when hand-sown, depending on the variety. The semi-dwarf varieties are particularly suitable for tropical Africa because of their short growing time. When a mechanical planter is used, a seed rate of 100–120 kg/ha is adequate. A higher seed rate of 150–180 kg/ha gives an increase in yield when there is reduced tillering in areas with only a short cool dry period. When it is not possible to drill wheat in rows, seeds should be broadcast evenly and thoroughly raked into the soil. Medium-sized and small-scale farmers who sow their seed by broadcasting should obtain yields which are not significantly less than those achieved by drilling the same weight of seed in 20 cm rows. With adequate soil moisture, germination of seeds should be evident in four to five days after planting.

In Nigeria, planting of wheat is usually at the beginning of the cool weather between October and November. This is also the general practice in temperate areas of the northern hemisphere, where sowing usually takes place in October. Sowing later than the end of November carries the risk of yield losses, primarily due to high temperatures during the period of seed setting and grain filling.

In many countries the advantages of including leguminous crops in a rotation with cereals are fully recognised. The growing of wheat on the same piece of land year after year is not encouraged. Some of the common rotation systems followed are: wheat–legume–wheat–maize; wheat–cotton–legume–wheat; wheat–rice–legume–wheat; and sorghum–wheat–rice–wheat.

Fertiliser application

Wheat responds well to fertiliser applications, particularly nitrogen, phosphate and potash. Calcium ammonium nitrate or urea applied at 120–150 kg N/ha are reported to give good yields. A yield increase of up to 100% has often been reported following nitrogen fertilisation of wheat in Nigeria.

Where wheat is grown in rotation with a legume, the rate of nitrogen can be reduced by about a third. It is generally recommended that nitrogen fertilisers should be applied in two doses. The first, constituting about two-thirds of the total, should be applied at sowing and the balance should be top-dressed about three to four weeks after sowing or at tillering. However, recent studies in areas with fine-textured soils in Nigeria have shown that one full dose applied at planting is as effective as a split of the same amount.

Phosphates should be applied at planting and incorporated into the soil. About 25–45 kg of P/ha, applied either as single superphosphate, triple superphosphate or ammonium phosphate, should give a good return. In general, for medium to coarse-textured soils with a pH range of 6.5–7.5, the application of phosphates by broadcasting is suggested. Where soil pH is around 5.5, banding gives the best results, regardless of the texture of the soil.

Irrigation

The scheduling of irrigation is an important factor in irrigated wheat production. The practice of irrigation scheduling is site specific and takes into account the soil texture, the water-holding capacity of the soil, the infiltration rate of the soil, the evaporation demands of the area, the extent of depletion of soil moisture at a given stage of crop growth and the socio-economic objectives of achieving maximum or optimum yield. With shallow water table conditions, irrigation scheduling is a difficult task. Research in the Sudan zone of Nigeria suggests that when the water table is shallower than 60 cm, irrigation is not necessary, except for germination and nitrogenous fertiliser mobilisation. In general, applying 4–5 cm of water at weekly intervals is suggested for the medium heavy loams in West Africa, given that sufficient safeguards are provided for adequate drainage.

Weed control

When all other agronomic practices are properly and adequately manipulated, not more than 10% loss in the grain yield of wheat can be attributed to weed infestation. In other words, when the crop is sown at the correct time and with appropriate population densities, fertiliser rates and irrigation, the presence and effect of weeds is not highly significant.

Herbicides such as bentazon and chlortoluron applied about three to four weeks after seed-drilling will effectively control broad leaved weeds and sedges. Propanil used in combination with bentazon and applied post-emergence (14–21 days after seeding) is also effective. Where necessary this should be followed by supplementary hoe-weeding.

Pests and diseases

An important pest of wheat is the stem borer, *Sesamia calamistis*. Burning of crop residues after harvest and the use of resistant varieties are the most effective control measures. Termites, *Microtermes lepidus*, are rarely serious pests and where they pose a threat, the use of aldrin dust is suggested.

Wheat, like other cereals, is attacked by a number of diseases caused by fungal, viral and bacterial pathogens. Amongst these are the stem and leaf rusts and smuts and the streak and mosaic virus diseases. The stem rust, *Puccinia graminis tritici*, causes considerable damage. The semi-dwarf varieties are resistant to some strains of this rust. Root rot is associated with various *Helminthosporium* species. The occurrence, severity and the consequent yield loss varies according to the growth conditions as well as the cultivar used. The use of resistant varieties and a seed dressing are suggested control measures.

Harvesting and storage

Wheat should be harvested as soon as the stalks turn golden yellow and the grains feel dry and thresh easily. The use of a combine harvester is recommended for large fields (Figure 6.15). Where this is not feasible, the crop could be cut with sickles, tied in sheaves and left to dry for a few days. As soon as the sheaves are dry, they should be threshed preferably on a platform or concrete floor, and carefully winnowed to separate the grains from the chaff. The grains are then put in clean sacks and stored in a cool place.

Barley

Barley, *Hordeum vulgare*, is an important cereal crop used primarily as a source of malt in the

Figure 6.15 Harvesting wheat with a modern combine harvester.

brewing industry. Although it is predominantly grown in temperate regions, barley is also grown in the tropics for use in brewing. It is also used for making a variety of foods and non-alcoholic beverages. It is currently widely grown in North Africa, but the bulk of the production comes from Europe and America. Although barley is presently not an important crop in tropical environments, it is increasingly grown in irrigated areas, particularly in West Africa.

Types and ecological adaptation

There are three cultivated species of the genus *Hordeum*, which are classified according to the fertility of their lateral spikelets:

Hordeum vulgare (six-rowed barley) has lateral florets which are fertile, with sizes varying from slightly to markedly smaller than the central floret. With three fertile florets attached to each side of the rachis, six rows of kernels appear on the spike.

Hordeum distichum (two-rowed barley) has two sterile lateral spikelets, each with two rows of kernels.

Hordeum irregulare (irregular barley) has fertile central florets which have been reduced to rachillae

in some cases and these are distributed randomly on the spike. The remainder of the floret may be sterile or fertile.

Like wheat, barley is basically a temperate crop adapted to a wide range of conditions. However, production is best in areas with cool temperatures and where the growing period lasts about five months. The crop is generally more drought and heat tolerant than wheat and grows well in areas with 400—500 mm of rain. Hot winds tend to induce premature ripening of the grain and high humidity may impair proper maturation of the crop.

Light sandy loams or well-drained medium-heavy loams are preferred. A soil with a porosity of about 60% or more is generally ideal. The crop is quite tolerant of soil alkalinity and does well in soils considered to be too saline for wheat, although it does not tolerate wet, poorly structured soils as well as wheat.

Planting

Barley is usually cultivated on the flat. Seeds are sown either broadcast or drilled in rows 20—25 cm apart. When broadcast, seeds are worked into the

a)

b)

Figure 6.16 (a) and (b) Gravity-fed irrigation of barley.

soil by ploughing and harrowing. Drilled seeds result in a better germination rate than is achieved by broadcasting.

Irrigation scheduling at 6–10 day intervals should produce a good crop yield. The frequency of irrigation is usually influenced by the general slope of the land (Figure 6.16).

Fertiliser application

About 50 kg N/ha is considered adequate for a high barley yield. Excess nitrogen and sulphur adversely affect the brewing quality. The use of ammonium sulphate is therefore considered unsuitable. Where possible, urea, ammonium phosphate or calcium ammonium nitrate should be used as sources of nitrogen, especially if single superphosphate is applied as a source of phosphorus, since the sulphur contained is usually adequate for barley. About 20–25 kg P/ha should be adequate to give a good return.

Nitrogenous fertilisers should normally be split-applied. The first dose, constituting two thirds of the total, should be applied at planting and the remaining third top-dressed three to four weeks later. Phosphate fertilisers should be applied at seedbed preparation and should be well incorporated into the soil. Potash response is rare, but where it is considered necessary about 20 kg K/ha

should be adequate. This should be applied at planting together with the nitrogen fertiliser.

Harvesting

Barley is harvested when the grain is completely ripe and the straw has turned golden yellow and brittle. The plants can be either machine harvested or manually cut with a sickle. In the absence of appropriate threshing machines, the dried barley stalks are beaten with sticks or pounded in mortars and the grains winnowed. Simple mechanical threshers and winnowers are presently in use by some small and medium-scale farmers in the tropics. Large-scale harvesters or combines are often engaged when the size of the farm permits their use.

The grain yield of rainfed barley varies between 600–800 kg/ha. The irrigated crop yield ranges from 2000–3500 kg/ha. Under favourable conditions of manuring and crop husbandary, the irrigated crop may yield 4–5 tonnes/ha.

Acha

Acha, *Digitaria exillis*, is an annual grass, 45–50 cm high when fully grown. It has very small, soft, yellowish grains which average about 2000 seeds per gram. Acha is a very hardy crop, well adapted to inherently poor and impoverished soils. However, it requires heavy annual rainfall of about 1000–1500 mm, distributed over a period of six to seven months. Other requirements include a fine soil tilth and a cool climate. In essence, acha is largely restricted to high altitude tropical areas, 1000–1400 m above sea level. Its cultivation is presently solely for local consumption. In Nigeria acha is an important crop on the Jos and Bauchi highlands as well as in the southern parts of Kaduna State.

Planting

The crop may be planted on the flat or on ridges, the more common practice being to grow acha on the flat. Yield is, however, lower on the flat because the seeds are usually broadcast and so weed control becomes a major problem later. The seeding rate ranges from 30–45 kg seed/ha.

Fertiliser application

Although acha withstands poor fertility reasonably well, it responds well to fertiliser application, particularly to nitrogen and phosphate. As the crop has not been the subject of extensive research, there are presently no data on fertiliser use. However, the general practice of farmers is to apply about 20 kg P/ha and twice that amount of nitrogen, which improves grain yields considerably. Excess nitrogen tends to cause stem lodging, which in turn appears to deter farmers from using farmyard manure. Whether or not such an inference is justified, lodging can lead to serious yield losses if rains persist after the crop has ripened, particularly when it is almost ready for harvest. When using fertilisers on acha, the phosphate should be broadcast and thoroughly worked into the soil before final seedbed preparation. Nitrogen should be applied in two doses. The first half is applied at planting by placement on the side of the ridges or rows, or by broadcasting just after the soil has been pulverised ready for planting. The second half of nitrogen should be top-dressed about four weeks later.

Weed control

The problem of weeds is one of the major yield-limiting factors in acha production. At present weed control is achieved mainly by hand weeding. This practice is obviously labour intensive and constitutes a serious drawback in the production of the crop. One way of overcoming this problem is to grow the crop on ridges or in rows to permit use of tractors or animal drawn equipment when earthing up the ridges, thereby effecting weed control. Another approach to controlling weeds in acha would be to use herbicides. However, at present no suitable herbicides have been identified.

Pests and diseases

No important pests or diseases have so far been associated with acha. This observation is probably more as a result of a lack of research than to an absolute absence of pests and diseases associated with the crop. The only observed cases of stem borer have not had a serious effect on yield. At present acha is of such low cash value that any chemical control of pests and diseases is unlikely to be economical. Cultural control methods such as burning of infected crop residues should still be followed. A pre-planting chemical seed dressing is suggested where possible.

Harvesting and storage

In traditional agriculture acha plants are cut close to ground level with a sickle or knife and the stalks are stacked, heads up, to dry. The dry stalks are then threshed by beating the heads with sticks to free the grains which are separated from the chaff by winnowing. The left-over straw is used for feeding livestock. It may also be burned and the ash extracted with water, to be used as an addition to soups and various traditional dishes. Acha grains are usually put in sacks and stored in barns built of mud. Problems with storage pests are rare because only small quantities of the produce are stored for short periods, often less than six months.

Sugarcane

Sugarcane, *Saccharum officinarum*, is one of only a few plants which store sugar and not starch. The storage organ is the fibrous stem. Originating in Asia, sugarcane has now spread to most parts of the tropics, including the West Indies, Cuba, Brazil, Australia and West Africa. Historically, sugarcane has played a major role in the development of tropical agriculture, notably in tropical America. Beet sugar has, over the years, continued to be a major substitute for cane sugar in Western Europe. Since the late 19th century, beet and cane sugar have alternately taken the lead in supplying world sugar requirements. According to FAO agricultural statistics, about 35−36 million tonnes of cane sugar are produced annually, from nearly 750 million tonnes of sugarcane, harvested from about 13 million hectares world-wide. Although essentially a tropical plant, sugarcane is grown, albeit with lower yields, as far outside the tropics as Spain, Louisiana in the USA and New South Wales in temperate Australia. Within the tropics it can, with sufficient humidity, be grown at altitudes up to 1000 m. However, the main producers still remain Java, Cuba, the West Indies and India, while countries such as Barbados, Fiji and Mauritius depend mainly on sugar exports as their source of foreign exchange.

Commercial production of cane sugar is a highly technical process requiring very expensive machinery. It is this heavy capital outlay that constitutes the major limiting factor in production. For this reason, much of the sugarcane grown in many countries is only for sale as chewing cane and for the production of a number of local foods. Sugarcane is still widely cultivated by peasants in small patches all over the tropics largely for chewing. On a slightly higher level of organisation, small areas of up to one hundred or more hectares are grown for the manufacture of crude brown sugar. To prepare this raw sugar, the canes are crushed in a small mill worked manually by animal traction or a small engine and the juice is then evaporated in open pans over a fire. The yield of this sugar per unit area of production is high because it contains most of the molasses, which is of dietary value but generally causes the price to be low.

Types and ecological adaptation

There are four cultivated species of the genus *Saccharum* namely; *S. officinarum*, *S. barberi*, *S. sinense* and *S. edule*. Of these, *S. officinarum* was probably domesticated from the wild species *S. robustum* in New Guinea, and then spread rapidly to India through Java and Malaysia. It is the most important and most widely cultivated species. The species *sinense* is probably a hybrid between *S. officinarum* and *S. spontaneum*, a wild species. *S. sinense* is cultivated in India and Asia and is important in breeding programmes. *S. edule* is confined to New Guinea and is of little commercial value in the sugarcane industry.

Sugarcane is categorised into two main types of plant:

1. The noble (thick) canes which grow to about 5 cm in diameter. They are high yielders of sugar which need fertile land and do poorly under conditions of prolonged drought. *S. officinarum* is an example of a noble cane.
2. The thin canes are much more hardy and are capable of withstanding poorer soils and longer dry seasons. However, they yield generally lower than the thick canes. *S. sinense* is a typical member of this group.

Sugarcane requires deep, fertile, heavy loams. Low-lying land with a rich clay loam is best. Heavy rainfall (1200−1500 mm per annum) is essential. In most parts of West Africa, where sugarcane cultivation is mainly by small-scale farmers, areas of operation are restricted to the seasonally flooded wetland (fadama) soils along river banks.

Planting

Viable seeds of sugarcane are rare because the pollen retains its viability for only a few hours. Cross-pollination is therefore difficult to achieve. Propagation is normally by stem cuttings or setts.

Ridges, about 1.25 metres apart, are first made on land completely cleared of weeds. Holes are then made to take the cuttings, which are usually about 12 cm long with about three to four nodes. The soil must be well-loosened to enable the roots to develop properly. It should also be treated with aldrin dust to prevent termite attack. The cuttings or setts are planted about 150−155 cm apart (Figure 6.17). The setts take root in about two weeks, and subsequently the top shoots normally grow into the new plants. Subsequent mechanical cultivation is devoted mainly to weeding. Setts which do not take root should be replaced within a month or less. The main shoots may be pinched out to enhance branching, but care must be exercised to ensure that this is not overdone.

Figure 6.17 Planting sugarcane in Nigeria. Cane stalks are thrown from the tractor-trailer, then cut into short lengths.

Fertiliser application

The two most important fertiliser requirements of sugarcane are nitrogen and phosphorus, although responses to such trace elements as copper and manganese have been reported. For canes grown from setts, fertilisers are best applied at planting. With ratoon canes, fertiliser application should follow post-harvest weeding. Calcium ammonium nitrate or urea at 150 kg N/ha and single or triple superphosphate at 30–40 kg P/ha are suggested. The fertilisers should be thoroughly worked into the soil just before laying the setts.

Weed control

Weeding is usually necessary only at the initial stage of the establishment of sugarcane. Once the canopy has formed weeds are shaded, so weed control is not a major factor in sugarcane production.

However, the herbicides diuron and dalapon are suggested for the early control of broad-leaved weeds and grasses respectively. Where the setts are planted on ridges, weeding should be carried out by earthing-up the ridges. One weeding and earthing-up about a month after planting is often enough to kill all the weeds.

Pests and diseases

The main pest of sugarcane is termites (white ants) which bore into the stems especially in the dry season. If the setts are treated with an insecticide such as dieldrin or aldrin, this effects satisfactory control.

Red rot is a fungal disease which causes the head to wither away. The use of resistant varieties and a suitable fungicide (applied to the cuttings before planting) will control red rot. Chlorotic streak, caused by a mosaic virus, is also a common disease of sugarcane. This is often not an easy disease to control. However, resistant varieties are now available. Smut, *Ustilago scitaminea*, may also cause considerable damage to sugarcane. Infected plants should be destroyed and as far as possible resistant varieties should be used.

Harvesting and processing

The first cutting of the cane is usually at about 12–14 months. In cooler climates, such as in Uganda, the first cutting is at about 18 months. In the major producing countries where cultivation is largely mechanised, sugarcane is grown on large estates of several thousands of hectares, often supporting a well equipped factory. All operations on a sugar estate are timed very carefully so as to obtain the maximum sugar content. Periodic sampling is done and appropriate analyses are carried out for this purpose. Sugar yield is often highest just before flowering and this is usually the best stage to cut. The ripe canes should be cut as close to the ground as possible because most of the sugar accumulates in the lower region of the stems.

For small-scale farmers, harvesting is a hand-and-cutlass operation, with one man cutting and stripping 2–3 tonnes of cane a day. With mechanical cultivation, 25–50 times as much work

can be done (Figure 6.18). Stripping consists of cutting off the top portion of the canes, then removing the leaves and sheaths from the remaining stem. The tops are often used to provide the planting material for the next crop. The leaves may be used as fodder, they may be burnt on the field or they may be chopped and turned into the soil. On some estates, the leaves are burnt before the cane is harvested, particularly where harvesting and haulage to the factory are completely mechanised. Canes being cut for the first time are often described as 'plant canes'. After harvest, the stumps are left to ratoon (allowed to grow from the same roots) into a new crop of stems for cutting. This process can be continued for three or more crops, but the yield from successive ratoons gradually declines. The practice varies in different countries, but rarely more than two or three ratoon crops are taken before replanting the field with new setts.

After harvest, the sugar content of the canes deteriorates rapidly. Therefore the sugar must be extracted immediately, preferably not later than 48 hours after cutting (Figure 6.19).

The processing of sugarcane to produce cane sugar is quite involved and may be briefly stated as follows. The canes are first cut into short lengths and the juice extracted by milling in rollers or crushers to squeeze the sap from the stems. The liquid extracted is then boiled to further

Figure 6.18 A mechanical cane harvester on a large sugar estate.

Figure 6.19   A train of 5 tonne carts taking harvested cane to the sugar mill for processing.

concentrate it and then cooled so that the syrup crytallises. At this stage a brown crystalline sugar is produced. This is then subjected to a series of processes of refinement and purification, including phosphotation, sulphitation, carbonation, filtration and concentration. After the sugar has been sufficiently purified, it is then grained and crystallised. The crystals are separated by centrifuging and the residue at this stage is known as molasses. The final sugar crystals are dried, cooled, graded and bagged.

CHAPTER 7

Oilseed crops

Oil is obtained from a range of crops. Table 7.1 gives a list of selected crops containing oil in commercial quantities. However, this chapter deals only with crops grown primarily for their oil content. This means that crops such as cotton and soyabean, which are grown mainly for their fibre and proteinaceous grains respectively, are treated elsewhere.

Table 7.1. Major oil-producing crops and their oil content

Oilseed crop	Oil content (%)
Groundnut	45–50
Cotton	20
Castor	35–55
Safflower	27–30
Niger	40–45
Sesame	43–57
Sunflower	25–50
Linseed	40–46
Oilseed rape	40
Soyabean	18
Oil palm	65–84
Coconut	58–63

Groundnut

Groundnut, *Arachis hypogaea*, also known as monkey-nut, peanut and goober pea, is believed to have originated in Bolivia, South America. It is now grown extensively in West Africa, India and China. It belongs to the family Leguminosae (Fabaceae) and therefore has root nodules containing nitrogen fixing bacteria. The crop is primarily grown for oil extraction, as the kernels contain 40–50% edible oil. It is also grown for human consumption, exported groundnuts being consumed fresh, as roasted salted nuts, in confectionaries and as peanut butter. The butter is a highly nutritious food containing about 29%

protein, 47% fat and 17% carbohydrate. As a subsistence crop, groundnuts are eaten fresh or boiled, or they are roasted and ground into a flour.

The cake produced after oil extraction is fed to livestock as a good source of protein and the parts of the plant remaining after harvesting are also used extensively as a feedstuff.

The groundnut plant is illustrated in Figure 7.1. It is an annual plant which derives its name from the unusual characteristic of underground

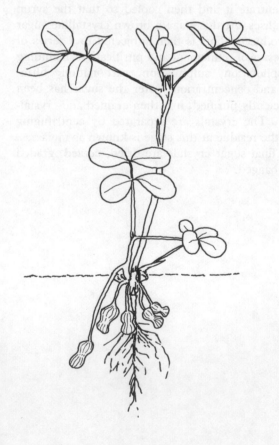

Figure 7.1 Groundnut plant of a bunch variety showing the pods developing underground.

seedpods, which are borne just below the soil surface. Each pod contains up to 6 seeds.

Types and ecological adaptation

Groundnuts can be grouped into two main varieties which are differentiated on the basis of their branching pattern along the main stem, being either alternately or sequentially branched. The distinguishing features of the two groups are compared in Table 7.2.

The groundnut crop is a major export crop in the moderate rainfall areas (500–1400 mm) of West Africa, lying between latitudes 8° and 13°N. For groundnut cultivation the rainfall must be well distributed throughout the growing season. The crop thrives best in bright sunshine and a warm environment. Groundnuts prefer light, well-drained, fertile soils, the best soil texture being a sandy loam. The crop is intolerant of extremely acid or alkaline soils. The cardinal temperatures for the various growth stages are shown in Table 7.3.

Table 7.2. Distinguishing characteristics of the two major groups of groundnuts

Sequentially branched	Alternately branched
Erect bunch growth habit	Runner or spreading bunch habit
Light-green foliage	Dark-green foliage
No seed dormancy	Seed dormancy present
Susceptible to leaf spot	Moderately resistant to leaf spot
Early-maturing (90–110 days)	Late-maturing (120–150 days)

Groundnut breeding and selection are carried out in many centres in the tropics. Selection is based on increased pod and kernel yields, high oil content and resistance to drought and diseases. New cultivars are frequently added to the list of

Table 7.4. Some recommended cultivars of groundnut in tropical Africa

Variety	Days to maturity	Other characteristics
RRB	100–110	High yielding; moderately tolerant to rosette
55–437 and Ex-Dakar	100–110	Small seeded; moderate to high yielding
RMP 12, RMP 91, M 2568 and 69–101	120–130	Rosette resistant; high yielding
MK 374	120–130	Improved local lines; moderate yielding

recommended varieties. Table 7.4 lists some of the varieties in use in various tropical countries. When choosing a suitable variety for a particular location it is important to match the cultivar with the length of the growing season in the area.

Planting

Whether to plant groundnuts on ridges or on the flat depends on local soil conditions, the availability of water and the possibility of soil erosion. In areas that are prone to waterlogging during the growing season it is advisable to plant the crop on ridges. The ridges should be spaced about 1 m apart to allow a closed canopy to form at the stage of peak leaf development. In areas with good drainage, the groundnut crop is often planted on the flat. In most savanna areas in West Africa where groundnut is a major crop, the particular rainfall regimes lead to the formation of soil surface crusts. As the production of pods is dependent on the fertilised pegs penetrating the soil, it is essential to till the soil to enhance pod formation.

The shelled groundnut seed is planted in holes

Table 7.3. Cardinal temperatures of the major growth stages of groundnut

Growth stage	Maximum temperature (°C)	Minimum temperature (°C)	Optimum temperature (°C)
Germination	35	12	24–30
Vegetative	38	15	28
Flowering and kernel development	33	18–20	25–30

Table 7.5. Effect of seed dressing on seedling establishment and pod yield of groundnut

	Dressed	Not dressed
Seedling establishment (%)	64	44
Pod yield (kg/ha)	2605	2171

on ridges or broadcast on the flat. Seeds should usually be sown at the beginning of the rains. The seed looses its viability soon after shelling so should be planted as soon as possible. Seed from the 'runner' varieties requires a period of dormancy before planting. Seed dressings are recommended to enhance the germination rate. In Nigeria, Aldrex-T, a combined insecticide and fungicide, is recommended at the rate of 300 g/100 kg of seed. Thiram + HCH (BHC) seed dressings can also be used. Table 7.5 illustrates the beneficial effects of seed dressing at Samaru, Nigeria.

Because groundnuts are grown for cash in areas of predominantly subsistence agriculture, the time of planting has traditionally been after other food crops are sown. This trend has persisted despite

Figure 7.3 A low population density of groundnut plants does not form a closed conopy, resulting in a lower yield.

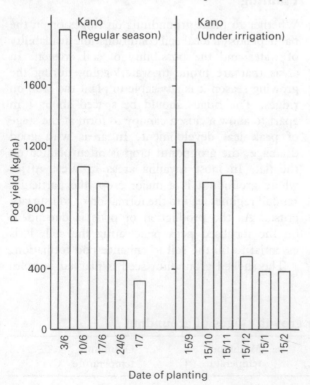

Figure 7.2 Pod yield of groundnuts in relation to the date of sowing at Kano, Nigeria.

overwhelming evidence that early planting increases yields. The relative advantages of planting early are demonstrated in Figure 7.2a using data from Kano, Nigeria. Figure 7.2b illustrates similar results during the dry season, with water being provided by irrigation. While the positive response to early planting during the rainy season may be due to optimal use of water, the response during the dry season depends on the stage of crop growth which is exposed to the cool temperatures between December and February.

Sowing density varies in different areas where groundnuts are grown, but all estimates show that peasant farmers normally sow at too low a density. In West Africa, groundnuts are planted at densities of about 47 000 plants/ha (a spacing of 91 × 23 cm for individual plants). In mixed cropping, the population is often as low as 28 000 plants/ha. At these levels, the plants do not provide a closed canopy and fail to fully utilise available soil moisture and solar radiation (Figure 7.3).

The relationship of groundnut population densities to vegetative growth, development and dry matter production is illustrated in Figure 7.4. Growth and branching of individual plants are

Table 7.6. Mean effect of population density on groundnut haulm and pod yields and shelling percentage at Dambatta, Nigeria, 1973–76

Population (plants/ha)	Haulm yield (kg/ha)	Pod yield (kg/ha)	Pod shelling quality (%)
57 000	1 266	752	54.7
86 000	1 503	822	54.0
115 000	1 543	805	57.0
172 000	1 814	914	58.7

reduced at high population densities, but more dry matter is produced per unit area of land, as closely spaced plants develop larger leaf areas earlier in the season, making better use of sunlight. Economically, the optimum population density for high kernal yields in the savanna zones of West Africa has been found to be as high as 86 000 plants/ha using ridges. Table 7.6 gives the mean effect of population density on groundnut haulm

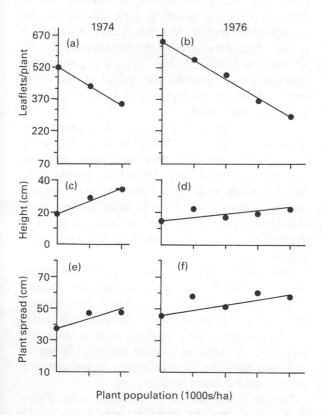

Figure 7.4 The effects of population density on (a and b) the number of leaflets per plant, (c and d) plant height and (e and f) plant spread.

and pod yields as well as shelling percentage at Dambatta, Nigeria, during 1973–76.

Fertiliser application

The nutrient requirements of groundnuts are primarily a function of the variety used, the soil nutrient content, the climate or ecological location and the level of crop husbandry practised. Requirements of groundnuts for the major nutrients are given in Table 7.7. In general, groundnut production is relatively less sensitive to fertiliser application than other field crops, mainly because groundnut is very efficient in obtaining nutrients from the soil and so exploiting residual fertilisers left from previous crops in a rotation. In this way the crop can become dangerously exhaustive if grown continuously, even though it is a legume and fixes nitrogen (Table 7.8).

Groundnuts have been reported to respond to fertiliser applications of nitrogen, phosphorus, potassium, calcium, magnesium, sulphur and the

Table 7.7. Amounts of major nutrients taken up by a crop of groundnuts yielding about 1200 kg/ha of kernels

Nutrient element	Quantity taken up (kg/ha)
Nitrogen	72
Phosphorus	7
Potassium	42
Calcium	15
Magnesium	9
Sulphur	6

Source: Balasubramanian, U. and L.A. Nnadi, 1977. Crop residue management and soil productivity in savanna areas of Nigeria. FAO Conference Paper, Buea, Camaroon.

Table 7.8. Effect of preceding crop on crop yield (kg/ha) in a rotation including groundnut, Malawi

Preceding crop	Maize	Cotton	Groundnut
Maize	2684	806 ★	149 ★
Cotton	2925	676	241 ★
Groundnut	3687 ★	815 ★	58

★ Significant difference, p = 0.05
Source: Webster and Wilson, 1980

Table 7.9. Mean pod yield of groundnuts in response to phosphorus and potassium fertilisers at eight sites in the Sudan and Guinea savanna zones of Nigeria

Phosphorus (kg/ha)	Potassium (kg/ha)	Sudan	Guinea
		(kg pods/ha)	
0	0	723	1209
18	0	919	1491
36	0	1028	1639
54	0	1165	1814
0	25	836	1369
18	25	1061	1680
36	25	1172	1884
54	25	1341	2072
0	50	939	1502
18	50	1131	1806
36	50	1267	2020
54	50	1434	2084

micronutrients boron and molybdenum, at different growing locations. Most locations show a positive response to phosphorus and sometimes potassium (Table 7.9). A starter dose of a low-level nitrogenous fertiliser has been shown to increase haulm size, even though its contribution to grain yield may be uneconomical or negative. The current practice in West Africa is for applications of 54 kg of P_2O_5 plus 50 kg of K_2O/ha to all soils in the Sudan savanna zone, and 54 kg P_2O_5 plus 25 kg K_2O/ha in the northern and southern Guinea savanna areas.

Weed control

At subsistence level, weed control in groundnut is by the use of a hoe. The practice of ridging also aids in burying weeds. It is always advisable to weed at least twice after seedling emergence. In commercial production, however, most weed control is through the use of pre- and post-emergence herbicides. Pre-planting weed control is usually combined with seedbed preparation. The number of applications will depend on the nature of the soil, the level of weed infestation and the time available to the farmer. The object should be to kill weeds and at the same time produce a reasonably loose seedbed. Excessive pulverisation of the soil should be avoided, although a fine seedbed will be required if soil-acting herbicides are to be used.

For pre-emergence application, metolachlor at 1.5 kg /ha (Dual 500 EC at 3 litres/ha) or alachlor at 1.5 kg /ha have been found to be effective. A post-emergence application of bentazone at 1.5 kg/ha (Basagran 48% EC at 3 litres/ha), when the weeds are at the four to five leaf stage, also gives adequate control.

Pests and diseases

Although various pests commonly occur in groundnut crops, their populations rarely increase to harmful levels. Millipedes, nematodes and thrips can occasionally cause damage. A serious problem in groundnut cultivation is the occurrence in large numbers of the aphid, *Aphis craccivora*, which is the vector of the groundnut rosette virus. This viral complex can prevent pod formation in young plants. Other symptoms of the rosette virus are chlorotic leaves, mosaic symptoms and deformed shoots, the distorted growth of the main stem giving rise to the rosette appearance. Suggested control measures for rosette include the use of resistant varieties, planting early and at the recommended spacing, the destruction of volunteer plants which may harbour the vectors over the dry season, and rotation. Close spacing is particularly important as it inhibits the spread of the aphid vector. At wider spacings insecticides are necessary to control the aphids.

Leaf spot diseases caused by *Cercospora* spp. occur mostly in warm, humid weather. They can be controlled by seed dressings, early planting, rotation and good crop sanitation. Spray treatments of Dithane M-45 at 6 g/litre of water applied weekly commencing 3 weeks after sowing, have also been reported to effectively control leaf spot diseases and significantly increase pod and haulm yields.

Groundnut blight, caused by *Sclerotium rolfsii*, occurs most frequently in wet weather, but is rarely of economic importance. Control is best effected by removal of infected plants. Bacterial wilt caused by *Pseudomonas solanacearum*, attacks and destroys the root system, causing the plant to wither and eventually die. Control measures include crop rotation and good crop sanitation.

Storage pests such as groundnut beetle, *Trogoderma* beetle and groundnut bruchid can attack the stored nuts. Keeping nuts in clean stores, prompt shipment and the use of insecticides are all control measures for storage pests.

A fungus, *Aspergillus flavus*, may also spread in the pods during storage. It produces highly toxic, carcinogenic substances called aflatoxins. It occurs mainly in humid conditions and enters the pods through cracks or termite holes. Careful harvesting to reduce pod damage and immediate drying to a moisture content of 10% are the suggested control measures.

Harvesting and storage

The time of harvesting for groundnut crops depends on the cultivar, as different varieties mature at different times. However, planting is usually timed to ensure harvesting during a dry period soon after the rains have ended, but while the soil is still moist. The whole plants, including the pods, are carefully lifted, the earth is shaken off and they are turned upside down and left to dry in the sun for 2−3 days. Drying can also be done in wind rows, in stacks or on racks (Figure 7.5). Care is taken not to damage the pods and drying should be as rapid as possible to prevent the spread of fungal infections.

After 2−3 days the pods should be removed from the plants, even if this necessitates subsequent drying on trays and mats. Early picking of pods reduces termite damage in the field. When drying the pods they should not be more than 4 cm deep on free standing trays or 2.5 cm deep in contact with a solid surface.

Nuts can be stored in the shell or after being shelled. The preferred storage method is in the shell, using lindane dust at the rate of 110 g/50 kg of nuts (2.2 kg/tonne) to control storage pests. Nuts for export are often shelled, either by beating the shells with sticks to release the nuts, by hand shelling or by hand operated shelling machines.

Castor

The castor plant, *Ricinus communis*, is thought to have originated in north Africa. It is now cultivated extensively throughout the tropics and subtropics, especially in Brazil, India and the USSR. It is a member of the family Euphorbiaceae. The oil ex-

Figure 7.5 Sacks of groundnuts stacked in the sun, Kano, Nigeria.

tracted from the castor bean is used to a limited extent for domestic purposes, such as in the production of soap and hair oil, for medicinal purposes as a purgative, and as an illuminant. However the main uses of castor oil are industrial, where it is used as a lubricant, in hydraulic fluids and in the production of paints and other synthetic materials. Silkworms, raised for the production of silk, are fed on the leaves of the castor plant. The cake remaining after oil extraction from the seed contains a toxic substance so cannot be used as a livestock feed, but is used as a good source of organic manure. Table 7.10 gives the nitrogen and phosphorous content of the cake.

Table 7.10. Nitrogen and phosphorus content of castor cake

Type of cake	Content (%)	
	N	P
Decorticated castor cake	6−7	2.3
Undecorticated castor cake	3−4	1.8

The castor plant is a perennial plant which can grow to a height of 12 m. Dwarf annual varieties have been developed recently which only grow to a height of 1−2 m and can be harvested 5−6 months after planting. The leaves of the castor plant are alternate and palmate and the flowers are in panicles with the female flowers developing in the upper part and the male flowers developing in the lower part (Figure 7.6). The perennial castor plant produces a very deep, extensive root system whereas the annual varieties have relatively shallow roots.

Types and ecological adaptation

The perennial and annual varieties of castor have different cultural requirements. Perennial castor with its extensive root system, thrives in dry areas with an annual rainfall between 500−850 mm. It cannot withstand waterlogging or heavy rainfall, especially at flowering. The annual varieties require higher rainfall for establishment, but also cannot tolerate heavy rainfall during flowering as this results in poor seed set.

Most perennial castor varieties shatter readily. They are susceptible to various pests and diseases, especially in wet weather. The thickness of their

Figure 7.6 The castor plant has palmate leaves and spiny fruit.

seed is highly variable, which prevents the use of mechanical hullers. Annual varieties are non-shattering but are more susceptible to pests and diseases. They produce relatively uniform seed which can be decorticated using mechanical hullers. Recommended varieties of annual castor are Baker 44, Baker 22 and Lynn.

Castor grows well in most well-drained soils. It does not tolerate heavy clay soils or alkaline conditions, but can survive in a slightly acidic soil.

Planting

For light soils, the land is ploughed at the onset of the rainy season. The soil is harrowed to crush clods. The harrowing is repeated once or twice in clay loams.

Castor is planted, as are other cash crops, after most food crops are established. Yields obtained from different planting dates in experiments in India show that early planting (June) yields better than late (August) planting (Table 7.11).

Table 7.11. Mean yields obtained from different dates of planting of castor at Himayatsagar, India

Planting time	Yield of dry fruit (% of July planting)
June	124
July	100
August	83
September	37

Source: Kulkarni, 1959.

Castor is usually sown in furrows or on ridges at a depth of about 10 cm. The seed rate varies depending on the size of the seed and whether the crop is annual or perennial. Medium-sized seeds are sown at a rate of about 20 kg/ha spaced at 90 × 90 cm. Wider spacings result in more profuse branching.

Fertiliser application

Highly fertile soils with a high nitrogen content are not desirable for optimum seed yields in castor. Such soils produce excessive vegetative growth and very tall plants, which are difficult to harvest. An application of about 30 kg N, 60 kg P_2O_5 and 45−60 kg K_2O/ha is considered sufficient on most soils. The nutrients may be applied in the form of mineral fertilisers or, even more beneficially, as farmyard manure.

Weed control

Castor usually requires relatively little attention after the canopy is fully established. In the early stages of growth, the young plants must be protected from competition by weeds. If the soil has been well cultivated at planting, two weedings are adequate, at about three and seven weeks after emergence. After the plants have attained a height of about one metre and have covered up the interspaces, removal of weeds around each plant can be done by hand-pulling. In mixed cropping, the cultivation given to the main crop also benefits castor.

Pests and diseases

Castor is very susceptible to pest damage. In some places the perennial castor crop is scattered over a large area, therefore inhibiting the spread of pests and diseases. However, annual castor is always cultivated in fields and as such is far more susceptible to pest and disease attack. Many insects damage the foliage, but the most serious yield losses are caused by various sucking insects which damage the flowers and growing points, especially in wet weather. Those causing the most damage are *Lygus* bugs and *Helopeltis* spp. Control is achieved by using insecticidal sprays every two weeks during the flowering period.

The castor crop has no diseases of economic importance, although various fungal infections do occur.

Harvesting and storage

Annual, non-shattering varieties of castor can be harvested when all the fruits are mature and fully dry, and when the leaves have fallen from the plants. For these varieties mechanical harvesting has been developed using converted corn pickers, which strip the fruits from the plant and hull and clean the seed. The perennial varieties which shatter readily, however, should be harvested before the seeds reach full maturity, when one or two fruits on the spike are dry. Harvesting of these varieties is by hand picking.

In most tropical countries, threshing is done by beating the dried fruits with a stick. After threshing, the seeds are winnowed. Castor seed is very hard and does not require much care during storage. No insect or pest attacks the seed. The seeds are usually decorticated for oil extraction using specially designed mechanical hullers. Castor seed does not exhibit dormancy. Germination commences within 70 hours and, with adequate moisture, is usually completed in 7−9 days.

Benniseed

Benniseed, *Sesamum indicum*, also known as sesame, or simsim was one of the first oilseed crops cultivated by man. It belongs to the family Pedaliaceae. Benniseed is an erect, annual herb 0.5−3 m high with a fetid smell. It takes 80−180 days to mature. Its stems are longitudinally furrowed and densely hairy. Varieties are classified as either shattering or non-shattering according to whether or not the seed capsules open on drying. The plant has a

Figure 7.7 A sesame plant showing axillary flowers and seed pods.

characteristic tap root system with a surface mat of lateral feeding roots. The flowers are axillary with 1–3 flowers per leaf axil (Figure 7.7).

The crop is thought to have originated in India or Ethiopia. It is, however, now cultivated in almost all tropical countries, alone or in mixtures with other crops such as cotton, maize, guinea corn or millet. The production figures of some tropical countries are given in Table 7.12.

The seed is utilised for the production of a high quality, odourless oil. The oil is used in the manufacture of compound cooking fats, margarine and salad oils. It is also used as a fixative in the perfumery industry, in various cosmetics, as a carrier for fat-soluble substances in pharmaceuticals such as penicillin, and in insecticides. The oil-cake can

Table 7.12. Total production figures and yield per hectare of sesame in six tropical countries

Country	Total production (1000 tonnes)	Yield (kg/ha)
India	426	187
Sudan	199	352
Mexico	246	1079
Ethiopia	37	—
Nigeria	17	314
Uganda	30	342

Source: FAO production yearbook, 1973–74.

be used as a poultry and livestock feed and as a fertiliser. The cake contains about 35% protein.

Types and ecological adaptation

Benniseed is grown in tropical, subtropical and warm temperate regions. It thrives in warm temperate regions during the summer and in tropical lowlands under semi-arid conditions at elevations below 1220 m. Optimum temperatures for growth are between 20 and 24°C during the early vegetative stages and about 27°C during growth and fruiting. The crop requires a rainy season of about 5 months and at least 500 mm total rainfall.

White, large seeds are generally preferred for human consumption even though brown seeds are reported to contain more oil and are therefore preferred by the oil-extracting industry. The variety used for planting should be high-yielding and resistant to diseases. Recommended varieties in Nigeria include Yandev 55, Pbtil No 1, 65B-28, E-8 and 69B-392.

Sesame can be grown on many types of soil, but it thrives best on well-drained, fertile soils with a sandy or loamy texture and neutral reaction. It is, however, not a poor-land crop. The site selected should be fairly flat.

Planting

The seedbed is best prepared immediately after the first rain of the season. The soil should be disc-ploughed and harrowed. If the seed is to be broadcast, a flat seedbed should be prepared. Otherwise ridges should be made if the seed is to be planted in rows.

Sesame should be planted soon after the rains

138

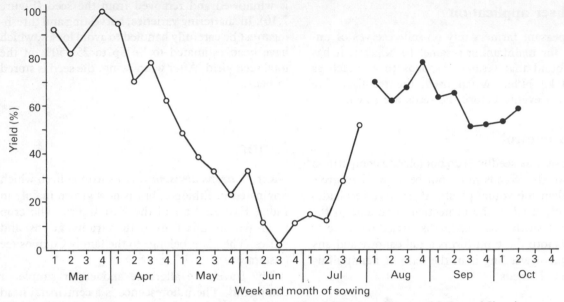

Figure 7.8 The effect of sowing date on the average yield of sesame, at Mokwa, Nigeria.

are well established. Where the rainy season extends over 180 days, two short season crops may be grown. The yields resulting from experiments conducted at Mokwa, Nigeria, using different sowing dates are shown in Figure 7.8. The effect of sowing date on the length of the growing season is also shown in Figure 7.9.

The crop tolerates a wide range of planting densities in different countries and locations (Table 7.13). In general, a narrow spacing and a high population density, up to 225 000 plants/ha, increase total yield, even though the yield and the number and length of capsules from individual plants may be reduced. When the seeds are broadcast, between 2–17 kg of seed/ha is mixed with sand or manure before spreading. This is to enhance uniform seed distribution.

Table 7.13. Common spacing of sesame in some tropical countries

Country	Spacing	Population density (plants/ha)
India	20 × 45 cm	107 000
Uganda	30 × 30 cm	108 000
Uganda	8 × 61 cm	215 000
Sudan	15 × 81 cm	81 000
Tanzania	28 × 71 cm	50 000
Nigeria	30 × 30 cm	108 000
Nigeria	23 × 23 cm	190 000
Nigeria	12.5 × 25 cm	300 000

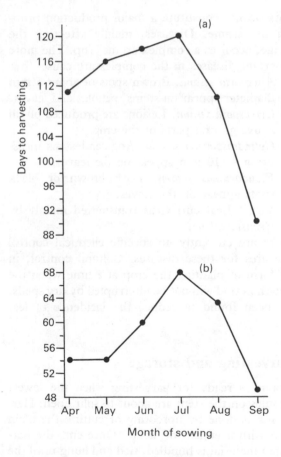

Figure 7.9 Effect of sowing date on (a) number of days to harvesting and (b) number of days to flowering in sesame grown in Nigeria.

Fertiliser application

Most peasant farmers rely on soil reserves of nutrients for maintaining sesame. In Nigeria, it has been found that sesame responds to as much as 25–30 kg N/ha. When used, the fertilisers are broadcast evenly before the seeds are sown.

Weed control

Young sesame seedlings do not tolerate competition from weeds. Weeds must not be allowed to grow taller than the young plants. Presently, herbicides are rarely used in the protection of sesame in the tropics. Usually hoe-weeding is carried out once or twice before the crop forms a full canopy and any subsequent weeding is in the form of handpulling of isolated weeds.

Pests and diseases

Pests do not constitute a major production problem in sesame. Diseases, mainly affecting the foliage, occur as a complex on the crop. The more important diseases in the complex are caused by:
(a) *Alternaria sesami*. Brown spots of about 5 mm diameter appear on stems, petioles and leaves.
(b) *Cercospora sesami*. Lesions are produced on all above ground parts of the crop.
(c) *Cylindrosporium sesami*. Angular lesions measuring 4–10 mm appear on the leaves.
(d) *Pseudomonas sesami*. Dark brown or black spots appear on the leaves.
(e) Virus. Leaf curl virus transmitted mainly by *Bemisia tabaci*.

There are currently no specific chemical control measures for these diseases. Cultural control, in the form of planting the crop at a time when the growth period will not be interrupted by dry spells, has been found to reduce the incidence of leaf curl.

Harvesting and storage

Sesame is ready for harvesting when the lowest capsules on the stem are about to split open. Harvesting is done by uprooting or cutting the main stem with a cutlass or sickle. Once cut, the harvested material is bundled, tied and hung until the capsules are dry and ready for threshing. Threshing is done by lightly beating the dry shoots or branches with a stick over a mat. All foreign matter is winnowed and removed from the seed (Figure 7.10). In shattering varieties, harvesting and threshing must be carefully handled to avoid losses, which have been estimated to be up to 25–50% of the total seed yield. After winnowing, the seed is stored in bags.

Niger

Niger, *Guizotia abyssinica*, is a shrubby herb which originated in Ethiopia, but is now grown mainly in India, East Africa and the East Indies. The crop is grown in mixtures with various grains and pulses. The plant belongs to the family Compositae (Asteraceae).

The leaves are alternate and either simple or compound. The inflorescence is a centripetal head of many tubular flowers, while the fruit is an achene, dry and indehiscent. The seed is used mainly for oil extraction. The best quality grades of oil are used for edible purposes, while the inferior grades are used in the soap and paint industries. The oil-cake, which is very rich in albuminoids, may be used as cattle feed.

Types and ecological adaptation

Niger is a warm season crop requiring a moderate rainfall of about 1000 mm. It does not tolerate waterlogging and prefers loamy or light sandy soils. The site selected for growing niger seed should be free-draining. The seed yield on light, poor soils is relatively low. Local varieties are considered to be best adapted to local conditions.

Planting

The seedbed must be well ploughed and harrowed. The seeds can be broadcast or drilled. About 5–10 kg of seed is required per hectare. The seed is drilled 30 × 30 cm and seedlings are thinned to leave a single plant per hole. The thinning allows adequate space for each plant to develop vigorously.

Fertiliser application

Niger is a hardy plant which survives on poor soils with adequate moisture. However low levels of nitrogen and phosphorus fertilisers are required to

Figure 7.10 Women winnowing sesame seeds at the Rural Research Station at Dongui, Chad.

increase yields. A rate of 25 kg each of N and P fertiliser per hectare is reported to be satisfactory on most soils.

Weed control

As niger seed is most often mixed with various grains and pulses, the crop benefits from the routine weeding given to the mixed canopy.

Pests and diseases

The major insect pest is the caterpillar of a moth, *Perigea capensis*, which feeds on the foliage. At low levels of attack, the pest is controlled by hand-picking and destroying the caterpillars. In more serious attacks, plants should be sprayed with 7% HCH (BHC).

A small, black, soft-bodied aphid, *Dactynotus carthami*, often becomes an important pest of niger in certain locations. The aphids suck the sap from tender plant parts causing stunted growth and reduced seed production. They rarely need to be controlled but, if necessary, they can be sprayed with nicotine sulphate solution.

Mildew caused by *Sphaerotheca* spp. and leaf-spot caused by *Cercospora guizoticola* are the main diseases of niger. However, their effects seldom justify any control measures.

Harvesting and storage

Once the fruits are mature, plants are cut with a sickle at the base of the stem and are stacked in a threshing yard for about a week. The harvest is spread to expose the branches to the sun. When dry, the seeds are retrieved by beating the stalks with sticks. The clean seeds are winnowed and sieved from the threshed material and are bagged for storage.

Sunflower

Sunflower, *Helianthus annuus*, belongs to the family Compositae (Asteraceae). It originated in Mexico, but is now grown throughout temperate regions as well as in the tropics and subtropics. It is an annual which can grow up to 5 m in height, although some dwarf varieties are only 0.6 m high.

Figure 7.11 Sunflower plants showing large yellow flowers borne terminally.

The yellow or orange flowers are up to 0.3 m in diameter (Figure 7.11). The plants usually produce a single stem and the flowers are borne terminally.

Sunflower seeds contain a valuable edible oil, high in polyunsaturated fatty acids. About 90% of sunflower seed production is utilised for oil extraction. The oil is mostly used to produce margarine, although some is used in the production of soap, paint and varnishes. Sunflower seeds are also eaten whole, in various confectionaries and are sold as birdseed.

Types and ecological adaptation

The sunflower plant produces shallow, lateral feeding roots and a deep taproot, which probably accounts for the plants' high drought tolerance. It thrives in areas with a rainfall of 300–800 mm, An adequate water supply is important during the flowering period, although high rainfall at this time results in a poor fruit set. While the seed

ripens, dry warm weather is necessary. Strong winds can cause the crop to lodge.

Sunflowers can be grown in most types of soil with moderate fertility. The crop does particularly well on light, rich, calcareous or alluvial soils with a pH between 6 and 7.5.

Varieties of sunflower are selected on the basis of their oil content and their yields. Seed colour also varies, and when selecting a variety for use in confectionaries or as birdseed, those with pale seeds are preferred.

Planting

The sunflower crop can be sown broadcast or in rows using mechanical seed drills. Flowering takes place 3–4 months after planting and the time to full maturity is 3½ months. Planting must take place at such a time as to allow adequate available moisture during flowering, but dry, warm weather during maturation. When the crop is rainfed a population density of 40 000 plants/ha is sufficient. For this a seed rate of about 10 kg/ha is used and drilled at a spacing of about 75 cm × 30 cm. In irrigated areas population densities of up to 80 000 plants/ha can be supported.

Fertiliser application

A good crop of sunflowers removes between 90–115 kg N, 17–20 kg P_2O_5 and 220–280 kg K_2O/ha from the soil. Pre-planting applications of super-phosphate at 36 kg P_2O_5/ha have been shown to increase yields, although too much nitrogen causes excessive vegetative growth, induces lodging and reduces the oil content of the seeds.

Weed control

Weeding twice while the plants are young gives adequate weed control. At the same time the plants should be earthed up to prevent lodging in strong winds. When the tall varieties are about 1 m high they produce sufficient shade to suppress weed growth.

Pests and diseases

As the crop is ripening, birds feeding on the seeds can present a severe problem. Timely harvesting and drying with the flower heads inverted significantly reduce bird damage.

Sunflowers are susceptible to several diseases. Rust, caused by *Puccinia helianthis*, can cause severe damage to leaves and a considerable reduction in yield. Leaf-spot, caused by *Septoria helianthis*, stem rot and head rot, caused by *Sclerotinia* spp., and grey mould, caused by *Botrytis cinerea* can also be a problem. Control measures for these diseases involve a crop rotation, with the sunflower crop being grown not more than once in 4 years. A rotation including cereals and grasses helps to prevent disease build-up. Planting resistant varieties and destroying all crop residues after harvesting are also good cultural practices.

Harvesting and storage

Most varieties of sunflower shatter readily so timely harvesting is essential. Dwarf varieties can be mechanically harvested, but tall varieties must be harvested by hand, usually in several stages as the whole crop does not ripen simultaneously. When the disc florets turn brown the flower heads are cut of and sun dried in stacks or inverted on individual spikes. The latter method reduces the amount of bird damage and promotes rapid drying. The moisture content should be reduced to 9–14%, depending on the proposed length of storage. Threshing is usually done by hand, using sticks to beat the dried flower heads. The properly dried seeds are bagged and transported for oil extraction.

Safflower

Safflower, *Carthamus tinctorius*, is thought to have originated in southern Asia. It is a member of the family Compositae (Asteraceae). The plant is an erect annual herb, 0.3–1 m in height, with serrate, spiny leaves and globular flower heads. The flowers are yellowish to orange-red and produce smooth four-angled achenes. The seeds yield a valuable drying oil which is the most unsaturated of all edible vegetable oils. It is mostly processed for human consumption, but is also used in the paint industry.

Ecological adaptation

The crop is cultivated in the drier regions of the tropics and in areas with a dry subtropical or Mediterranean climate. It tolerates a wide range of temperatures. Seeds normally require at least 4°C for germination while the seedling and bud stages can tolerate minimum temperatures of −7 and 0°C, respectively. The optimum temperature range for growth is 20–30°C. Safflower is cultivated both as a rainfed and an irrigated crop, requiring a minimum of 400–450 mm of available water for good seed yields. Even though the crop is drought tolerant, a rainfall regime of between 600–1000 mm is preferred for optimal growth and production. The crop prefers light, alluvial loams or medium-heavy soils, which must be deep and well-drained.

Planting

Thorough ploughing and harrowing of the soil is required. The seeds are drilled 40–50 cm apart at a rate of 15–20 kg/ha if sown in rows, or 40–50 kg/ha when broadcast. When naturally rich soils are utilised, heavy fertilisation is considered unnecessary as it produces profuse vegetative growth and reduced grain yields. In the absence of specific research data, it is suggested that safflower should be fertilised with nitrogen and phosphorus at rates similar to those used for other small grain crops. The crop should be kept weed-free.

Pests and diseases

Grey mould, *Botrytis cinerea*, attacks safflower heads in high-humidity areas. *Alternaria* leaf-spot occurs in areas with frequent rains, heavy dews and on irrigated lands. Rust, *Puccinia carthami*, is associated with warm temperatures and high humidity, particularly at the end of the rainy season. Rust-resistant varieties are available and should be used. To effectively check the occurrence of diseases, it is advisable to rotate the safflower crop with other non-susceptible crops. The safflower fly, *Acanthiophilus helianthi*, and soil nematodes can also cause considerable damage.

Harvesting and storage

As safflower varieties are non-shattering the crop is normally harvested when most leaves have turned brown and the seed rubs freely from the least mature heads. Handling the spiny varieties during harvesting can be difficult. Harvesting involves cutting the plants at ground level or uprooting the entire plants, which are then heaped for a few days to dry. The dried harvest is threshed by beating with sticks and cleaned by winnowing. For safe storage, threshed safflower seed should have a moisture content not exceeding 8–10%.

Oil palm

The oil palm, *Elaeis guineensis*, is a member of the family Palmae. Two types of oil are produced from the fruit; palm oil obtained from the pericarp and palm kernel oil. World production figures for palm oil and kernels are given in Table 7.14. Both oils are used in the manufacture of margarine, compound cooking fats and soap. Oil content of the pulp averages about 56%. Palm kernels yield 46–48%. Analysis of palm kernel cake indicates a content of 13–17% crude protein, 6–9% fat and 47–60% carbohydrate.

Types and ecological adaptation

Commercial production of palm oil is presently based on various varieties of *Elaeis guineensis*, although there are interests in *Elaeis melanococca* for breeding purposes. *E. melanococca* hybridises readily with *E. guineensis* and is dominant for such characters as large leaves, higher oil content and a shorter habit (Figure 7.12).

There are three distinct colour varieties of *E. guineensis*. The variety *virescens* has green immature fruits, which ripen to bright orange. The variety *nigrescens*, which is more common, has black fruits which ripen to deep red. The third variety is white (unpigmented) at both the immature and mature stages.

The oil palm is suited to lowland tropical climates where there is plenty of rain (at least 1300 mm) and an abundance of sunshine. It requires warm temperatures and must be grown at low elevations. A mean annual temperature of about 21°C is required for fruit production, the optimum temperature for growth being between 24–27°C. The best soils are rich alluvial or sandy loams with a clay subsoil, in which a good response to fertiliser can be obtained. The oil palm tolerates temporary flooding at the late seedling stage. It also tolerates acid soils.

The best land for the cultivation of oil palm is freshly cleared virgin forest because of the inherent fertility. A recently cleared bush follow or land that has been farmed can also be used. The land should be well-drained and the location must enjoy plenty of sunshine. Flat or gently undulating land is ideal.

On the basis of their genetic forms, commercial oil palms can be grouped into three types, depending on their shell thickness: the dura, the pisifera and the tenera types. Palms which are homozygous for the shell-less character produce fruits lacking shells and are called the pisifera type. They usually produce sterile females and their bunches normally abort, although some are able to set seed and are occasionally fertile. The dura type has a thick hard shell with a fibre ring enclosing the kernel. Pisiferas are used as pollen parents and are crossed with dura mother palms to

Table 7.14. Production of palm oil and palm kernel oil (1000 tonnes).

	Palm kernel				Palm oil			
	1965	1975	1980	1985	1965	1975	1980	1985
Nigeria	461	295	345	370 F	530	640	675	770
Benin	55	73	70	75 F	43	22	28	37
Brazil	170	240	266	250 F	na	7	16	22
Costa Rica	20	7	10	7 F	8	22	23	24
Honduras	0.9	2.9	3	3 F	1.5	9	11	15
Malaysia	350	na	1040	1210	1480	na	2575	4130
Indonesia	35	82	120	214	163	6	650	1148
Africa (total)	776	681	725	781	992	1278	1365	1474
World (total)	1062	1360	1812	2659	335	3263	5080	7578

Note: na = data not available.
Source: FAO production yearbook, 1965–85.

Figure 7.12 A short variety of oil palm grown in West Africa.

produce progeny which bear fruit with reduced shell thickness, enhanced mesocarp yield and increased content of palm oil and palm kernel oil. The hybrids, referred to as teneras, form the bulk of modern planting material in commercial estates. The tenera fruit characteristics are thus intermediate between those of dura and pisifera.

Planting

In commercial production, oil palm seeds or seedlings are mostly obtained from specialised government or private organisations involved in the genetic improvement of the crop. Germinated nuts or bare root seedlings are planted into black polythene bags filled with free-draining loamy soil. The seed must be planted on its side with the shoot side upwards, or a twisted seedling will result. Seedlings remain in the nursery for 6−9 months. One month after establishing the seedling, an N−P−K compound fertiliser is added at the rate of 5 g/polythene bag in the first month, rising to 50 g/bag in 6−9 months (Figure 7.13).

Weed control in the nursery must be thorough, and can be either by hand pulling or a combination of hand pulling and the use of a pre-emergence herbicide such as diuron. When planting out, the seedlings must be carefully selected, and any stunted or deformed seedlings should be discarded. Transplanting of seedlings into the field should start as soon as the first rains fall and should be completed just as the rains have become established. Later plantings may result in the death of young palms within one year of transplanting.

In areas where 'grass cutters' and other rodents are found, it is necessary to protect the young seedlings from these animals. A collar of 2.5 cm wire mesh 30 cm in diameter and 45 cm high or alternatively a fence of split bamboo should be placed around each plant at the time of transplanting. The collar should be inserted into the soil and securely pegged down.

Oil palms are planted at a 9.0 m triangular spacing, thus giving a population of 150 trees/ha. The area betwen the seedlings may be intercropped with annual food crops for the first three years.

145

Figure 7.13 Oil palm seedlings growing in polythene bags in a nursery.

Fertiliser application

For maximum yield and productivity, the application of fertilisers is essential for oil palm. Fertilisers should be broadcast on a clean-weeded circle, 7.5 cm away from each seedlings. For the first 6 months after transplanting to the permanent site, each plant should receive a mixture of 0.25 kg each of ammonium sulphate and potassium sulphate. The rate should be increased to 0.5 kg each during the first year of yielding. From the second year of yielding onwards, 0.66 kg of potassium sulphate only must be applied to each plant every year. In areas where symptoms of magnesium deficiency occur, 0.6 kg of magnesium sulphate should also be applied. The best time of the year for the application is as soon as the rains are well established.

Weed control

Leguminous cover crops may be established between the rows of oil palms to check the growth of

weeds. However, in the absence of cover crops the field should be slashed about three times during the year. Weeding should be clean around the individual trees to enhance the effectiveness of applied fertilisers and to allow easy identification and collection of fallen fruit, the presence of which indicates bunch maturity. Weeding clean around the palms also helps to reduce the risk of fire.

The use of chemicals to control weeds in oil palm plantations is becoming widespread. The use of terbuthylazine + paraquat (4.0 + 1.0 kg/ha) applied post-emeragence three times (for young palms) and twice (for older trees) is reported to be effective in checking weed growth.

Pests and diseases

The few important pests of the oil palm include the leaf miner and various boring beetles, the most serious of which is the rhinoceros beetle, *Oryctes rhinoceros*. This pest feeds on the leaves and the growing points and can thereby kill the entire

Figure 7.14 Harvested bunches of palm nuts ready for oil extraction.

keeping the seedlings adequately watered all the time can reduce the incidence.

Harvesting and storage

Given the proper care and management, oil palms can give a harvestable crop by the fourth year from planting. The right time to harvest bunches is when the fruits become loose and can be dislodged with relative ease. In practice, the farmer normally estimates the number of loose fruits on the ground or in the frond axil below the bunch as a guide when deciding whether or not to harvest. Experience has shown that if there are about ten loose fruits at the base of the palm, then the bunch is sufficiently ripe to harvest. If harvested earlier than this, the fruits may be too young and may contain little oil. If harvested too late, the free fatty acid content of the oil may affect its overall quality.

Bunches from young palms should be harvested with a special palm-harvesting chisel or a sharp pointed cutlass. Only dead leaves should be pruned off. With taller palms, a hooked Malayan knife should be used (Figure 7.14).

The yield of oil palm is highly variable depending on environment, management, planting material and the fertilisers used. Palms growing in environments with an even annual rainfall and a suitable soil will yield twice or three times as many fruits as palms on similar soil but where there are dry periods and inadequate applications of fertilisers. At the onset of bearing (in the fourth year), the yield should be about 1 tonne/ha. This should increase progressively to about 20 tonnes/ha of fruit after about 10 years, after which there is a gradual decline with age.

The palm oil is extracted by machine, although in the village setting it is more commonly done by hand, by pounding in a mortar and manually squeezing out the oil, or by boiling the fruits and skimming off the oil. Kernel oil is also obtained by soaking the kernels in water, crushing them, steeping then in hot water and skimming off the oil which rises to the surface. In traditional practice, the kernels are toasted in an open pot and as the oil oozes out it is collected. The residue left, known as the palm kernel cake, is used in animal feeds. Processed in these ways, palm oil and palm kernel oil may be stored for over one year. The free fatty acid content of the oil increases with storage.

plant. It may be controlled by spraying with insecticides as well as by observing proper field sanitation.

Anthracnose can be a problem during the nursery stage. Brown spots appear on the leaves and damage can be severe in overcrowded conditions. It can be controlled by proper spacing in the nursery and by spraying with Dithane M-45 or captan. Blast is a more serious nursery disease, causing decay and collapse of the root system so that the leaves become soft and yellow and the plants soon die. It is most severe during the early part of the dry season. There is no effective control of the disease, but

Coconut

Coconut, *Cocos nucifera*, belongs to the family Palmae. It is the sole species in the genus. The coconut palm supplies a variety of the needs of inhabitants of the wetter tropics. The nuts are used for food and drinks and the copra (the dried endocarp) and coconut oil are both used in the production of margarine and compound cooking fats. Production figures of coconut and copra across the world are given in Table 7.15. The unopened flowering spathe, when wounded, yields a delicious sap.

Types and ecological adaptation

Coconuts thrive in the humid tropics and do not survive outside the tropics. For good yields, the annual rainfall should exceed 1200 mm and the length of the dry season should be minimal. In coastal locations and on small islands with steep inland hills, where rainfall is less than ideal, coconuts still do well on the seepage water from inland.

High levels of sunshine are necessary for good yields. The preferred temperature range is 27–32°C. The crop can be grown successfully at a fair distance from the sea, but the majority of plants occur in coastal regions. Coconut palms grow best in deep, light, free-draining soils which encourage healthy root development. The plant also does well in heavier soils, provided they are also well-drained. It tolerates a higher degree of salinity than most other crops.

There are a number of distinct natural varieties,

the most important of which are the dwarf forms referred to as var. *nana*, and tall forms referred to as var. *typica*. Dwarf coconuts show distinct characteristics, with a degree of self-pollination. The tall varieties are mostly cross-pollinated and exhibit a gradation of characters. There are three basic nut colours: pale yellow, green and orange, all of which hybridise to produce a range of colours.

Dwarf coconuts are of little commercial interest and are seldom planted on a large scale because the plants are short-lived and yield little copra. However, they produce excellent immature nuts for fresh fruit. Selection for commercial production of copra is made from the nuts of high-yielding mother trees of the tall types. The seeds must be quick to germinate and vigorous in growth and they must also be as round as possible. These characters show a high correlation with the production of good yields.

Planting

The coconut is propagated only from seed. Nuts of high-yielding palms are selected. If liquid cannot be heard when the nut is shaken, it should be discarded. Coconuts exhibit a short dormancy of a few weeks after attaining maturity. The usual method of raising nursery plants is to plant mature nuts (fruits) in loose sandy soil beds about 30 cm wide and 15–20 cm deep. The nuts are laid horizontally in shallow troughs about 15 cm apart and lightly covered with soil. Coconuts may be planted with or without the husk. Removal of the husk is known to accelerate germination. There is no need

Table 7.15. Production figures for coconut and copra (1000 tonnes)

	Coconut				Copra			
	1965	1975	1980	1985	1965	1975	1980	1985
Tanzania	na	300	310	320 F	na	27	29	29 F
Ghana	191	311	300	108	3	17	10	7 F
Nigeria	200	90	90	90 F	6	9	10	10 F
Indonesia	4699	6942	10 900	10 754 F	482	885	991	1160
India	4999	4331	4500	4550 F	266	314	387	380
Mexico	770	960	710	655 F	168	145	120	120
Africa	1299	1549	1601	1567	89	158	167	198
World	26 636	30 896	35 422	34 661	3296	4368	4552	4507

Source: FAO production yearbook.

Note: na = Data not available.

F = FAO estimate.

to shade the nursery, but at the appearance of shoots, the soil should be watered regularly.

Seedlings are ready for transplanting to the field after about 5−6 months, at which time they have 6−10 large, thick, pencil-like roots which can be pulled upwards without causing them harm. The first seed leaves appear a few weeks after planting. At transplanting, the seedlings should have at least three leaves. Transplanting of seedlings is best done at the beginning of the rainy season. Seedlings selected for planting should only be those that are healthy.

Coconuts are usually planted on a square pattern of 10 × 10 m for tall palms grown on light and easily permeable soils, and 8 × 8 m on relatively poor soils. Malayan dwarf and other dwarf palms grown on good soils are spaced at 6 × 6 m, while on poor soils the spacing is reduced to 5 × 5 m. Coconut palms may also be grown in a triangular arrangement. In general, smallholders adopt a wider spacing than is suggested and interplant with cashews or mangoes. Planting annual crops between the rows, especially early in the life of the coconut plantation, is also a common practice, even at close spacings. Coconuts also provide protection against strong winds when they are intercropped with a wind sensitive crop such as banana.

Fertiliser application

Fertiliser application during the early seedling stage is important to encourage vigorous growth, early bearing and high initial yields. Potash is the main nutrient required, while nitrogen is necessary for leaf development and early flowering. Response to phosphate occurs only where soils are markedly deficient. A typical compound fertiliser mixture recommended per coconut palm is 0.2−0.3 kg of N, 0.2−0.3 kg of P_2O_5 and 0.5−0.7 kg of K_2O, applied in split doses at the beginning of the rains. The fertilisers are placed in a band encircling the tree, up to 1.5−2 m from the trunk. Farmyard manure and compost applications have also been found to be beneficial. Magnesium deficiency sometimes occurs and is corrected by foliar sprays.

Weed control

It is necessary to clean weed around each tree during the first few years of growth. In mature trees, weed and bush growth must be checked by regular slashing between the rows of plants. Intercropping with food crops, or the use of cover crops, also help to control weed growth.

Pests and diseases

The rhinoceros beetle, *Oryctes rhinoceros*, is the most serious pest in coconut plantations. The large, black adult beetles eat the new unopened leaves and can eventually destroy the growing point. As coconut palms cannot form side branches, destruction of the growing point results in the death of the palm. The bore holes formed by the beetles in the terminal buds can be readily identified and the beetles can be destroyed by forcing a strong wire down the holes. Good crop sanitation is an effective control measure. This involves the destruction of all decaying coconut palms, which are breeding sites for rhinoceros beetles. Control has also been achieved by the deliberate introduction of a virus disease to destroy the beetles. This method has been particularly successful in confined environments, such as the South Pacific Islands.

Coreid bugs, *Pseudotheraptus wayii*, can cause considerable reductions in yields. The bugs attack the young nuts, which subsequently either fall prematurely or produce deformed mature coconuts with a low copra content. Control can be achieved by spraying insecticides and by good crop sanitation.

Various fungal wilt diseases occur in coconut, caused by the fungi *Fusarium* spp. and *Helminthosporium* spp. One important wilt disease is the Cape St. Paul wilt. The symptoms of this disease include nut dropping, discolouration of the inflorescence and drying out of the lower leaves from the tips.

Harvesting and storage

With tall palms, the first harvest is usually taken 6−8 years after planting. It takes a further 5−8 years to reach the stage of full maturity. The time taken from setting to full maturity of fruits is 12 months. Since one inflorescence is produced every month, harvesting is done throughout the year. For best yields of copra (the white kernel dried for storage) and for the highest oil content, the nuts must be fully mature. Ripe nuts are also essential for the manufacture of dessicated coconut (the

149

a tonne of copra varies with the cultivar; in tall palms, it is from 3000–6000 and in dwarf palms from 6000–8000 nuts. The copra : nut ratio varies from 0.2–0.4 in tall palms, the copra being about 32% of the weight of the husked nut. The average yield varies from 2500–7500 nuts/ha. Copra contains 65–70% oil.

Both the copra and the oil extracted from it can be stored for over one year. However, storing copra for too long may increase the free fatty acid content of its oil.

Shea butter

The shea butter tree, *Butyrospermum parkii*, belongs to the family Sapotaceae, which includes other trees and shrubs of economic importance. The family is characterised by a milky latex in the stems and branches. The ovaries of the fruit possess several locules, with a solitary ovule in each. The shea butter tree occurs wild in the West African savanna and in Central Africa. The tree grows to a height of 9–12 m and has a bark that is rough and fissured, giving it the appearance of crocodile leather.

Shea butter fruits yield an edible oil which is locally used in the same way as palm oil for cooking, anointing and illumination. Industrial uses of shea butter oil are in the manufacture of soap and candles and in confectionery. The refined oil is edible and is utilised in the preparation of butter substitutes.

Types and ecological adaptation

Different varieties of the plant are known to occur throughout Africa, some having long fruits and others with more rounded berries. The trees of the southern parts of Sudan are believed to belong to the botanical variety *niloticum*.

The shea butter tree occurs naturally on lateritic slopes in the savanna zone of West Africa and throughout equatorial Africa where the rainfall is not too high, 1000 mm or less. The tree grows in areas with annual temperatures of 24–32°C, with a minimum of 21°C and a maximum of 36°C. Shea butter is the major local vegetable fat in the drier parts of the equatorial belt of Central Africa, where oil palms do not grow.

When the seed germinates it sends down a long

Figure 7.15 Coconuts are harvested by climbing up the trunk of the palm and cutting the base of suitable bunches.

grated and sun-dried white flesh, without the brown skin). Nuts for coir (the brown fibre round the nut) production are best picked at 11 months old. Drinking nuts are picked earlier, at about 7 months.

The nuts are harvested by climbing up the trunk in order to cut the base of suitable bunches. If the trees are short, knives or long poles may be used to harvest nuts from the ground. Cutting steps on the trunk of the tree should be discouraged as it harms the trees (Figure 7.15).

Yields vary greatly with the environment, the age of the plantation and the standard of cultivation. The number of coconuts required to produce taproot before producing any leaves. This type of growth has probably developed as a survival mechanism against fires and drought. The trees are deciduous, shedding their leaves in the dry season to conserve moisture. The sweet-scented white flowers are produced before the leaves, at the end of the dry season.

Virtually no work has been undertaken on the agronomy of the shea butter tree, which almost always grows wild. The fruits often ripen on the tree and fall to the ground, where the seeds eventually germinate. Germination of the seeds and vegetative growth are extremely slow. The tree attains a height of only 7–10 cm after two to three years. This slow rate of vegetative growth is probably a consequence of the diversion of resources towards the development of the long taproot system, which sustains the developing seedling.

Planting

The seed may be planted at the stake or grown in a nursery and transplanted later to the field. In the latter case, the nuts are sown in the nursery at 20 × 15 cm intervals in deep soil. The first transplantation takes place during the second dry season after planting and is followed by a second transplanting two years later. The seedlings are planted in their final position the following year, that is four and half years after sowing. The field spacing is at 1.5–2.0 m intervals in rows 8 m apart, or 8 m triangles.

No data on the fertiliser requirements of the shea butter tree are available. Protection against weeds reduces competition for water and nutrients.

Pests and diseases

Various pests attack the shoots, leaves and fruits. The adults of the beetle *Curimosphena senegalensis* bore into all parts of the growing shoot and lay their eggs. The larvae also cause damage to the leaves and young fruits. Control may be achieved by spraying with a systemic insecticide. A leaf-spot disease caused by *Phyllosticta* spp. produces round zonate spots on the older leaves. Spraying with Dithane M-45 controls the disease.

Harvesting

The tree begins to bear after 12–15 years and reaches full bearing capacity after 20 years. The shea butter fruit takes from four to six months to mature. The seed (nut) is embedded in a pulp, which is sweet and edible when ripe. The white kernel of the nut is rich in oil (up to 50%), which is solid below 28°C.

The harvesting of fruit from wild trees entails the collecting of freshly fallen fruits, any fruits with germinating seeds being discarded. The kernels are sun-dried or kiln-dried in preparation for oil extraction.

Other oilseed crops
Linseed

Linseed, *Linum usitatissimum*, is a cool-season crop that may be grown during the dry season in tropical environments. The crop is adapted to medium or heavy-textured soils. It grows best at temperatures below 21°C. It is frequently grown at high altitudes (1800–2500 m) in the tropics. Linseed belongs to the family Linaceae.

An important factor in determining the yield of linseed in the tropics is the time of sowing. Experiments conducted at several locations in the Sudan savanna zone of Nigeria show that planting soon after the onset of the dry season gives maximum yields (Table 7.16).

Table 7.16. Mean yield of irrigated linseed sown on different dates in the Sudan savanna zone of Nigeria

Sowing date	Seed yield (kg/ha)
October 14	1963
November 4	1346
November 25	950
December 16	847
January 6	441
January 27	206

The crop responds to the application of nitrogen but not phosphate fertilisers. Between 50 and 90 kg N/ha is required for optimum yields. Rust (*Melampsora lini*), wilt (*Fusarium oxysporum*), leaf spot (*Alternaria lini*) grey mould (*Botrytis cinerea*) and *Sphaerella linorum* cause appreciable losses. The main form of control is to use resistant varieties.

Linseed plants are cut as close to the ground as

possible. The cut plants are allowed to dry, after which they are threshed, either manually or using appropriate mechanical threshers. Clean seed is obtained by winnowing and sieving. Linseed contains 40–60% of a drying oil.

Neem

Neem, *Azadirachta indica*, is a member of the family Meliaceae. It is a large, glabrous, evergreen tree, 12–18 m high, with a straight trunk and long spreading branches. It originated in the open scrub forests of the dry zone of Burma but now occurs thoughout tropical Africa and Asia. Neem is an important shade tree in dry areas, such as in northern Nigeria. It is also a useful browse tree and has been found to be partially vermifugic for goats and other browsing animals in East Africa. The seeds yield an oil which is mainly used for lighting, soap making and medicinal purposes. The oil content is as high as 40–45%.

The tree tolerates maximum shade temperatures as high as 49°C. It thrives best in drier climates, with a rainfall between 450 and 750 mm. It is moderately salt resistant and can be grown on a wide range of soils.

Seedlings can be easily raised in nurseries and then transplanted, but direct sowings have proved even more successful than transplanting. Seeds should be fully ripe and sown soon after collection. Seedlings raised in nursery beds are ready for transplanting when they reach a height of 8–10 cm.

Neem fruits are beaten down from the tree and then swept up from the ground, which is cleaned beforehand. The fruits are sun-dried and decorticated. Well dried seeds can be stored for 4–6 months in well ventilated rooms.

Other potentially important oilseed crops which are yet to assume economic importance in the tropics include:
(a) African walnut (*Coula edulis*) with about 30% oil.
(b) Jojoba (*Simmondsia chinensis*), with about 50% oil.
(c) Soursop (*Annona muricata*), with about 24% oil.
(d) Sweetsop (*Annona squamosa*), with about 15% oil.

CHAPTER 8

Grain legumes

The grain legumes include all the cultivated plants or wild species belonging to the family Leguminosae (Fabaceae). The botanical characteristics of this group of food plants are common to the whole family. The Leguminosae is a large family made up of 600 genera and 13 000 species. The grain legumes are distinguished in this family only by virtue of having their seeds used for food. They include various beans and peas as well as the lupins and bambarra groundnuts.

The three subfamilies of the Leguminosae are sometimes classified as families: the Mimosaceae, Ceasalpiniaceae and Papillionaceae. The first two families are mainly trees and shrubs while all the grain legumes of agricultural importance are members of the family Papillionaceae.

Grain legumes have been recognised as being important in human and animal diets since the beginning of agriculture. The importance of grain legumes is largely due to the high nutritional value of their seeds (Table 8.1). They form a good source of protein, especially in areas where there is a shortage of animal protein.

The protein content of the leaves and ripe seeds of most cultivated grain legumes is 4–10% and 18–40%, respectively. Estimates of the quantities of legume grains consumed in Africa range from 11.0 g per head per day in Liberia and Madagascar to 160 g per head per day in Chad. On average, the consumption in most countries is 30–80 g per head per day. Some grain legume crops are also used in soil conservation, where they provide a mat over the soil surface and are routinely grazed.

Cowpea

The cowpea, *Vigna unguiculata*, is an annual plant. The different varieties have spreading, semi-upright or erect growth habits with purple, pink, white, blue or yellow flowers. Most varieties have pods which hang downwards, but in a few varieties the

Table 8.1. Nutritive value of important grain legumes (per 100 g edible portion)

Common name	Botanical name	Calories	Water	Protein	Fat	Fibre g	Carbohydrate (g by difference)	Ash
Groundnut	*Arachis hypogaea*	546	5	25.6	43.3	3.3	23.4	2.5
Pigeon pea	*Cajanus cajan*	343	11	20.9	1.7	8.0	62.9	3.5
Chickpea	*Cicer arietinum*	358	11	20.1	4.5	4.9	61.5	2.9
Horse gram	*Dolichos uniflorus*	338	10	22.0	0.5	5.3	66.3	0.9
Soyabean	*Glycine max*	335	8	38.0	18.0	4.8	31.3	4.7
Lathyrus pea	*Lathyrus sativus*	293	10	25.0	1.0	1.5	61.0	3.0
Lentil	*Lens culinaris*	346	11	24.2	1.8	3.1	60.8	2.2
Lima bean	*Phaseolus lunatus*	341	11	19.7	1.1	4.4	64.8	3.4
Kidney bean	*Phaseolus vulgaris*	341	11	22.1	1.7	4.2	61.4	3.8
Pea	*Pisum sativum*	346	11	22.5	1.8	5.5	62.1	5.5
Broad bean	*Vicia faba*	343	11	23.4	2.0	7.8	60.2	3.4
Mung bean	*Vigna aureus*	340	11	23.9	1.3	4.2	60.4	3.4
Cowpea	*Vigna unguiculata*	342	11	23.4	1.8	4.3	60.3	3.5
Bambarra nuts	*Voandzeia subterranea*	367	10	35.0	18.0	4.5	32.5	na

Note: na = Data not available.
Source: Anon., 1975.

pods point sideways or upwards. The colour of the seeds varies from white, cream-coloured, purple, brown, mottled brown to black. Cowpeas are mainly short-day to day-neutral plants with well developed root systems which sometimes have nodules (Figure 8.1).

Types and ecological adaptation

The cultivated cowpeas can be divided into three species, namely, *Vigna unguiculata*, *Vigna sesquipedalis* and *Vigna catjang*. The cowpea, or black-eyed bean, refers specifically to the most widely grown cultivar of African origin, *V. unguiculata*. The other two, *V. sesquipedalis* and *V. catjang* originated in India and the Far East and differ from the African forms in length of pod and habit. *Vigna sesquipedalis* is a strongly climbing form with pods up to 60 cm long, whereas *Vigna catjang* is an erect bushy plant with up-tilted pods from 6 to 8 cm long.

Cowpeas are found throughout the savanna zones from the forest margin to sub-desert conditions, at altitudes below 1500 m. The major producing countries include India, Niger, Burkina Faso, Uganda and Senegal. In high rainfall areas, cowpeas are attacked by a wide range of insect pests and diseases which cause very low yields unless they are controlled.

Cowpeas need warm conditions (Table 8.2), and thrive from the equator to 30° north and south of the equator. The crop is drought resistant and is intolerant of waterlogging. Cowpeas perform best in soils with a pH between 6.6 and 7.0.

Cowpeas are grown on an extremely wide range of soil types, varying from sands to the heaviest 'cracking' clays. Because the crop is sensitive to waterlogging, the site must be free-draining.

Planting

Cowpeas normally require soils that have been well cultivated, including the use of a rotary hoe where possible. Recent studies suggest, however, that zero tillage or partial tillage may be as effective as conventional tillage if weeds and soil pests are adequately controlled, and soil structure is reasonable.

The choice of planting on ridges or on the flat depends on soil conditions, the need for water conservation, the dangers of erosion and whether or not the required planting density can be

Figure 8.1 Cowpea variety with pods which hang downwards, growing in Ghana.

accommodated using available ridging implements. Ridging is advised in areas which have a marked water surplus during the growing season, but the ridges should not be so widely separated as to prevent a closed canopy from forming at peak leaf development. Ridging should also be used where there is a need to conserve limited rainfall, especially on sloping lands.

All grain legumes are propagated by seed. The general principle of using large, plump, healthy seeds is applicable. The choice of variety suitable for any given location is arrived at with a thorough understanding of the environment, the potential yield capability of the seed and the taste of the

Table 8.2. Temperature requirements for different growth stages of cowpeas

Growth stage	Maximum	Minimum	Optimum
		(°C)	
Germination	42	12	25−35
Vegetative growth	40	10	25−35
Flowering and grain development	35	18	25−30

154

consumers. In general, people in the tropics prefer large, rough, white or brown-seeded varieties for human consumption.

In addition to the appearance of the seed, breeders are interested in varieties that are adapted to the seasonal and variable water supply in the tropics. Hardy, drought-resistant, high temperature tolerant and sun-loving crops are preferred. The use of a variety is more widespread if it can also tolerate relatively poor, nutrient deficient soils.

There are local differences in the variety recommended for use in most tropical countries where cowpeas are grown because a variety which yields well in one locality does not necessarily perform well in others. Table 8.3 gives a list of common, high-yielding well-adapted varieties utilised in some countries in the tropics.

The time to plant cowpeas at a particular location depends on the availability of water, the length of the growing period of the variety used, its sensitivity to temperature and photoperiod and the seasonal fluctuations of temperature and daylength at the site. Variations in temperature and day-length at any location within the tropics are relatively small compared with those in temperate regions. Variations in temperature may significantly affect cropping time at high altitudes and during the dry season. Daylength variations often necessitate precise timing in the planting of wet season crops of daylength-sensitive cultivars, if these cultivars are not to remain on the field for too long before flowering.

The main determinants of planting time in most parts of the tropics are, however, a compromise between the time when water becomes available on a stable or continuous basis and the need for dry weather at crop maturity. Since most grain legumes are traditionally intercropped, the other crop in the mixture, most often a cereal, is planted as soon as the rains are established, while the sowing of the grain legume is delayed for a further three to four weeks.

Various monocropping experiments involving cowpeas have indicated that the crop can benefit from early planting (Table 8.4). However, even though late planting reduces grain yield, the later the planting date the lower the incidence of most cowpea diseases. The yield advantages of early planting of cowpea may, therefore, be lost if diseases such as scab, septoria leaf spot or brown blotch are not adequately controlled.

Table 8.3. Some improved cowpea varieties cultivated in tropical Africa

Cultivar	Origin
Black eye	Botswana
Rhenostar	Botswana
Bambey-21	Senegal
Gorom Local	Burkina Faso
IAR 341	Nigeria
IAR 48	Nigeria
IAR 355	Nigeria
IAR 1696	Nigeria
Mougne	Senegal
TN 88−63	Niger
TN 13−78	Niger
KN−1	Burkina Faso
TVx 1999−01F	Burkina Faso
VITA-4	Burkina Faso
TVx 1948−01F	Burkina Faso
TVx 3236	Burkina Faso
Kpodiguegue	Benin

Table 8.4. Effect of time of planting on grain yield (kg/ha) of cowpea at three locations in Nigeria

Sowing date	Ibadan (7°23′N, 3°35′E)	Daudawa (11°38′N, 07°09′E)	Kano (12°03′N, 08°32′E)
April 19	1804	—	—
May 4	1277	—	—
June 15	—	1578	1546
June 30	—	1031	1750
July 15	—	860	1260
July 30	—	814	953
August 15	—	701	650
August 30	—	828	368

The response of cowpeas to spacing depends on the characteristics of the variety being tested. However, spacing has no significant effect on seed yields of varieties with a spreading habit. Spacing cowpeas at 15 × 90 cm to 30 × 60 cm achieves populations densities that are considered optimum for most varieties. Experiments conducted at Kano, Nigeria, show that seed yields of both upright (Acc. 261) and spreading (Acc. 588/2) varieties may be expected to increase when the population density increases from 43 000 to 86 000 plants/ha. However, no significant seed yield benefit could be expected from a further increase in population to 172 000 plants/ha (Table 8.5). In fact, the seed yield of the spreading variety is reduced at such close spacing.

Table 8.5. Mean grain yield of cowpea as affected by variety and population density at Kano, Nigeria

| Population density (plants/ha) | Yield (kg/ha) | |
	Acc. 588/2	Acc. 261
43 000	1416	483
86 000	1480	702
172 000	1220	766

Fertiliser application

Most cowpeas have active nodules on their roots which fix a substantial proportion of the nitrogen required by the crop. The innoculation of most cultivars in current production with *Rhizobium* bacteria is largely unnecessary, as local strains of the bacteria have been shown to be highly effective, particularly in tropical Africa.

The cowpea crop is normally expected to make use of any residual fertiliser from the cereal crop in a mixed canopy. As a sole crop, however, it has been shown to respond to phosphates and sometimes to nitrogen and potassium (Table 8.6). Other mineral nutrients such as sulphur and molybdenum, that may be required for optimum performance of the symbiotic bacteria, are assumed to be supplied in adequate quantities by the phosphatic fertiliser.

Weed control

While some of the most vigorous varieties of cowpea can compete successfully with grass weeds, clean weeding is an important factor in the reduc-

Table 8.6. Main effects of nitrogen, phosphorus, potassium and molybdenum fertilisers on grain yield of cowpea in Nigeria

Fertilizer treatment	Kano	Samaru (kg/ha)
kg N/ha		
0	1374	1120
29	1420	1134
kg P$_2$O$_5$/ha		
11	1223	1135
22	1576	1191
kg K$_2$O/ha		
0	1426	1051
53	1373	1203
kg Mo/ha		
0	1383	1173
0.03	1415	1081

tion of leaf diseases and nutrient competition early in the growth of the crop, and in the reduction of the competition for moisture later in the growing season.

Weeds offer serious competition to cowpeas, especially in the first 30 days of growth, and must be controlled for optimum performance. Conventional cultivation before planting controls weeds for about 20 days only and some form of weeding is necessary if the yield is not to suffer. In most parts of the tropics cowpeas are either weeded using hoes or using animal-drawn and to a lesser extent tractor-drawn implements. Herbicides are not widely used except on large-scale farms. In Nigeria, a combination of metolachlor and Metabrom at 1.0 + 1.0 kg/ha, metolachlor and diuron at 1.5 + 0.8 kg/ha or norflurazon and diuron at 0.5 + 0.8 kg/ha have been found effective as pre-emergent herbicides for cowpea crops.

Pests and diseases

Cowpeas are susceptible to a large number of insect pests, especially when varieties are grown outside the conditions for which they have been selected. Some important pests of cowpea, with the time of their peak activity in relation to the growth cycle of the crop, are illustrated in Figure 8.2. The current recommendation for the control of insect pests on cowpeas in the savanna zones of

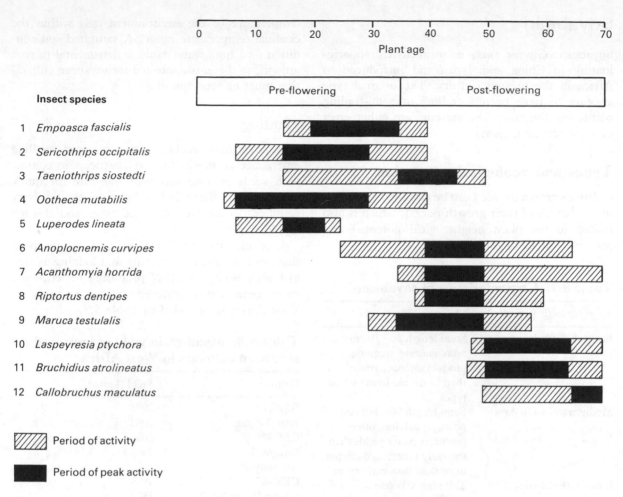

Figure 8.2 Diagram to show the time of occurrence of cowpea pests and their peak activity, in West Africa.

Nigeria consists of four applications of cypermethrin (100 g/ha) at weekly intervals, followed by two applications of a mixture of cypermethrin and dimethoate (50 g + 500 g/ha) also at weekly intervals. The cypermethrin controls thrips while the dimethoate controls coreids.

Diseases reported on cowpea include root-rot caused by *Sclerotium rolfsii*; wilt caused by *Colletotrichum lindemuthianum*; rust caused by *Euromyces vignae*; leaf-spot caused by *Erisiphe polygoni* and *Cercospora* spp.; scab caused by *Elsinoe* spp.; and a viral mosaic. These diseases are best kept under control by the choice of a good cultivar. Most cowpea diseases can be controlled with bi-weekly applications of a mixture of Benlate and Dithane M-45 at a rate of 2.5 litres/ha ultra low volume or 2.5 litres in 100 litres of water/ha very low volume, beginning six to seven weeks after planting.

Harvesting and storage

Because of the indeterminate growth habit of most varieties of cowpea, the pods tend to mature in succession along the flowering stem. This factor, as well as the possibility of shattering, necessitates the harvesting of mature pods as they become dry. When fully dried, the pods are threshed to release the seeds. This is achieved using appropriate threshing machines or by beating lightly with a stick. The seeds can then be transferred into airtight containers for a short period of storage. For longer periods of storage, 1.25 g of lindane (0.5% dust) per kg of seed should be mixed with the seeds before sealing in airtight containers. If the seeds are to be used for replanting, then a higher concentration of dust could be applied to ensure complete freedom from insect damage.

Soyabean

Soyabean, *Glycine max*, is an ancient crop originating in China and Japan and introduced to Africa in the present century. The original type appears to have been a trailing or semi-trailing plant, but the cultivated varieties are either erect or have reclining stems.

Types and ecological adaptation

Cultivars grown for seed can be classified according to the length of their growth period, which is also linked to the plant height, yield potential and protein content (Table 8.7).

Table 8.7. Characteristics of soyabeans

Type (growth period)	Characteristics
Early (85–95 days)	Stem length 30–70 cm; homogeneous ripening; production more reliable than in the medium and late types
Medium (95–110 days)	Stem length 50–110 cm; prone to lodging; often produces better yields than the early types; seeds richer in protein than early types
Late (110–125 days)	Tall plants; prone to lodging; under favourable conditions very good producers; seeds have high protein content

Soyabeans are tolerant of a wide range of climatic conditions, but cultivars introduced into tropical regions from their normal temperate environment frequently show a shortening of their growth period and reduced productivity. The soyabean plant is photoperiod sensitive. The varieties are never neutral, but possess varying degrees of long-day or short-day dependence.

The crop is relatively drought and high-temperature tolerant. Germination is slow below a temperature of 15°C (minimum temperature for germination is 11°C) while the optimum temperature ranges between 30 and 36°C, depending on the cultivar. Plant growth is best between 27 and 32°C.

Soyabeans will grow on most soil types if their nutrient and water requirements are met and the temperature of the environment falls within the cardinal temperature range. A saturated soil condition or a high water table is detrimental to root growth, so that a site selected for soyabean cultivation must be free-draining.

Planting

Well-pulverised seedbeds promote rapid seedling emergence as moist, fine soil particles in contact with seeds provide sufficient moisture for quick emergence. Tillage before seeding should bury residues, loosen the ploughed layer and destroy weeds.

It is advisable to choose varieties of soyabean that are resistant to shattering and lodging as well as high yielding. A list of promising varieties and their mean yields achieved at four locations in West Africa is provided in Table 8.8.

Table 8.8. Mean grain yield of improved soyabean cultivars in West Africa

Cultivar	Yield (kg/ha)
Bossier	2098
Imp. Pelican	2082
CES 486	2080
Samsoy-1	2000
Samsony-2	2000
CES 407	1929
Chung Hsing No. 1	1851
Chippewa 64	1827
Kent	1816
JGm 393 (S)	1661
Clark 63	1586
Amsoy	1445
Shelby	1425
Malayan	1214
35s/58	1127
Hale 3	1372

Source: ITTA Annual Reports, 1982–83.

Most grain legume plants have nodules on their roots containing bacteria which fix a major part of the nitrogen required by the crop. Soyabeans are an exception, requiring the introduction of special strains of *Rhizobium* bacteria. The *Rhizobium* strains may come from the wild *Glycine javanica* or may be applied to the seed at planting. A list of *Rhizobium* strains that can be used and their host plants is provided in Appendix 5.

Optimum planting dates are determined by

moisture supply and distribution as well as the variety used, which can be early, medium or late maturing. Table 8.9 gives the results of sowing date experiments conducted at Yandev and Samaru in the Guinea savanna of Nigeria.

Table 8.9. Mean grain yield (kg/ha) of soyabean as affected by sowing date at two locations in Nigeria

Sowing date	Yandev (7°23′N, 9°01′E)	Samaru (11°11′N, 07°38′E)
June 1	1250	1521
June 15	1460	1150
June 29	1736	1134
July 13	1763	1047
July 27	720	653
August 10	– *	387

* Sowing date not included.

Varieties differ in their ability to produce good yields with different row widths and planting densities. The early-maturing varieties tend to yield more when planted in narrow rows, but this consideration must be balanced against the need for a row wide enough to allow the use of weeding implements. Lodging may also be a problem with narrow rows, especially if the plant population density is too high. The current recommendation in Nigeria is for a spacing of 30 × 15 cm (about 220 000 plants/ha), although studies at IITA in Ibadan have shown that some varieties continue to have increased yields at populations as high as 555 000 plants/ha (Figure 8.3). Soyabeans intercropped with maize are shown in Figure 8.4.

Fertiliser application

Soyabeans are not considered to respond well to the direct application of fertilisers. However, the

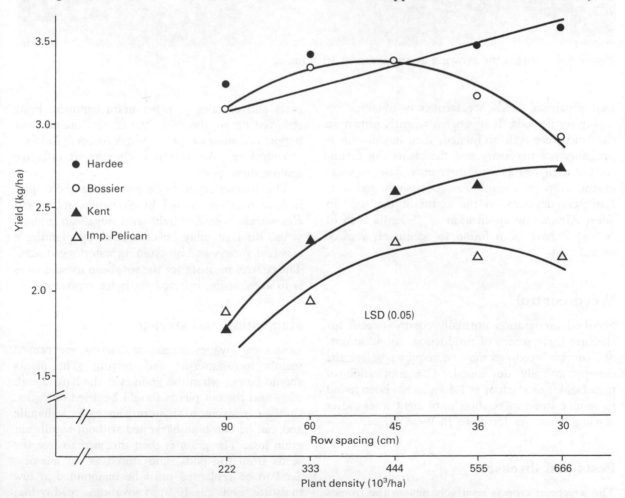

Figure 8.3 Effect of row spacing on seed yields of four soyabean cultivars.

Figure 8.4 Soyabeans growing as an intercrop with maize.

best soyabean yields are frequently obtained on highly fertile soils. If soyabeans recently grown on the land were well nodulated, then inoculation is probably not necessary and the plants can manufacture their nitrogen requirements. The response of the crop to phosphorus, potassium and zinc fertilisers depends on the status of the soil. In West Africa, the application of 45 kg/ha each of N and P have been found to adequately replace inoculation.

Weed control

Seedbed preparation normally controls weeds for the first three weeks of production. An additional one or two weedings may be required before the canopy is fully developed. The application of metolachlor or alachlor at 1.5 kg/ha has been found to reduce weed infestation until eight weeks after sowing at various locations in West Africa.

Pests and diseases

The soyabean crop is relatively new to the tropics and is not, therefore, afflicted with as many insect

pests and diseases as other grain legumes. Pests reported on soyabeans in the tropics include leaf hoppers (*Empoasca* spp.) and pod borers (*Laspeyresia ptychora*). Azodrin and Surecide are effective against these pests.

The diseases reported on soyabean in the tropics include root rot caused by *Sclerotium rolfsii* and *Rhizoctonia solani*, blight and soyabean mosaic virus. Root rot may be controlled by planting a resistant variety and by avoiding waterlogged sites. The control measure for the soyabean mosaic virus is to avoid using infected seeds for replanting.

Harvesting and storage

Losses of soyabean grain at harvest are caused mainly by shattering and lodging. The plants should be cut when the grain is at the hard dough stage and the cut plants should be dried on racks. Cultivars resistant to shattering are easier to handle and can be combine harvested without significant grain loss. The grain is then threshed to free the seeds from the pods, and stored as for cowpea. Seed to be replanted must be maintained at low moisture levels (9–10%) as soyabeans readily lose their viability if adequate care is not taken.

Figure 8.5 A fruiting branch of a pigeon pea plant.

Pigeon pea

Pigeon pea, *Cajanus cajan*, belongs to the family Leguminosae (Fabaceae). The crop, which is a native of Africa, is known by the common names of pigeon pea, red gram, Congo bean, dhal, yellow dhal and Angola pea. The plant is an erect, branching shrubby, glandular, pubescent annual (or short-lived perennial) with a very well developed taproot (Figure 8.5).

Pigeon pea is an important crop in India (Figure 8.6). It is found throughout Africa although it is important only in limited areas, including the Lango district of Uganda. Other than providing a proteinaceous food crop, pigeon peas are also used for erosion control and as a nurse crop for cocoa, coffee, tea and various other tree crops.

Types and ecological adaptation

The pigeon pea plant is one of the most deep-rooted and drought resistant of all legumes. It survives well even on the poorest soils and in the

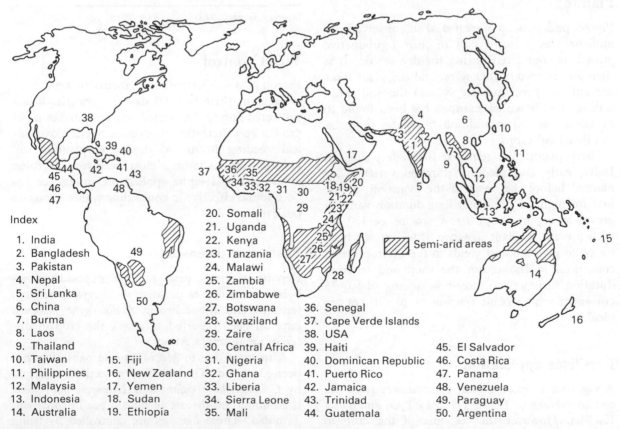

Index

1. India
2. Bangladesh
3. Pakistan
4. Nepal
5. Sri Lanka
6. China
7. Burma
8. Laos
9. Thailand
10. Taiwan
11. Philippines
12. Malaysia
13. Indonesia
14. Australia

15. Fiji
16. New Zealand
17. Yemen
18. Sudan
19. Ethiopia

20. Somali
21. Uganda
22. Kenya
23. Tanzania
24. Malawi
25. Zambia
26. Zimbabwe
27. Botswana
28. Swaziland
29. Zaire
30. Central Africa
31. Nigeria
32. Ghana
33. Liberia
34. Sierra Leone
35. Mali

36. Senegal
37. Cape Verde Islands
38. USA
39. Haiti
40. Dominican Republic
41. Puerto Rico
42. Jamaica
43. Trinidad
44. Guatemala

45. El Salvador
46. Costa Rica
47. Panama
48. Venezuela
49. Paraguay
50. Argentina

Semi-arid areas

Figure 8.6 Countries where pigeon pea crops are grown.

driest conditions. Pigeon peas are susceptible to damage by cool temperatures, excessively saline soils and waterlogging. They perform well on alkaline and moderately saline soils, up to pH 8, but also seem well adapted to a soil pH as low as 5. In India, the crop is grown in areas with 500–1500 mm of rainfall and a temperature range between 20°C and 40°C.

Breeding and selection of pigeon pea seeds has been carried out with the objective of producing high yielding, photo-insensitive, nutritious varieties. The International Crop Research Institute for the Semi-Arid Tropics (ICRISAT) in India has produced a range of varieties adapted to the different lengths of growing season found in India. Some of these varieties are also disease and insect resistant.

The crop is highly adaptable to different soil conditions, but thrives best on deep loams, free from excessive amounts of soluble salts. The site must be free-draining as the crop is very sensitive to poor aeration in the soil.

Planting

Pigeon peas depend a great deal on conserved soil moisture, as a major part of their reproductive growth is completed during the dry season. It is therefore necessary that water and the plant roots are able to penetrate deeply into the soil. Sub-soiling of soils with a hardpan has been found to be advantageous, in addition to regular tillage as in other field crops.

Early planting is required for high yields. In India, early and medium duration varieties are planted before the onset of the monsoon, in the first fortnight of June. For long duration varieties grown as a pure crop, a row spacing of 120 cm with a plant to plant distance of 60 cm has been found to give the best yields in the environmental conditions in India. For the short and medium duration varieties, an inter-row spacing of 50–75 cm and an inter-plant spacing of 20–30 cm are ideal.

Fertiliser application

A vigorous crop of pigeon pea removes from the soil an average of 132 kg N, 25 kg P_2O_5 and 64 kg K_2O/ha. However the response of the crop to direct applications of nitrogenous fertilisers is generally negligible or negative, because the nitrogen required by the crop for growth and development is amply supplied by fixation in the root nodules. pigeon peas show a very positive response to applications of organic manures and phosphate fertilisers (Table 8.10). About 60–100 kg of P_2O_5/ha is considered to be adequate on most soils. Pigeon peas are very susceptible to zinc deficiencies. A soil application of 2–4 ppm zinc or a foliar spray of 0.5% zinc sulphate with 0.25% lime have both proved effective in controlling this deficiency.

Table 8.10. The affect of phosphate fertilisers on the mean grain yield of pigeon pea at New Delhi, India

Level of P_2O_5 (kg/ha)	Grain yield (kg/ha)
0	1640
25	1910
50	2010
75	2080
100	2220

Source: Singh (1973).

Weed control

Pigeon peas are particularly sensitive to weed competition in their first 60 days of growth. When protected during this period, the crop makes rapid growth and thereafter suppresses weeds. Mechanical weeding 20 and 45 days after planting will keep pigeon pea farms almost weed free. Nitrofen at the rate of 1 kg/ha applied pre-emergence has been found effective in controlling weeds in pigeon pea crops.

Pests and diseases

Reports of insect pests on pigeon pea have not been extensive. Scale insects (*Icerya purchasi*), termites and the eelworm *Meloidogyne javanica* have all been reported to attack the crop in different areas across Africa.

A root rot due to *Macrophomina phaseolina* has been identified. The leaves are sometimes attacked by *Colletotrichum cajani* and *Uromyces* spp. Wilt is an important disease of pigeon pea in India and Trinidad. These diseases are controlled by using resistant varieties.

Harvesting and storage

The crop ripens irregularly and may be picked green for canning. The pods shatter when dry, so the final drying must be done after harvest. Threshing is carried out to obtain the seeds, which are stored in clean bags before milling.

Chickpea

Chickpea, *Cicer arietinum*, also referred to as gram, hommes, garbanzo beans, Bengal gram and Egyptian pea, is not known to grow in the wild. It is the world's third most important pulse crop and the most important legume grown in India. Chickpeas are also cultivated in the Middle East, the Mediterranean and Ethiopia (Figure 8.7).

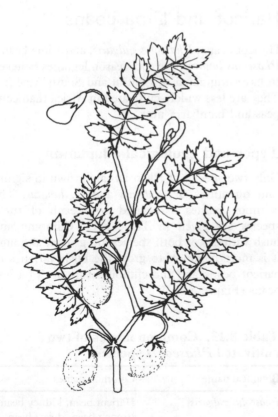

Figure 8.7 A branch of a chickpea plant showning mature pods.

Types and ecological adaptation

Chickpeas are cultivated during the cool season in the tropics. They are primarily grown in low rainfall areas but give good yields under irrigation. Excessive rains soon after sowing or at flowering can harm the crop. They are grown on soils ranging from light sands to heavy loams and are sensitive to saline and alkaline soils as well as soils with a high pH.

Most of the research work evaluating chickpea germplasm and breeding new chickpea varieties has been undertaken in India. ICRISAT currently has over 9000 germplasm lines from which varieties that are high-yielding, early or late-maturing or resistant to diseases have been isolated.

Like pigeon pea, the chickpea crop is adaptable to different soil conditions, but thrives best on deep loamy soils which are free-draining and free from excess amounts of soluble salts.

Planting

The tillage requirements of the crop vary according to soil type. On heavy textured soils, a rough seedbed is recommended. The depth of planting is an important factor for rainfed chickpeas as it affects both the initial germination and the subsequent mortality. A planting depth of 10 cm is recommended for better germination and reduced seedling mortality.

The date of planting is considered to be the most important factor in determining the performance of chickpeas in the tropics (Table 8.11). In India, current recommendations are for chickpeas to be planted from 15–30 October in the semi-arid north Indian plains, and from 1–15 October on the peninsular, after the monsoon rains are over.

Table 8.11. The effect of sowing date on the grain yield (kg/ha) of three chickpea varieties at Pantnagar, India

Sowing date	G-130	H-355 (kg/ha)	T-3
October 1	752	367	390
October 15	968	417	830
October 30	1260	445	695
November 15	2146	1661	1885
November 30	2399	1787	2176

Source: Anon., 1975.

Chickpeas readily adjust to the available space. Generally, yields have been higher with a 30 cm row spacing than with 45 or 60 cm rows, for a fixed plant population density. Seed rates of 75–100 kg/ha are sufficient for most varieties of chickpea.

Fertiliser application

A starter dose of nitrogen at a rate of 15–25 kg/ha and phosphate fertilisers at a rate of 50–75 kg P_2O_5/ha are required for good growth and yields (Table 8.12).

Table 8.12. The effect of fertilisers on the grain yield of chickpeas in India

Fertiliser treatment	Grain yield (kg/ha)
kg N/ha	
0	1011
22.5	1164
kg P_2O_5/ha	
0	929
45	1158
90	1183

Source: Anon, 1975.

Weed control

Mechanical weeding at 30, 45 and 60 days after planting or two weedings at 30 and 60 days give grain yields equivalent to completely weed-free conditions. Prometryne at 0.25 kg/ha or alachlor at 1.0 kg/ha are effective pre-emergence herbicides. Spraying of the herbicides MCPB or 2,4-DB at 0.75 kg/ha four weeks after sowing controls weeds between the young seedlings.

Pests and diseases

The most serious and widely reported insect pests of chickpea are the pod borer (*Heliothis armigera*) in the field, and beetles (*Callosobruchus* spp.) in the store. Insecticides may be used to control both insects, but where possible resistant varieties should be used.

Wilt, caused by *Rhizoctonia* and *Fusarium* spp., blight (*Ascochyta* spp.) and rust (*Uromyces* spp.) cause major yield losses. The application of a fungicide as a seed dressing has been tried as a control

measure. However the long-term objective is to develop and use varieties that are resistant to these diseases.

Harvesting and storage

The chickpea seed is mature about seven weeks after the opening of its flower. At this time, the moisture content of the seed is still about 70%. In harvesting, the plants are cut about 30 cm above the ground when about 50% of the pods are dry. The pods are allowed to dry completely before threshing the branches to release the seeds.

Cultivars differ in their storage requirements. At room temperatures, germination capacity is retained for between one and five years for most varieties. In closed glass jars, some varieties have been found to be 90% viable after nine years.

Haricot and Lima beans

Haricot beans, *Phaseolus vulgaris*, and Lima beans, *Phaseolus lunatus*, are cool season legumes believed to have originated in Central and South America. They are less widely cultivated in Africa than cowpeas and bambarra nuts.

Types and ecological adaptation

Only two species of *Phaseolus* are grown in significant quantities, *P. vulgaris* and *P. lunatus*. The common names associated with each of these species are given in Table 8.13. Climbing and bush varieties of both species are grown, although it is more common to grow the bush varieties of haricot beans and the climbing varieties of Lima beans (Figure 8.8).

Table 8.13. Common names of two cultivated *Phaseolus* species

Botanical name	Common names
Phaseolus vulgaris	Haricot bean, kidney bean, French bean, dwarf bean, princess bean, string bean, navy bean, pinto bean and snap bean
Phaseolus lunatus	Lima bean and Madagascar bean

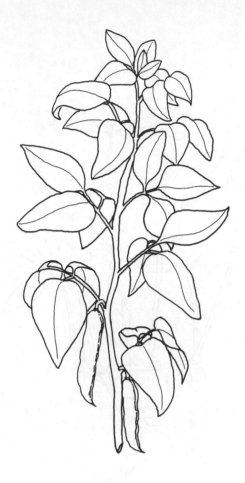

Figure 8.8 A fruiting branch of a Lima bean plant.

with a pH between 6 and 7. Lima beans will grow at higher altitudes than haricot beans.

Planting

A combined insecticidal and fungicidal seed-dressing is required. Aldrex-T, as recommended for groundnuts, is suitable.

Normal field planting for the bush varieties of haricot beans is a row spacing of 40−50 cm, by 8−12 cm in the row. Depth of sowing should be 3−6 cm, depending on soil type, the deeper planting being in light soils. Deeper planting is necessary because of the greater probability of dry surface soil in light soils whereas shallow planting increases the likelihood of successful emergence in heavier soils. Lima beans are planted in rows 75−100 cm apart, with an intra-row spacing of 10−50 cm, depending on variety.

Fertiliser application

Unlike other legumes, *Phaseolus* species, especially the bush haricots, respond to nitrogenous fertilisers, even though they form root nodules like other legumes. Nitrogen can be applied in amounts up to 20 kg/ha.

Weed control

Phaseolus species need a soil that has been well cultivated, but they do not like a loose soil as they lack a vigorous root system and in windy conditions the bush varieties may rock, thus aggravating the problems of stem and collar rot. Cultivation on the flat is preferred to cultivation on ridges and minimum disturbance of the soil around the beans is desirable. Herbicides, such as dalapon, can effectively be substituted for hand weeding.

Pests and diseases

Phaseolus species share the wide range of insect pests found on cowpeas and groundnuts. In addition, haricot and Lima beans are more susceptible to soil-borne pests and diseases than either cowpea or groundnut crops. A seed dressing of aldrin applied at 22 g of 40% wettable powder per 100 kg of seed controls most soil-borne pests and root diseases. For leaf diseases and wilts, the selection of resistant varieties is more appropriate.

Being cool season crops, the plants are better adapted to the moist highland areas of the tropics, for example in Africa they grow well in East Africa and Madagascar. Haricot bean is the most important pulse crop grown throughout tropical America, but is of little importance in India and tropical Asia. The plants require high temperatures (10−30°C) during germination and growth, but cannot stand waterlogged or saline soil conditions.

The climbing varieties tend to mature later than the bush types. Canadian Wonder is the most adaptable variety of haricot bean grown in Africa, while a range of Lima bean varieties (for example TP1-9, TP1-10, TP1-17, TP1-79 and TP1-187) have been found to give good yields in Nigeria.

Both beans require deep, free-draining soils

165

Harvesting and storage

When harvesting haricot beans, whole plants are pulled out of the ground as soon as approximately half the pods are ripe. The plants are tied in bunches and hung to dry on frames. After threshing, the seeds can be preserved using 1.25 g of lindane per kg of seed.

Lima beans ripen irregularly and should be harvested before the pods are completely brittle, particularly if the variety shatters. The pods are then dried on trays and threshed.

Other grain legumes

There are a number of other grain legumes that are grown in the tropics which are not of great economic importance. These leguminous crops may become more important in the future as more research is directed towards adapting them to the tropical environment. Some of these other legumes are briefly discussed here.

Bambarra nut

Bambarra nut, *Voandzeia subterranea*, also variously known as Congo goober, earth pea, Kaffir pea and Madagascar groundnut, is a native of the Sudan zone of Africa. The crop is grown throughout the African continent. It is a high-temperature crop which has a high tolerance to drought. It thrives well on the poorest and most sandy soils and gives good yields even in areas of uncertain rainfall (Figure 8.9).

In order to ensure a good crop, a close spacing of 20 × 50 cm is suggested. The depth of planting should be 3–6 cm. Innoculation of bambarra seed with *Rhizobium* is necessary when planting a new field. A fertiliser application of 15 kg N and 30 kg P_2O_5/ha is reported to be adequate on most soils. Although the crop is normally planted on ridges, ridging may not actually be required because of the light nature of the soils on which bambarra nuts are usually grown.

The bambarra nut crop is one of the most pest and disease-free legume crops and is subjected to less damage to its pods in the field and in storage than any other major grain legume, although it can suffer from bruchid beetle damage. The crop

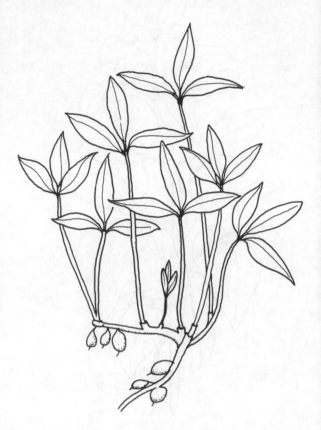

Figure 8.9 A bambarra groundnut plant showing the development of subterranean pods.

is normally harvested in much the same way as groundnut.

Jack bean

Jack bean, *Canavalia ensiformis*, is widely cutivated throughout Africa, although it is grown more extensively in the West Indies and Brazil. The young pods are used as a vegetable and the seeds are sometimes eaten after long cooking. The crop is planted at the beginning of the rains at a spacing of 30–40 cm square. The seed requirement is 25–30 kg/ha and it is planted about 2 cm deep. When jack beans are three-quarters grown, they are a useful vegetable. The seeds are harvested as the pods become ripe.

Sword bean

Sword beans, *Canavalia gladiata*, are thought to have originated in India or Africa. The plant usually has a twining or climbing habit, but semi-erect cultivars are known. The production of sword bean is similar to that of jack bean.

Dolichos bean

Dolichos bean, *Lablab niger*, also referred to as hyacinth bean, is a native of India. Few reports of its culture in Africa are available, except that it has been grown to a limited extent in Angola and Malawi. In India, the dolichos bean crop is sown at the beginning of the rains and is used as a hay crop during the rainy season. The seeds are normally harvested about three months after the end of the rains. The ecological requirements of the dolichos bean crop are similar to those of other legumes such as groundnut and cowpea. Once established, dolichos bean is capable of sustaining growth on light sandy soils.

The planting distance varies from 25×10 cm to 80×30 cm, depending on the cultivar and the location. Many cultivars are very vigorous and are capable of weed suppression once they are well established. Dolichos bean has been reported to respond to both phosphate and potash fertilisers. A basal dressing of about 25 kg P_2O_5/ha is suggested. The crop does not require innoculation with *Rhizobium*, especially when it is grown on soils previously cropped with cowpeas. In most cultivars the pods mature in succession up the flowering stem. However, in some of the very early upright varieties all the pods tend to ripen together. The pods are harvested as they become ripe and the dry pods are threshed. The grain is then stored in the same way as for cowpea.

Winged bean

The winged bean, *Psophocarpus tetragonolobus*, has been cultivated for generations in the humid tropics of South and Southeast Asia. The crop is grown in backyard gardens in many countries, but in Papua New Guinea, Burma and Thailand it is cultivated as a field crop. In Nigeria winged bean has only recently been introduced and it is now successfully grown in various parts of the country.

For good growth, winged bean requires ample, well distributed moisture. It is cultivated in areas with an annual rainfall of 700–4000 mm. However, it thrives best in hot humid areas with an annual rainfall of 2500 mm or more. It is tolerant of a wide range of soils and thrives well even in soils with low organic matter and in sandy loams or heavy clays. It withstands high temperatures but almost never survives frosts. Temperature is as important as day-length in controlling flowering. Day temperatures higher than 32°C or lower than 18°C inhibit flowering even under otherwise suitable short-day conditions.

Under subsistence farming conditions, winged bean is grown mainly in small plots either in pure or mixed stands. Planting is carried out at the beginning of the rains. The recommended spacing is about 1 m between rows and 25–70 cm between plants. Planting density is in the order of 125 000–150 000 plants/ha. Seeds are planted at a depth of 2–3 cm and stakes or a trellis 1.2–2 m tall are generally used as staking the crop increases yields two fold. Supports can be made of bamboo, wire, string or rope. The plants may also trail on living supports when grown in mixtures with cereals, especially maize, which serves to support the beans.

When grown in mixed garden cultivation, winged bean is usually free of pests and diseases. Nevertheless, the crop is known to be susceptible to a number of pests and diseases. The bean borer, *Maruca testulalis*, attacks the flowers. A number of leaf-eating insects also attack the leaves, and nematodes attack the roots. The most important fungal disease is the 'false rust' caused by *Synchytrium psophocarpi*. Others are leaf-spot, powdery mildew and collar-rot. Nematodes can be important pests.

Immature pods, dry seeds and tubers are all edible. Estimates of the yield levels of the various parts are shown in Table 8.14.

Table 8.14. Yields (kg/ha) of the various parts of winged beans

Edible part	Yield range (kg/ha)
Immature pods	34–36 000
Dry seeds	2–5000
Tubers	5–12 000

Lentil

Lentil, *Lens esculenta*, is an annual grain legume grown widely in the Near-East and India. It is cultivated as a cold-season crop throughout the subtropics and is adapted to a wide range of soils. In the tropics it is most successful as an irrigated dry-season crop, at elevations over 1000 m.

Yam bean

The yam bean, *Phaseolus adenanthus*, is grown for its tubers. It is cultivated in Liberia, but no reports are available on its cultivation in other parts of Africa. Other leguminous crops which yield tubers include the Mexican yam bean (*Pachyrrhizus erosus*), the African yam bean (*Sphenostylis stenocarpa*), the winged bean (*Psophocarpus tetragonolobus*) and one *Vigna* species, *Vigna vexillata*.

Pea

The pea, *Pisum arvense* or *Pisum sativum*, is a cold-tolerant legume which can also be grown in many tropical environments. The types of pea commonly cultivated are short-day and dwarf cultivars which are adapted to lowland conditions under dry-season irrigation.

Green gram

Green gram, *Vigna mungo*, is a native of India, but has been grown in parts of Africa. The ecological requirements of the crop are similar to those for cowpea. The plants are, however, smaller than cowpea plants and can be planted more densely. In Nigeria, dry seed yields of over 1300 kg/ha have been recorded using varieties such as TVra 4, TVra 31 and TVra 87.

CHAPTER 9

Fibre crops

Fibre crops comprise a diverse group of plants all of which produce fibres of sufficient length, strength, and durability to be utilised in the manufacture of cloth, cordage and numerous other related articles. Fibres used for cloth and similar goods are classed as 'soft fibres'. Those employed principally for heavy cordage and ropes are termed 'hard fibres'. Cotton, *Gossypium* spp., which belongs to the soft fibres, is the most important fibre crop in the world. Other soft fibres are kapok (*Ceiba pentandra*), ramie (*Boehmeria nivea*), kenaf (*Hibiscus* spp.), jute (*Corchorus capsularis*), hemp (*Cannabis sativa*), and flax (*Linum usitatissimum*). Included in the group of hard fibres are manila hemp (*Musa textilis*) and sisal (*Agave* spp.). The world output of fibres is shown in Table 9.1 for the period 1978–80.

Also discussed in this chapter is rubber (*Hevea brasiliensis*) which, although it is not a fibre crop, is included here for convenience.

Cotton

Cotton, *Gossypium* spp., is a tropical crop which requires a long warm dry season for ripening and drying the lint-carrying fruit, referred to as the boll. It does well in areas with a mean annual rainfall of 750–1200 mm. In the hot humid tropics the production of cotton suffers from the problems of inadequate solar radiation, pests and diseases. Total production of cotton lint in 1983–84 was about 15 million metric tonnes from nearly 32 million hectares of land world-wide. Of the total world production, the USSR contributes about 50%, Indo-China 20%, North America 14% and Africa 8%. The rest of the world's cotton lint output comes mostly from Europe and South America (Figure 9.1).

Types and ecological adaptation

The genus *Gossypium* includes a number of wild and domesticated species. The wild and lintless diploid species probably evolved in southern Africa and spread from there to the arid and semi-arid regions of Arabia, Southeast Asia, Australia, the United States of America and parts of West and East Africa. Such cotton types have no real economic value because they are lintless. They produce small dehiscent capsules which contain small seeds covered to varying degrees with short and dark-coloured surface hairs (fuzz) which cannot be spun to make yarn.

The linted diploid species comprise most of the Indian and Asiatic cottons, including *Gossypium herbaceum and G. arboreum*. The tetraploid species are by far the most important cotton types. Two

Table 9.1. World production of major fibre crops, 1982–1984

| Crop | Total production (million tonnes) | | | Land area cultivated (million ha) | | | Grain yield (tonnes/ha) | | |
	1982	1983	1984	1982	1983	1984	1982	1983	1984
Cotton	14834	14380	17794	na	na	na	na	na	na
Sisal	436	351	389	580	521	538	750	674	725
Jute	3745	3895	4157	2314	2333	2308	1618	1669	1801
Hemp	201	187	255	367	358	391	546	522	575

Source: FAO production yearbook, 1984.
na = data not available.

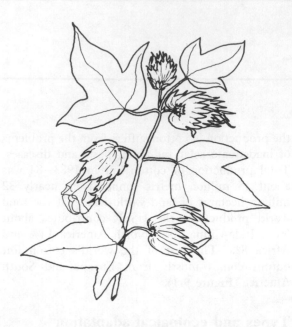

Figure 9.1 A flowering branch of a cotton plant, *Gossypium* sp., showing an unopened boll (fruit).

members of this group, *G. hirsutum* and *G. barbadense*, today dominate the world cotton industry.

From the commercial point of view, the most important differentiating parameter in the cotton industry is the staple length of the lint hairs on the seed. For *Gossypium hirsutum* (the American upland-type cotton), the staple length varies from 25–28 mm. It is widely grown in the semi-arid tropics, notably in West and East Africa and India. The Samaru Allen variety (which at one time was the best known cotton in West Africa and from which most present-day cultivars in the region have developed) was the earliest short-staple upland type to be introduced to the savanna zone of Nigeria.

The species *Gossypium barbadense* yields lint of longer staple length. The 'sea island' cotton grown in the southern United States of America the Egyptian cotton cultivated completely under irrigation and the Ishan cotton of southern Nigeria are all members of this group. They yield lustrous and strong lint, with an average staple length of about 50 mm. Commercially, the long staple cottons are usually highly priced since they are more suited to the manufacture of high-quality textiles.

Cotton grows well on a wide range of soil types, but prefers well-drained and deep, heavy loams which are high in natural fertility. However, the crop is able to withstand poor soils more than most annual crops. It is quite sensitive to soil acidity, doing best in slightly acid (pH 6.5–6.9) to slightly alkaline soils (pH 7.0–7.5). Soils with a pH below 5.5 should be limed before being cropped to cotton.

Planting

Cotton may be planted on the flat or on ridges, depending on the preference of the grower. In West Africa, the crop is mostly grown in the northern Guinea savanna zone and is usually the first crop to be planted on newly cleared land. In general, cotton is relatively insensitive to cropping sequence such that it can be grown on the same piece of land for several years without an appreciable decline in yield. However, such a practice should not be encouraged since it may lead to the build-up of disease.

Four to six seeds should be planted about 2–3 cm deep and 45 cm apart on 90 cm ridges. The young seedlings should be thinned to two plants per stand about three weeks after germination. In general, thinning is best done after good rain. Research results have shown that it is not worthwhile re-seeding to fill missing stands, unless the number of empty stands is scored at more than 10%. A population density of about 150 000 plants/ha has been shown to be optimal in the northern Guinea savanna of Nigeria (Table 9.2). In general, germination occurs within a week, while flowering

Table 9.2. Seed cotton yields as affected by nitrogen level and population density at two locations in the savanna zone of Nigeria

Treatment	Yield (kg/ha)	
	Tumu	*Gusau*
kg N/ha		
0	1325	924
26	1505	1129
52	1745	1181
78	1707	1205
Plants/ha		
50 000	1487	1059
100 000	1580	1042
150 000	1645	1226

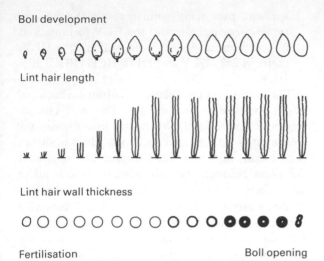

Boll development

Lint hair length

Lint hair wall thickness

Fertilisation   Boll opening

| | | | | | | | | | | |
0   5   10   15   20   25   30   35   40   45   50

◄──────── Days from fertilisation ────────►

Figure 9.2 Boll and lint development in cotton, *Gossypium* spp. (after Kochhar, 1986).

commences seven to eight weeks after sowing. Boll maturation and opening occurs over a period, four to seven months after sowing (Figure 9.2).

Fertiliser application

The benefits derived from fertiliser applications to cotton are to a large extent dependent on the efficiency of pest control measures. Cotton generally responds very well to nitrogen and phosphate. However, the application of too much nitrogen can result in the crop developing excessive vegetative growth at the expense of lint production.

For a well-managed cotton crop, the application of 50 kg N and 20–40 kg P/ha should be adequate (Table 9.2). Boron deficiency in cotton is a common occurrence in many of the producing countries, particularly the United States of America, Australia and West Africa. In such cases, boronated superphosphate (containing about 0.5% borax) may be used so as to supply both the boron and phosphate, although there is the potential danger of persistent use leading to an excess of boron. The phosphate fertiliser should be broadcast and worked into the soil before planting or before the ridges are made. The fertiliser can alternatively also be applied at planting by placing it in holes 10 cm away from the seeds.

The application of nitrogen should be carried out in two equal doses. The first after sowing by placement, and the remaining half top-dressed seven to nine weeks after sowing. Potash deficiency is not as common as nitrogen or phosphorus deficiency. However, where there is a definite need for potassium, about 30 kg K/ha should be adequate.

Weed control

Early weeding of cotton fields is essential to enable the seedlings to establish, particularly since cotton is sensitive to weed infestation. Where ridges are used, these should be moulded at each weeding either manually using hand hoes, or by mechanical means using ox-drawn or tractor-drawn implements. A pre-emergent herbicide such as diuron, applied in combination with either norflurazon, fluridone or alachlor and followed by a supplementary hoe weeding or ridge moulding should effect satisfactory weed control. Where herbicides are used, only one supplementary weeding is usually required before the bolls reach maturity.

Pests and diseases

Cotton is the favourite host for a wide variety of insect pests which attack the vegetative and the flowering and fruiting parts of the plant. Pest attack is generally not serious during the vegetative stages prior to flowering, but post-flowering damage can be devastating if control measures are not taken. Cotton planted at the correct time and then well tended benefits tremendously from insecticide application. The important insect pests of cotton are:
Cotton stainers, *Dysdercus* spp., are piercing and sucking insects which feed on the juices of unripe cotton seeds. When the bolls finally open, the lint is often stained yellow or even completely destroyed. The damage caused by these insects is not mechanical, but is the result of certain fungi and bacteria which are introduced into the bolls and transmitted from one boll to another. The bugs are usually conspicuous, being red, black, or white in colour.
Bollworms, which are mostly members of the order Lepidoptera, cause damage to cotton by boring through and eating out the inside of the bolls. The damage may cause the immature boll to die off or not to open properly. The lint may also become badly stained and worthless.
Cotton jassids, *Empoasca* spp., are sucking bugs which only cause damage to the leaves. Varieties

which have hairy leaves tend to be free from attack by these insects.

Cotton aphids, *Aphis gossypii*, are soft-bodied insects with piercing and sucking mouth parts. They feed on the terminal buds and leaves, resulting in stunted growth and curling of the leaves. They produce honeydew which spreads over the leaves and interferes with the photosynthesis of the plant.

Control measures against insect pests on cotton include the use of resistant varieties. Experience has shown that those varieties with hairy leaves also provide an effective control measure and that crop rotations help to prevent the build-up of pests on the same site. When a crop other than cotton is planted on the site, the hibernating pests are deprived of their usual host plant, which soon leads to their extermination. A cotton—sorghum—groundnut rotation is widely practised in many of the cotton producing areas of Africa.

A valuable cultural method of checking insect pests on cotton is the strict observance of the closed season. This is the practice whereby all cotton plant material is destroyed by burning at the end of the harvest and all further sowings are prohibited by law until the following season. This is to ensure that there is no suitable host for the cotton pests and diseases, many of which have few alternative hosts. The insects are thereby starved out or their life-cycle broken. In practice, all of the season's cotton plants are cut, heaped and burnt. At the beginning of the following rainy season, seedlings arising from the diseased bolls or waste seeds should be uprooted and any shoots on old root stumps should be cut off.

A number of insecticides are now available for use in controlling pests on cotton. Among the more widely used chemicals are:
1. A formulation containing 20% DDT and 2% HCH (BHC).
2. A wettable powder containing 85% carbaryl currently available in West Africa as Vetox 85, Sevin 85 or Dicarbam 85.
3. Dicampethrin pyrethroid, which is a formulation derived from pyrethrum. It is not only an effective insecticide but is also relatively safe to use.
4. The development of ultra low volume (ULV) spraying has shown that 3 litres per hectare of an oil-based insecticide formulation can be applied using a battery-operated spinning disc sprayer, producing similar results to conventional spraying. Some of the ULV formulations in common use include endosulfan 25% ULV, carbaryl 25% ULV and DDT/HCH (BHC) 20% ULV.

The two important diseases of cotton are bacterial blight and damping-off disease. The main characteristics of bacterial blight, caused by *Xanthomonas malvacearum*, are angular leaf-spots, dead branches (black arm) and boll rot. Bacterial blight attacks the plant through the leaf stomata. It can affect more than 50% of the seedlings grown from untreated seeds and then secondary spread through a cotton field occurs up to eight weeks after germination. Control measures include seed dressing with an appropriate insecticide. In some West African countries this is done at Government-owned ginneries and the treated seeds are supplied to the farmers free. Another effective control measure is the destruction of all cotton plant debris during the closed season. Resistant varieties should be used where available.

Damping-off disease is caused by the fungi *Fusarium aspergillus* and *Penicillium pythium*. The fungi live in the soil where they attack susceptible seedlings, although sometimes the seeds are sown already infected with fungi. When the seedlings are attacked, the cells just below the soil level become waterlogged, resulting in damping-off and death of the plants. The use of crop rotations and resistant varieties provide two of the best control measures.

A third disease, *Alternaria* leaf-spot, is also caused by a fungus. The disease mainly attacks the cotton seedlings, causing premature senescence followed by defoliation. It is not as important in its effect as bacterial blight and damping-off. Control is largely by the use of resistant varieties and by the adoption of a crop rotation.

Harvesting and storage

The first bolls are formed about 10—11 weeks after planting, and flowering may continue up to 16 weeks after planting. The bolls mature and open over a period of 4—7 months after planting. In most of the semi-arid tropics, this is from November to December. The crop is ready for harvest when the mature bolls split open, showing tightly-packed seed cotton.

In most of the developing countries, cotton is

Figure 9.3 Mechanical cotton picker harvesting cotton in Mississippi, USA.

picked by hand. Leaves and other unwanted vegetative or diseased parts are carefully discarded and sometimes the seed-cotton is graded at the time of picking. Two or three pickings are usually necessary because the bolls mature at different times. The large labour requirement for picking is one of the important economic considerations limiting the production of cotton at the level of the small-scale farmer. In countries like the United States of America, picking is done by machine, and the plants are first defoliated with a chemical spray to induce uniform ripening of the bolls. By so doing, there is little chance of contamination from plant debris. Once the harvested seed-cotton is cleared of all dirt and diseased material, it is put into sacks ready for ginning. Having separated the cotton seed from the lint, the latter is then compressed into bales and sold to the textile mills or stored to be sold later or for export (Figure 9.3).

The cotton seeds remaining after the lint has been removed are usually crushed and the cotton seed oil (20%) is removed. This oil is used for the manufacture of margarine and edible vegetable oil. The residue from the crushed seeds is made into cotton-seed cake for livestock feed.

Sisal

Sisal, *Agave sisalana*, originated in Central and South America, being a native of the Yucatan Peninsula of Mexico. It is the most important of the group of hard hemp fibres, which includes Manila hemp and New Zealand flax. The world's largest production of sisal comes from East Africa, notably Tanzania, Kenya and Mozambique. In North and Central America, Mexico remains the leading producer. A considerable amount also comes from the West Indies and India, with smaller amounts coming from West Africa, particularly Togo and Senegal.

Sisal is essentially a commercial crop which is hardly ever grown by small-scale farmers, except as hedges. Like sugarcane, sisal requires large scale production to justify the use of the expensive machinery required. Heavy capital investment is necessary because of the waiting period before harvestable maturity and because of the need for a factory.

The greatest demand for sisal is for use as binder twine, but it also makes ropes, sacks and bags of various types as well as marine cordage.

Types and ecological adaptation

Besides *Agave sisalana*, which produces the bulk of the commercial sisal crop, there are three other closely related species which are also of commercial importance. These include *A. fourcroydes*, Yucatan sisal, which produces the henequen hemp of Mexico; *A. cantala*, the Maguey sisal of the Philippines; and *A. letonae* which produces the Salvadorean henequen hemp of San Salvador. *Agave fourcroydes* is cultivated principally on the Yucatan Peninsula of Mexico and in Cuba. Henequen plants closely resemble sisal in appearance, but have bluish-grey leaves with strongly spiny margins and stout tip spines. The fibres tend to be coarse and of reddish yellow tinge. It is used almost entirely to make twine.

Sisal prefers a hot climate and grows virtually throughout the humid and sub-humid tropics, notably the lowland tropics, preferring a rainfall of not less than 1000 to 1500 mm per annum, well distributed over the year. The more even the distribution of rainfall, the higher the fibre quality and the better the opportunity for continuous leaf harvesting. The crop generally prefers a medium or light soil.

Fertiliser application

A superphosphate fertiliser applied at about 25 kg P/ha/annum together with about twice that rate of nitrogen, applied as calcium ammonium nitrate or urea, should ensure a good crop. Lime is desirable where the pH of the soil falls below 6.5. The yield deteriorates over the year, and more rapidly under continuous cultivation owing to nutrient depletion. The productive capacity of a soil is an exhaustible index, particularly when large quantities of green matter are removed each year. Potash shortages are often reflected by a bending-over of the older leaves which are normally stiff and straight. Nitrogen and phosphate deficiency can easily be corrected, but boron deficiency, which causes leaf cracking and is a common occurrence in sisal plantations, is more difficult to control.

In some sisal-producing countries such as Java and Sumatra, the use of decortication wastes, including both solids and liquids, especially when fortified with lime, has proved effective as a fertiliser supplement. However, the large bulk of such wastes limits their wide use, except on large plantations with economic means of transportation.

Planting

Sisal is propagated by means of bulbils, which appear on the flower stalk, or by suckers growing around the base of the plant. Bulbils are generally preferred since a single plant produces up to 4000 bulbils compared with less than a quarter of this number of suckers. Only the large bulbils should be selected for planting. Bulbils are first planted in nursery beds about 25−30 cm apart in 50 cm rows. They are allowed to grow to about 40 cm or until they are about 9−12 months old when they are transplanted to the field. At this time, the bulbils are well rooted. Transplanting should preferably be done at the beginning of the rainy season. A recommended planting pattern in the field is a series of double rows 60 cm apart with a 2.5 m alley between each pair, and the plants set out about 75 cm apart within the rows, giving a population density of about 2500 plants/ha. Alternatively, the plants are spaced 1 m apart in 3 m rows.

In general, the field should be kept weed-free. Where rows are about 3 m apart, crops such as beans or other annual legumes can be grown between them to help suppress weed growth and to check erosion.

Pests and diseases

Sisal is a very strong plant which is well protected with a leathery epidermis and, in some species, sharp spines. It is rarely attacked by insect pests or disease organisms. There is a sisal weevil which only very occasionally reaches pest proportions.

Figure 9.5 Sun-drying sisal is a lengthy, space-consuming process.

Figure 9.4 Harvesting sisal in Sudan.

Harvesting and processing

With good crop management and high soil fertility, cutting of the leaves could begin in 18–24 months after planting. In practice, however, the first harvesting commences after the plants have been in the field for 24–36 months. At this time about 50 leaves, each weighing up to 1 kg, may be cut from one sisal plant each year. Harvesting starts with the removal of the ripest (lower) leaves and this continues periodically over the next four years. On average, two cuttings are made annually for the first four years. Thereafter, only one cutting is made each year until the flower stalks begin to develop. A total of about 300 leaves may be harvested during the economic life of each plant, thus giving a total of 500–600 tonnes of fibre/ha. Over the normal eight-year harvest period, the average annual yield is thus about 67 leaves which is about 2.25 tonnes of fibre/ha. Sisal dies after producing an enormous inflorescence and bearing a crop of vegetatively produced bulbils. The age at which

175

this takes place varies, but ranges from 8–10 years. Obviously, the length of the period between maturity and flowering is a prime factor in profit-making and has, therefore, become an important genetic parameter in varietal selection. It appears to be influenced by climate (Figure 9.4).

The fibre is extracted from the leaf tissues by a process of decortication. In modern factories this is done entirely by machinery. The estate factory must, of necessity, have a good water supply. The leaves are first crushed by rollers, decorticated, and then washed before sun-drying the fibres. Proper drying of the fibres is very important because the quality of the fibres produced depends largely on their moisture content. Artifical drying has been found to result in generally better grades of fibre than the space-consuming, lengthy sun-drying process (Figure 9.5). The dry fibres are then machine-combed and sorted into various grades, largely on the basis of the previous separation of the leaves into size groups in the field.

Jute

Jute, *Corchorus* spp., is one of the best known and most widely used raw materials of plant origin, there being few industries which do not employ the fibre in one way or another. It is the usual material for the manufacture of sacks. The leading producers of jute are Pakistan, which produces about half of the world's supply, India, Brazil, China and Nepal. A few countries in Africa are also producers of jute, but the quantities are relatively small (Figure 9.6).

The main features which make jute valuable and popular include its long and readily spun fibres, its durability which makes the fibre particularly suitable for packing and for weaving light ropes and carpets, its relatively low cost and its wide climatic adaptability.

Types and ecological adaptation

Jute is obtained from two closely related species, *Corchorus capsularis* and *C. olitorius*, both of which originated in Indo-Burma and south China, but have become naturalised in many parts of the tropics. The more widely cultivated species is *C. capsularis* which thrives best in conditions of flooding on alluvial soils. It is widely cultivated in

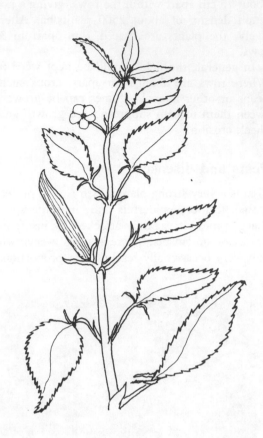

Figure 9.6   A fruiting branch of *Corchorus olitorius*, long-fruited jute.

the central and eastern parts of Bengal, principally in the Dacca region of India. *Corchorus olitorius* is usually grown on higher land and cannot withstand flooded conditions. The fibres of *C. olitorius* fetch a slightly lower price than those of *C. capsularis*.

Jute requires a hot, humid or sub-humid tropical environment. It's cultivation in Pakistan and India is essentially confined to the hot, moist, fertile plains of the delta formed by the Brahmaputra and the Ganges rivers flowing through Bengal and adjacent provinces. The rainfall during the growing season should range from 1500–2500 mm and the mean daily temperature should be 24–32°C. Indeed, so widely adapted is jute that even though it is cultivated over vast areas and under such diverse upland and lowland conditions in India and Pakistan, the local inequalities in rainfall distribution tend to average out. As a result, complete crop failures hardly ever occur.

Planting

Jute is usually planted on the flat. When sown broadcast, the suggested seeding rate is 20–25 kg/ha for *C. capsularis* and 10–15 kg/ha for *C. olitorius*. Germination takes place within a week. The seedlings are thinned to 10–15 cm apart after emergence. When sown in rows 60 cm apart, the seedlings should be thinned to 7–10 cm apart in the row in order to cause rapid unbranched growth. The plants normally attain a height of about 3 metres in 4–6 months. As soon as plants begin to bloom, they are ready for harvest.

Fertiliser application

It is suggested that for good fibre yield and quality, jute requires 25–30 kg N/ha and 10–15 kg P/ha. Soils with a pH below 5.5. should normally be limed. Jute is usually grown on heavy clayey loams and potash deficiency in such soils is not common.

Weed control

Close spacing of jute plants is desirable for two reasons, to prevent branching and to suppress weed growth. The operation of moulding up the ridges, either manually or by tractor, also helps to control weed growth. When sown broadcast on the flat, the easiest way of controlling weeds is with the use of herbicides. Diuron plus alachlor or norflurazon applied pre-emergence have been found to effectively check weed growth in jute.

Pests and diseases

Problems with nematodes and fungal pathogens are normally important considerations in jute production. Stem lesions, caused by *Fusarium* spp., tend to reduce the quality of the fibre. Root diseases caused by *Pythium* spp. are also of importance, as are root knot nematodes which can only be controlled by employing an effective crop rotation. Insect pests associated with jute and kenaf are numerous and include certain species of *Dysdercus*, *Hippopsis* and *Agrotis* as well as the cotton aphid, *Aphis gossypii*. The use of DDT, toxaphene and parathion is suggested to control these pests.

Harvesting and processing

When harvesting jute, the stems are cut at ground level, tied into bundles and stacked for a few days to shed their leaves. Thereafter, the jute is retted in water tanks, ponds or pools for 10–15 days. Most of the jute produced in Pakistan and India is retted in shallow parts of the Ganges river. At the time of retting the water temperature is usually around 24–27°C, a range well suited to rapid bacterial activity on the stalks. When cold weather sets in, the retting period may extend to 50 days. When fully retted, the fibres are stripped, washed and dried ready for baling or for direct sale to fibre product manufacturers. Jute production is highly labour-intensive, which makes India and Pakistan leading producers in view of their large and relatively cheap labour forces.

Kenaf

Kenaf, *Hibiscus* spp., and certain other members of the same genus have long been cultivated in many parts of the tropics. Its origin is not certain, but it most probably originated in Africa and has been widely cultivated since in Asia, especially India and Pakistan. The fibre from kenaf is widely used for rope making and in the manufacture of sacks. Although kenaf is commercially less important than jute, its increasing level of production has enabled it to come into prominence as a commercial substitute for jute.

Types and ecological adaptation

The various known types of kenaf exhibit wide variations in their plant habit and other characteristics associated with the period of their vegetative growth. The two most important species are *Hibiscus cannabinus* and *H. sabdariffa* (Figure 9.7).

Kenaf, as well as other species of *Hibiscus* grown for their fibre, thrives best in warm, moist climates. Deep, well-drained, fertile loams are most suited to the production of high yields for good quality fibre. Adequate moisture is needed throughout the growing period. Where the rainfall pattern is bimodal, production is best at locations which remain moist for most of the year, such as fadamas, or where facilities for supplementary irrigation are

Figure 9.7 Hibiscus cannabinus, the most widely cultivated species of kenaf, showing a cultivar with cordate leaves.

available. Kenaf is much more tolerant than jute of poor soil conditions.

Planting

The time of planting is usually determined by the photoperiodic requirements of the variety in use. The seedbed preparation for kenaf is a critical cultural operation. The seeds are usually drilled 3.5–5 cm apart for fibre production and up to 15–18 cm for seeds on rows made 15–20 cm apart. In order to encourage branching in a crop intended for seeds, row widths may be up to 30–40 cm. The seed rate therefore depends on the purpose for which the crop is grown. For seed production, a seeding rate of 10–15 kg/ha is considered adequate, for fibre production, 25–35 kg seed/ha is suggested. Because the young plants grow rather rapidly, there is usually no need for subsequent cultivation, especially if the crop is intended for fibre. But where the interplant spacing is relatively wide, one hoe-weeding should be adequate to check weed growth.

Pests and diseases

In light-textured sandy loams where root knot nematodes may be a problem, crop rotation is strongly advocated as this provides the only practical control measure. Kenaf is attacked by a range of insect pests, including some species of *Hippopsis*, *Anomis*, *Agrotis*, *Dysdercus* (cotton stainer) and *Aphis gossypii* (cotton aphid). The use of DDT and toxaphene in appropriate doses should effectively check these pests.

A number of leaf and stem diseases occur, caused by *Pseudomonas* sp., *Phytophthora* sp., *Fusarium* sp. and *Pythium* sp.

Harvesting and processing

For fibre production, best yields are obtained when the kenaf is harvested just prior to or at the beginning of flowering. The stems are cut at ground level and subjected to retting. Manual harvesting is highly labour-intensive and therefore expensive. The use of mechanical equipment, as for example a combined reaper-binder harvester, enhances the economic production of kenaf. Once the ribboning of the stems is accomplished, the ribbons must be retted using either the water or chamber method. Chamber retting is an aerobic process which is free of any odour. It takes only 2–3 days to complete. In contrast, water retting is an anaerobic process which may require up to two weeks to complete. In general, the time taken for retting depends on the temperature of the water. After retting the fibre is washed in clean water, to remove all waste and woody materials, before drying.

Silk cotton

Silk cotton, *Ceiba pentandra*, also referred to as kapok, is a member of the family Bombacaceae. This family, which contains about 20 genera and 150 species of trees in the tropics, is closely related to the Malvaceae. Within the genus *Ceiba*, there are nine species of large deciduous trees of which eight are of tropical American origin. One of the species, *C. pentandra*, extends through the tropics to West Africa and Southeast Asia (Figure 9.8).

The fruit of kapok contains numerous black seeds embedded in tight balls of silky floss derived not from the testa of the seed, but from epidermal cells of the fruit wall. The floss varies in colour from grey to light-yellow and pure white. Its quality depends largely on the handling process. Java produces 90% of the world's demand for kapok lint. Wherever the tree occurs, its floss is used for a variety of purposes, including stuffing pillows, mattresses and saddles. Being inflammable in the dry state kapok lint is used in India in the manufacture of fireworks. Up to 25% of the seed is a non-drying oil used for lubrication, soap making and, when purified, for cooking purposes.

Types and ecological adaptation

There are three important varieties of *Ceiba pentandra*:

(a) *C. pentandra* var. *caribea* which occurs wild in the American tropics and in evergreen, moist, semi-deciduous and gallery forests in West Africa. It is a gigantic tree, reaching 70 m in height and is one of the tallest trees in Africa. The trunk is unforked and spiny with large buttresses. When mature, the fruits (capsules) are dehiscent and release grey to white coloured floss.

(b) *C. pentandra* var. *pentandra* is the cultivated kapok of West Africa and Asia. Ecologically, this variety shows a wide range of tolerance and can be grown under both forest and savanna conditions. It is a tree of moderate height, up to 30 m. The trunk is unbranched and usually spineless with small or no buttresses. The fruits are usually indehiscent, producing kapok which is usually white.

(c) *C. pentandra* var. *guineensis* grows wild in the savanna woodlands of West Africa. It is seldom more than 18 m in height. The trunks are spineless and without buttresses. The fruit is dehiscent and produces grey coloured floss.

Kapok thrives best at elevations below 5000 m. It grows in a wide range of conditions, but for high production it requires abundant rainfall (1250–1500 mm annually) during the vegetative period of growth, and a drier period for flowering and fruiting. Fruits are not set when night temperatures fall below 20°C. For best results, kapok should be planted on rich, deep and permeable soils which are free from waterlogging. Situations. in which silk cotton trees are exposed should be avoided as the trees are easily damaged by wind. Any of the three types referred to earlier may be grown. However, var. *pentandra* appears to be the most favourable because of its high range of tolerance and moderate height which makes harvesting easy.

Figure 9.8 The silk cotton tree, *Ceiba pentandra*, is the source of kapok lint.

Planting

Kapok is usually propagated by seed when grown on a plantation scale. Seeds are sown in a nursery and transplanted in the field at 8–10 months old after removing the crown and leaving 1.3 m of stem. Kapok is also easily propagated by means of cuttings which should be 5–7 cm in diameter and 1.3 m in length, from wood that is two to three years old.

The crop is normally not grown under irrigation, therefore planting should be done at the beginning of the rains, early enough for the plants to become established before the dry season. For successful germination, it is necessary to remove all the floss from the seed. Seeds may also be treated with a dilute solution of acetic acid to accelerate germination. Seedlings and cuttings are spaced about 6 m apart in the field. In the early stages of growth, intercropping with annual crops is encouraged as this does not hinder the growth of silk cotton trees.

Fertilisers are rarely used on a routine basis for silk cotton, but trees are known to benefit much from organic matter arising from leaf litter decomposition. Weeds should be kept under control by circle-weeding around individual trees and brushing in between rows.

Kapok appears to have few serious pests and diseases. In Ghana, it is an alternate host for the cocoa swollen shoot virus, but shows considerable resistance to it.

Harvesting and storage

The tree comes into bearing in the fourth or fifth year, producing about 100 pods which give about 500 g of cleaned floss per tree. It reaches full bearing during the seventh to tenth year when it yields 330–400 pods/tree/year, thus giving 1.6–1.8 kg of kapok. Properly managed, individual trees may yield up to 2.7–4.5 kg of kapok/year. Once established a normal tree continues to bear for 60 years or more. The pod contains approximately 44% husk, 32% seeds, 17% floss and 7% placenta by weight. The average weight per pod is about 28 g.

The pods are harvested when fully ripe and, in the dehiscent types, before they open. Once harvested, the pods are hulled and the kapok is dried in the sun. The seeds are removed by beating with sticks or by using a suitable mechanical devise. The quality of the kapok lint is judged by its freedom from foreign matter, including seeds, its moisture content and its colour, smell and lustre. Well dried kapok lint stores for many years.

Rubber

Natural latex, containing rubber in large quantities occurs only in species of the families Moraceae, Euphorbiaceae, Apocynaceae and Compositae. Natural rubber has been most exploited among members of the family Euphorbiaceae, particularly the nine species of the genus *Hevea, Hevea brasiliensis, H. benthamiana, H. guianensis, H. pauciflora, H. spruceana, H. microphylla, H. nitida, H. rigidiflora* and *H. camporum*. The first three species yield commercially acceptable latex, but the remaining six are too resinous or even act as anti-coagulants. Today, almost all commercial rubber is produced from *Hevea brasiliensis* and, although the tree originated in tropical South America, the most important region of production now is Southeast Asia.

Despite the enormously increased production in recent years of synthetic rubber, the natural latex-producing *Hevea brasiliensis* is still one of the most important tropical crops. Since the discovery in 1839 of the vulcanisation process, whereby sulphur is added to toughen the product, more and more uses have been devised for the latex, including the manufacture of tyres, tubes, boats, shoes, raincoats and many other things. The rubber seed has a soft kernel which contains about 50% of a dark-red drying oil which finds use in the manufacture of soaps.

Types and ecological adaptation

The rubber tree is native to rainforest areas, but can be grown at different elevations and in regions with different amounts of rainfall. *H. brasiliensis* is a lowland type, requiring 1900–2500 mm annual rainfall and a temperature range of 24–32°C or warmer. However, excessive humidity interferes with tapping schedules and also increases the incidence of leaf diseases. The dry season produces a condition known as 'wintering', in which leaf-fall occurs and yields drop to very low levels for several weeks or months.

Strong winds can cause damage to rubber trees, but there are major clonal differences in susceptibility to wind damage. A great deal of rubber is grown on well-drained inland tropical soils. Under conditions of high humidity and rainfall such soils are to be preferred. Soils of higher water-holding capacity are preferred in sub-humid conditions. Rubber tolerates acidic soils and liming is not normally carried out unless the pH is found to be below 4.

The most successful aspect in rubber cultivation has been rubber breeding. While unselected seedlings yield about 0.5 tonnes of dry rubber/ha, the first cycle of clones doubles that yield at 1.1 tonnes/ha. Indeed, current standard clones and advanced materials in large-scale trials yield around 1.6 and 2.0 tonnes/ha, respectively.

The various rubber varieties in cultivation in the tropics have been bred for different purposes. For instance, 'GT 1' and 'PR 107' are relatively wind-resistant, while 'RRIM 501' is high-yielding, but has only a few large branches joined at acute angles, and is therefore more susceptible to wind damage. Other clones which have performed well include 'PRIM 623' and 'PRIM 628' in which yields of 2000 kg/ha or more can be obtained with correct planting, fertilising, weed control, and other management practices. Several institutions in different parts of the tropics are engaged in the breeding of rubber for high latex yields, good tapping capability and resistance to diseases, pests and adverse weather conditions.

Planting

Rubber may be planted from seed or as a budded material on a rootstock. Budded material gives a more uniform mature stand of trees. A polythene bag nursery or ground nursery may be adopted. Generally, rootstock seedlings and seedlings that are to be planted as unbudded trees are grown from clonal seed of a recommended clone. Seeds are planted in polythene bags of size 24 × 40 cm, filled with a suitable soil. A rather heavier soil, adequately drained, is more suitable than a very sandy soil. Organic manure may be mixed, if available, at the rate of about 150 g/bag. Seeds should be planted lying flat, at a depth of about 2 cm. If seeds are not planted flat, the main taproot may be twisted.

For a ground nursery, a site with a fairly heavy soil should be chosen and ground rock phosphate should be worked into the top 20 cm of the soil at a rate of 400 kg/ha along with ground magnesium limestone at a rate of 100 kg/ha. Mulching with grass or other materials is advantageous in a ground nursery. After about 4−6 weeks when germination has occurred, manuring with an N:P:K:Mg fertiliser at a ratio of about 18:10:12:2 should be carried out at the rate of 5 g per plant, every three weeks and up to three weeks before planting out in the field.

Budding of rubber seedlings is now carried out at the green wood stage, a few months after planting when the stem is about 15 mm thick. Patch budding is used for rubber, the buds being obtained from multiplication nurseries. A small panel of bark is removed leaving a flap at the top to hold the bud patch. The bud is then inserted and bound in place with transparent polythene (Figure 5.16). After a few weeks the bud union should have occurred. With care, a high percentage of success should be obtained.

After a few weeks, the stock is pruned back some centimetres above the bud-patch and the bud should begin to grow. At this stage, and until the budded scion has formed, a main shoot or any other buds developing on the stock stump should be pinched off. Budding should only be done when the plants are in a vigorous state of growth and not during adverse weather conditions. Buds from the maintenance nursery should only be taken during active growth and are best taken when the top whorl of leaves has just matured.

The budded seedlings remain in the nursery for about a year before they are transplanted to a permanent site. Transplanting must be done at the beginning of the rainy season. Rubber is normally not grown under irrigation so for proper establishment high rainfall must be available during its initial period of growth.

Seeds are not treated before sowing, but when sown fresh after collection the success of germination is very high. Stumped buddlings, which are normally used for replacement planting of buddlings which have failed in the field, are extracted from the ground, trimmed of side roots to form a club-like base, topped and the base is whitewashed for protection. Because of the large size and the high content of food reserves in such buddlings, they normally establish very quickly in the field.

Rubber is planted at densities between about

320 and 480 trees/ha. Generally, a higher density is more suitable for smallholders using family labour, and a lower density for estates in which hired labour is used. This is because the yield per tree is higher at the lower density and a hired tapper can tap a higher volume of latex per day. A higher density will give higher yields in the first few years of tapping. If interplanting of the rubber with legume crops such as cowpeas is intended, then an avenue planting arrangement of trees in 10 m rows, planted at 2−3 m within the rows, is recommended. Otherwise a more rectangular arrangement of about 4 × 6 m should be used, as this will give more rapid ground shading and will reduce the cost of weeding during establishment.

Fertiliser application

The rubber plant takes up some 92 kg N, 16 kg P, 70 kg K and 18 kg Mg/ha/year. To adequately fertilise young rubber trees on relatively poor soils, up to commencement of tapping, about 100 g of ground rock phosphate should be placed in the planting hole at the time of planting. Subsequently, fertilisation is best done at about two-monthly intervals along the weeding strips, using N:P:K:Mg mixtures in the following ratios: for the first three years, 15:15:10:2 and from the fourth to the sixth years, 18:10:12:2. Suggested application rates for the first six years are 15, 30, 100, 150, 200 and 250 g/12 months/tree.

For rich clay soils, nitrogen is the main fertiliser required. This may be supplied in the form of urea at rates up to 150 kg/ha/year, given at two monthly intervals. Sometimes a dressing of potassium chloride may also be required on clay soils. For wind-susceptible clones and windy locations, the quantity of urea should be halved. For mature rubber trees, magnesium fertilisation and liming are not required, unless deficiency is serious, as these encourage latex coagulation.

Weed control

It is common practice to plant leguminous cover crops when establishing young rubber plantations. The species most commonly used are *Pueraria phaseoloides*, *Centrosema pubescens* and *Calapogonium mucunoides*, the seeds of which are generally scarified to promote germination. Apart from controlling weeds by suppression, leguminous cover crops fix about 225 kg N/ha in their first year and consequently reduce fertiliser costs.

Traditionally, sodium arsenate has been the herbicide of choice for general weed control in rubber and is used for spraying along the rows of trees. In mature rubber plantations, the need for weeding is greatly reduced because of the dense canopy. In avenue planting, the weeding costs can be considerable. Intercropping with cash crops offers a viable solution, especially since the profit margins and the perennial nature of the rubber crop require that the expenditure on weeding be minimal.

Sodium arsenate is highly poisonous to mammals and so is considered to be dangerous for use as a herbicide. Paraquat (gramoxone), which is also highly toxic, has been used to some extent, as well as monosodium methyl arsenate (MSMA) and disodium methyl arsenate (DSMA), but these are more costly. For long term pre-emergence weed control in the weeding strip, diuron or simazine at a rate of 1 kg/ha may be used. Higher concentrations of diuron are toxic to rubber seedlings. If grass weeds predominate, dalapon at a rate of 10 kg/ha may be used as a directed spray. Glyphosate (Roundup), which is a much safer herbicide, may also be used to give a long period of control. In mature rubber the dense canopy greatly reduces weed growth and weed control normally consists of periodic spot weeding of invading weeds using the herbicides already referred to.

Pests and diseases

Rubber is not generally affected by insect pests at epidemic levels. Termites sometimes occur and are treated by spraying around the trunk with dieldrin or a similar insecticide. Cockchafer grubs sometimes occur in the soil. They may be controlled with endrin, poured into holes in the ground.

Rubber is, however, affected by a number of serious diseases. Of the major diseases, the South American leaf blight, caused by the fungus *Dothidella ulei*, which is a native of tropical America, is by far the most serious. Attempts to develop resistant clones have not been successful because of the production of new pathogenic strains by the fungus and the long period required to produce a new rubber clone. The disease flourishes in humid conditions when the relative humidity is over 95%. Good field sanitation may help reduce the incidence of the disease.

Leaf blight, caused by *Phytophthora* spp., may sometimes be serious. It is controlled by prophylactic spraying with fungicides or dusting with sulphur to control the spread when an outbreak occurs. Another leaf disease is caused by *Oidium heveae* which causes leaf-fall. It is serious at high elevations and under excessively humid conditions. It is also controlled by spraying or dusting.

Various root diseases occur in rubber of which *Ganoderma* spp. and *Fomes* spp. are the most serious. These spread from the stumps and large woody roots of both jungle and old rubber trees. It is not practical to remove jungle stumps during land preparation, but in replanting old rubber trees, stumps are commonly removed using heavy machinery. Isolation trenches are used to prevent spread from infected trees. The main roots and collar of *Genoderma*-infected trees may also be treated with a preparation of drazoxolon dissolved in grease.

Two tapping panel (cut) diseases of some importance in rubber are black stripe, caused by *Phytophthora palmivora*, which produces dark lines on the panel, and mouldy rot, caused by the fungus *Ceratocystis fimbriata*, which initially produces a darkening of newly-tapped bark, followed by a white mould. Both diseases are controlled by field hygiene and the removal of weeds from near the tree, as well as treatment with fungicides. Black stripe is usually treated with mercury fungicides or with captafol. Mouldy rot is best treated with a 1% solution of benlate. Affected trees should not be tapped until the disease symptoms have disappeared.

A physiological disease of the bark caused by over-tapping and stimulation is known as brown bast. The renewed bark becomes dry and lumpy. The disease is controlled by resting the trees and its occurrence is prevented by tapping at a lower intensity.

Tapping

Tapping is the most exacting operation in rubber production. Trees are opened for tapping when they reach a girth of 50 cm or at a height of 150 cm from the ground in the case of budded trees. Trees grown from seed have more tapered trunks and are usually opened for tapping at a height of 120 cm, when the circumference has reached 50 cm. Plants fulfilling these specifications would be between four and seven years old from planting. The tapping cut is made to run downward from left to right at an angle of 22–30° and at a height of about 50 cm from ground level. The most common method of tapping is that in which half the trunk is tapped on every second day. In many areas an instrument called a 'tapping template', made of metal or plastic, is placed against the tree to give a guide as to how the tapping cut (panel) should run.

In tapping, the actual cut is made with a special knife which has a curved tip. The cut is made such that only the bark, the phloem plus the outer cork of the tree is removed without damaging the cambium layer. If the cambium is damaged, regeneration of the bark in the tapped region cannot occur normally. When the tapping cut is made latex oozes out, runs along the edge of the panel and is collected in a container at the base of the tree. The flow soon diminishes so that every two days the tapper must remove a further 1.5 mm of bark at the lower edge of the panel. This is just enough to break the coagulated ends of the lactifers (latex ducts) and to cause renewed latex flow. In this way the tapping panel is gradually widened, but the rate of widening should not exceed 2 cm per month. Tapping is usually done between five and eight o'clock in the morning because latex flow is greatest at this time, since general physiological activity increases in daylight (Figure 9.9).

When the first tapping panel has become too wide, a new panel is opened up on the other side of the trunk while the first one is left to heal. Tapping should be discontinued during the dry season.

The stimulation of latex flow by the use of chemicals is now widely practised and generally yields more rubber per tapping by inhibiting the process of plugging at the ends of the latex vessels, which ordinarily limits latex yield. The most common substances in general use include 2,4–D and 2,4,5–T. The chemical is applied to an area of scraped bark below the cut at a concentration of about 1% mixed with a natural carrier such as palm oil. Stimulation is usually carried out twice yearly. Many other substances, including copper salts, also stimulate latex flow. All stimulants should be used with care and should not be regarded as a means of obtaining a higher total yield, but as a means of reducing the tapping task and the labour requirement for tapping.

Figure 9.9 Rubber tapping, showing the latex collecting in a bowl.

The yield of rubber depends not only on climatic and soil factors, but also on the genotype. Under favourable growth conditions, improved rubber clones yield from 600–2000 kg of dry rubber/ha/year from the first year to about the tenth year of tapping.

Processing of latex

The latex collected is bulked and diluted with water so that solid impurities are able to settle out. The liquid is then passed through a sieve and transferred to a coagulating tank where a coagulant, such as formic acid, acetic acid or fermented coconut water, is added. In this way the material coagulates into thick solid blocks which can be used to make different kinds of rubber. For example, to make sheet rubber the coagulated lump is passed through sets of rollers. The resulting sheets are dipped in paranitrophenol to prevent the growth of moulds. They are then dried by smoking for four days, and baled. Passing the coagulated lumps through a macerator chops them up into fine pieces,

which are used to form rough sheets referred to as crepe rubber. The sheets are cut to size, dried and baled. Crumb rubber and concentrated rubber are two other forms into which coagulated rubber is processed for industrial use.

CHAPTER 10

Root and tuber crops

A root crop is grown for its enlarged roots, whereas a tuber crop is grown for the swollen ends of its rhizomes or underground stems. With the possible exception of cereal crops, roots and tubers are the most important group of staple foods in the tropics. This is particularly true of the wet tropics where people depend on various root and tuber crops as their major source of starchy food. In addition to their food value to man, roots and tubers are also becoming increasingly important as livestock feed.

As well as the species which are indigenous to the tropics, several other root and tuber crops introduced from temperate environments are successfully cultivated, especially where part of the year is cool or where good soils occur at fairly high altitudes. The wide cultivation and popularity of root and tuber crops in tropical countries is influenced by the fact that they can be grown in a variety of soil types and are less prone to pest and disease attacks than other starchy crops, such as cereals. In addition, root and tuber crops require little cultivation, but still produce heavier yields than other field crops. They are easy to prepare for human consumption, and have the advantage of being available throughout the year.

One disadvantage of root and tuber crops which is the subject of research efforts in tropical and temperate countries, is their relatively poor nutritive value, particularly their protein content.

Root and tuber crops vary in their importance, depending on the locality, although on a world-wide basis the roots and tubers of greatest importance are cassava, yams, potatoes, cocoyams and carrots.

Cassava

Cassava, *Manihot esculenta* syn. *Manihot utilissima*, variously described as manioc or tapioca, is a native of South America. It is a dicotyledonous plant belonging to the family Euphorbiaceae, which characteristically contain latex-producing organs in their tissues. It has achieved great importance as a food crop. A third of the world's production comes from Africa, with Nigeria being among the leading producers. Cassava is mainly grown by peasant farmers, for whom it is often the primary staple. It is also a cash crop, being used to produce industrial starches, tapioca and livestock feeds. In West Africa, cassava is processed into 'gari'. The leaves are also eaten as pot herbs in various countries in Africa, Asia and South America (Figure 10.1).

Types and ecological adaptation

Various botanical classifications of cassava have been made. The two species widely recognised are a poisonous bitter species, *Manihot esculenta*, and a sweet non-poisonous species, *Manihot palmata* (syn. *M. dulcis*, *Fatropha dulcis* and *M. aipi*). This classification is based on the ability of the roots to produce hydrogen cyanide. Sweet or non-poisonous forms usually contain only small amounts of the cyanogenic glucosides or the enzyme linamarase which are abundant in the bitter or poisonous varieties. Cultivars are also distinguished on the basis of various morphological characteristics such as leaf shape and size, plant height and petiole colour as well as tuber shape, earliness of maturity and yield level.

In the genus *Manihot*, there are species adapted to a wide range of habits, from the swamp forest environment to semi-desert regions. However, the cultivated cassava is mostly grown in relatively humid regions of the tropics where temperatures range from 25–29°C. It grows in lowland regions and up to 1500m altitude but cannot withstand cold or frost. It is grown in areas with a rainfall of 500–5000 mm per annum. Except at planting, it can withstand prolonged periods of drought by

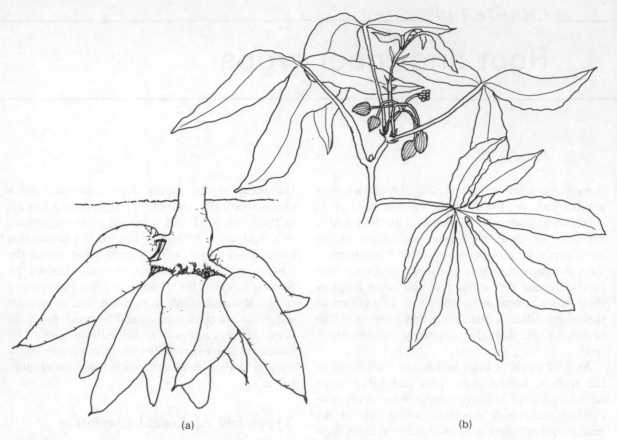

Figure 10.1 (a) Cassava tubers are harvested when needed as they deteriorate rapidly after harvesting. (b) A flowering branch of cassava, *Manihot esculenta*.

shedding its leaves, then rejuvenating quickly after rain.

Cassava thrives best in regions with average temperatures around 20°C, with small diurnal fluctuations and temperatures not falling below 10°C. It grows in a wide variety of soils, from heavy clays to light sandy loams. Heaviest yields are obtained on deep loose permeable soils with a high humus content. Tuber formation is under photoperiodic control. In short-day conditions, tuber formation occurs readily, but when daylength is greater than 10–12 hours, it is delayed and subsequent yields are lower. An abundance of starch is formed when the plant receives adequate amounts of sunlight, even though a good crop can be obtained under light shade.

There are currently many cultivars of cassava available for planting; some of these are genetically improved while others are local selections. The choice of variety should be guided by several factors, among which are high yield, stable productivity, disease resistance, quick canopy development, high dry matter content, level on the cassava flour index and amount of digestible crude fibre. The last three criteria describe the usefulness of the cassava crop for industrial, human and animal consumption, respectively. Popular varieties currently in cultivation are TMS 30572, 30555, 30337, 3000 58308 and 30211 in Nigeria; ROCASS 1, 2 and 3 in Sierra Leone; Kinuari (30085/28) and Kivuvu (30070/4) in Zaire; Gabon CIAM 76–6, 76–7, 76–13 and 76–33 in Gabon; Sey 14, 28, 32, 41 and 54 in Seychelles; CARICASS 1, 2 and 3 in Liberia; Kekabu, Jurai and Medan in Malaysia; IAC–7–127 in Brazil; Cambi, Pembero and Blanco in Argentina.

Planting

Cassava may be planted from seed, but this method is often used for breeding purposes only. The main mode of propagation is by stem cuttings which are planted and staked. The site for cassava cultivation should be well-drained, sandy loams being preferred. Planting can be done on ridges or on the flat, depending on the environment and the soil type and structure. On sandy soils, planting may be on the flat, whereas on heavy soils planting is usually on ridges or mounds. One advantage of ridge planting is the ease of harvesting the crop.

The planting time of cassava is very dependent on rainfall. In regions that have distinct wet and dry seasons, planting should be done as early as possible in the wet season. In this way the plants are able to make significant growth before the onset of the dry season. However, it is possible to obtain quite reasonable yields from plantings done at any time during the rainy season. In areas where cassava bacterial blight (CBB) is prevalent, it is recommended that planting should be done either early in the rainy season or towards the end of the rains. This timing allows the crop to become established before the peak of the rains, and hence the incidence of blight, thus giving the crop a head start on the disease. Where cassava is intercropped with another crop, it is often the practice to delay the planting of the cassava until the first crop is nearly ready for harvest.

In many countries, the protracted storage of planting material (cassava sticks) is neither practised nor recommended. The stem cuttings may be planted on the same day as they are harvested or they may be stored for 7–14 days prior to planting. The cuttings are normally tied in bundles and stored vertically, in the shade of trees or houses. The risk of desiccation may be minimised by covering the sticks with leaves. If planting is to be done during the dry season, the sticks may be treated with aldrin dust to protect them against termite attack.

The spacing used for cassava is dependent on the cultivar used and the growing conditions. Generally, cultivars with a spreading habit should be spaced further apart than those with an upright habit. As a general guide, a population of 10 000 plants/ha (1 m x 1 m spacing) gives a balance between high tuber yields and good tuber size, which is essential when tubers are marketed fresh and tiny tubers may not command a good market price. It is also important that plants should be spaced so as to ensure that the canopies of adjacent plants merge within the first two to three months after planting, to maximise both photosynthesis and weed exclusion.

Fertiliser application

By virtue of its deep and extensive root system, cassava is a good forager, able to reach less readily available soil nutrients. In many parts of the tropics, cassava is the last crop in a rotation sequence before the land is reverted to bush, yet under such conditons the crop still gives a good yield.

A major point to consider when fertilising cassava is that it has a relatively high requirement for potassium. When the level of potassium is low, the response to nitrogen and phosphorus fertilisers is generally poor. As a rule, cassava is fertilised with 90–120 kg K_2O/ha when the level of exchangeable potash is lower than 0.06%. The current recommendation in Nigeria is for applications of 30 kg N/ha, 15 kg P_2O_5/ha and 90 kg K_2O/ha. In Brazil, cassava is reported to respond reasonably well to phosphatic fertilisers, sometimes even better than it responds to nitrogen and potassium fertilisers.

Potassium sulphate is a better source of potassium than muriate of potash. Because of its chlorine content and its lack of sulphur, muriate of potash tends to depress cassava yield. Both urea and calcium ammonium nitrate are effective sources of nitrogen, while single superphosphate is a suitable source of phosphorus.

Band placement of fertilisers on one or both sides of each plant is preferred to broadcasting. Application of fertilisers directly to foliage is rarely practised, although when it is done, it is essentially to supplement rather than to replace soil applications. It is customary to apply most of the fertiliser at the time of planting, followed with a supplementary application two months later.

Weed control

Early weeding is important in cassava production. Keeping the crop weed-free for the first four months after planting gives an equivalent yield to that achieved with weed-free conditions throughout its life. This is the period of early canopy and tuber formation. At least two precisely timed weedings are needed to achieve optimum yields in cassava.

Several herbicides have been tested and found successful for weed control in cassava. Applications of diuron at 1.6 kg/ha are used on soils with up to 60% sand. Pre-emergence application of diuron at 2–4 kg/ha is safe for the cassava plant and provides good weed control until the cassava forms a close canopy. When cassava is intercropped with other food crops, especially maize, Primextra (atrazine plus metolachlor) at 2.3 kg/ha applied as a pre-emergence spray or Lasso at 0.5 kg/ha also applied pre-emergence, are effective.

Pests and diseases

The common grasshopper (*Zonocerus variegatus*) is an important pest of cassava which defoliates the plants, especially during the dry season. Insecticides of the lindane group provide effective treatment. In areas with distinct wet and dry seasons the cassava green spider mite (*Mononychellus tanajoa*) is often a serious pest during the dry season, but is less of a problem during the rainy season. The mite, which has been identified as a serious pest in Uganda, Colombia, Brazil and to a lesser extent in Tanzania, Kenya, Nigeria and Burundi, is found on buds, leaves and stems near the growing point. Another important insect pest is the cassava mealybug, *Phenacoccus manihoti*. Apart from the damage it causes, the cassava mealybug has also been implicated in the transmission of viruses or virus-like diseases. Native habitats of both the mealybug and the green spider mite have been explored and several natural enemies have been identified which feed on the pests. In addition, the breeding of resistant varieties is being carried out as a means of control. The cassava hornworm, *Erinyis ello*, causes damage during its larval stage, which feeds voraciously on the leaves. Biological control of the hornworm, which is a serious cassava pest in Brazil, has been possible by the use of its natural insect enemies, *Trichogramme spp.*, which parasitise the hornworm eggs, and wasps, *Polistes spp.*, which are predators of the hornworm larvae. Biological control may also be complemented with low levels of insecticides.

Cassava is also attacked by a host of fungal diseases, the important ones being brown leaf spot, anthracnose and *Phyllosticta* leaf spot. Brown leaf spot, which occurs quite commonly in South America, Asia and Africa is caused by the fungus *Cercospora henningsii*. Disease symptoms are the appearance of tan-coloured lesions with a distinct dark border on the upper leaf surface. The most favourable conditions for the development of leaf-spot occur during warm humid periods. The disease can be controlled by treating with copper fungicides or the use of resistant varieties.

Cassava anthracnose disease, caused by *Colletotricum* sp., is also an important stem disease in grassland savanna areas where soils are infertile and acidic. A severe attack results in defoliation. Control is by rigid sanitation and the use of resistant varieties.

The most important virus disease of cassava is the cassava mosaic virus which is very widespread and is found in all cassava growing areas, as it is easily carried over in infected cuttings. It is transmitted by a number of species of whitefly, including *Bemisia tabaci*. The symptoms are chlorosis and constriction of the lamina, together with leaf distortion and a reduction in size. Several control measures have been suggested, including careful selection of planting material from only disease-free setts, the control of the whitefly vector and the use of resistant varieties (Figure 10.2).

Figure 10.2 A cassava plant with chlorotic young leaves, a sign of cassava mosaic virus.

Harvesting and storage

It is best to harvest cassava at a time when the tubers are old enough to have stored sufficient starch, but not so old that they become very woody or fibrous. The exact time to harvest varies and depends very much on the cultivar. Some cultivars, particularly the sweet ones, are ready for harvesting in about seven months. Others require up to 18 months before they are ready for harvesting.

In practice, cassava plots are rarely harvested all at once or all at the recommended time of harvesting. The main reason for this is that the cassava tuber, once harvested, deteriorates rapidly. It cannot be kept in good condition for more than one or two days after harvesting. Therefore, the farmer harvests what he needs, leaving the remaining tubers unharvested and thereby storing them in the soil.

The yield level expected from a cassava crop depends on such factors as the cultivar used, cultural operations, fertiliser levels, the type of soil, the field spacing and the climatic conditions. Average yields are estimated at about 10 tonnes/ha on a world-wide basis, compared with a potential of 15–25 tonnes/ha.

Fresh cassava tubers, unlike yams, cannot be stored for more than a few days after harvesting, after which they begin to deteriorate rapidly. A bluish discolouration of the vascular bundles of the tuber develops, a symptom known as 'vascular streaking'. Streaking is caused by an enzymatic process and it fails to occur if the tuber has been dipped in warm water at 53°C for 45 minutes, stored in anaerobic conditions, submerged in water or refrigerated. In addition to vascular streaking, the quality of the starch in the tubers also deteriorates during storage. It has, however, been possible to store fresh cassava roots for up to eight weeks in boxes containing moist sawdust. The storage is preceded by curing the tubers at 30–35°C and 80–85% relative humidity. This type of storage is economical for small quantities of cassava to be used as a fresh vegetable, but may not be the answer for the storage of large amounts of cassava. Most cassava is therefore stored in processed forms. The major processed forms fall into four general categories; meal, flour, chips and starch. Meal forms include 'gari' and retted cassava, and account for the bulk of cassava consumed as human food in the tropics. Cassava chips and starch are mainly industrial products, accounting for most of the cassava in international trade.

Yam

Yams belong to the family Dioscoreaceae and the genus *Dioscorea*. This genus contains about 600 species out of which about ten are presently of economic importance. The genus is subdivided into sections within which the species fall. Most of the economically important yam species, *D. rotundata*, *D. alata*, *D. cayenensis*, *D. opposita* and *D. japonica* are in the same section (Enantiophyllum), which is characterised by vines twining to the right (clockwise) when viewed from the ground upwards. Species in other sections, *D. dumetorum* and *D. hispida* (Lasiophyton), *D. bulbifera* (Opsophyton), *D. esculenta* (Combilium), and *D. trifida* (Macrogynodium), twine to the left (Figure 10.3).

The regions of origin of the common species of yam are shown in Table 10.1.

Table 10.1 Regions of origin of common yam species

Botanical name	Area of origin
Dioscorea rotundata	West Africa
Dioscorea alata	Southeast Asia
Dioscorea cayenensis	West Africa
Dioscorea esculenta	Indo-China
Dioscorea bulbifera	Tropical Asia and West Africa
Dioscorea dumetorum	Tropical Africa
Dioscorea hispida	Tropical Asia
Dioscorea trifida	South America

Most of the yams produced in various parts of the world are consumed within the country of production, although appreciable quantities are shipped to developed countries for consumption by African and Caribbean ethnic minorities. The economic importance of the crop, therefore, lies in its use as a carbohydrate food for the producing region rather than in its ability to earn foreign exchange. Yams are consumed in different forms, which include boiled, pounded, mashed, fried, roasted and baked preparations. Fried yam balls, yam chips, yam flour and yam flakes are also popular. Yam peelings and waste are often fed to livestock, especially goats.

189

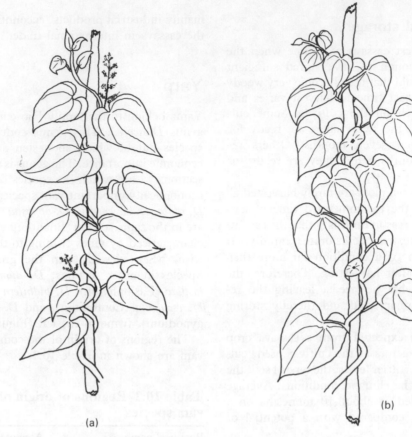

(a)

(b)

Figure 10.3 In (a) *Dioscorea alata*, the water yam, the vines twine to the right while in (b) *Dioscorea bulbifera*, the aerial yam, the vines twine to the left.

Types and ecological adaptation

The major characteristics of the various species of yam may be outlined as follows:

Dioscorea rotundata

Variously referred to as white yam and white guinea yam, *D. rotundata* has stems which are cylindrical in cross section and twine to the right. The stems usually bear spines and are long and wingless. The leaves are usually opposite in arrangement, simple, broadly cordate, acuminate and dark glossy-green. The tubers are cylindrical with rounded or pointed ends. The tuber skin is smooth and brown in colour, while the flesh is white and firm. The roots bear spines.

Dioscorea alata

The stem of *D. alata* (water yam) is square in cross section, winged and spineless. It twines to the right. It is usually green or purplish in colour, due to anthocyanin pigments. The leaves are vari-able in shape and colour, although they are generally light green in colour and large in size. They are ovate, deeply cordate and opposite in arrangement on the stem.

Dioscorea cayenensis

The stem of *D. cayenensis* (yellow yam, yellow guinea yam) is long and cylindrical and twines to the right. The stem bears spines especially at the base. The leaves are alternate in arrangement, simple, cordate and acuminate. They are dark green in colour and shiny above. The tuber flesh is pale yellow and firmer than the flesh of *D. alata*. In many respects, *D. cayenensis* and *D. rotundata* are similar, to the extent that some authorities regard the former to be a subgroup of the latter.

Dioscorea esculenta

The stems of *D. esculenta* (lesser yam, Chinese yam) are thin and cylindrical in cross section. They bear spines and twine to the left. The leaves are few, simple, broadly cordate, light green in colour and alternately arranged on the stem. The base of the leaf petiole is enlarged by three to four conspicuous spines. Tubers are small in size and produced in clusters by each plant. The tubers are cylindrical with round ends. They have smooth skins and are soft-textured and white-fleshed with few fibres. Their palatability is good as they taste slightly sweet. The roots bear spines.

Dioscorea dumetorum

The stems of *D. dumetorum* (bitter yam, trifoliate yam) are robust, spiny, cylindrical and hairy. They twine to the left. The leaves are trifoliate with ovate leaflets. The tubers are large and may be produced singly or in clusters. They are bitter and, in many cultivars, poisonous due to their high content of the alkaloid 'dioscorine' or its derivatives. The tubers may be detoxicated by extraction of the alkaloid, by soaking and boiling. The roots bear spines.

Dioscorea bulbifera

The stems of *D. bulbifera* (aerial yam) are cylindrical and spineless and twine to the left. The leaves are usually larger than those of other *Dioscorea* sp. They are simple, broadly ovate, widely cordate and are alternately arranged on the stem. The axil of each leaf contains two bud primordia and a bulbil primordium. It is the bulbil primordium that proliferates to give rise to shoots, roots and tubers when a vine cutting of *D. bulbifera* is planted. The subterranean tubers are much reduced or absent and are usually bitter and unpalatable. Aerial tubers (bulbils) are freshly angled and grey or brown in colour. The bulbils produced in the leaf axils are succulent and edible.

Dioscorea trifida

The stem of *D. trifida* (cush-cush yam) is rectangular in cross section, with angles sometimes extended to form wings. The vines twine to the left and, like *D. esculenta*, produce many individually small tubers per plant. The leaf is simple, but is deeply divided into three to five lobes. The petiole is winged. Neither leaf nor stem bear spines. The tuber flesh is white, yellow, pink or purple and is highly palatable.

Dioscorea opposita

The stems of *D. opposita* (cinnamon yam, Chinese yam) are cylindrical in cross section and spineless and twine to the right. The leaves are simple, accuminate and opposite in arrangement. The tubers are spindle-shaped and up to a metre long. Aerial tubers are formed in the leaf axils and are a convenient means of propagation, although such tubers take up to three years to yield a reasonable amount.

Dioscorea japonica

This species very closely resembles *D. opposita* in structure. It is used for food and in traditional medicine, and is limited to Japan and China.

Being essentially tropical crops, the edible yams cannot withstand frost. They do not do well in temperatures below 25°C, their optimum temperature range being between 25°C and 30°C. Ideally the crop requires at least 1000 mm of rainfall. Certain forest species, for example *D. cayenensis*, grow almost continuously throughout the year and cannot tolerate a rain-free period of more than two to three months. The savanna species *D. rotundata* and *D. alata* require a growing period of six to eight months to complete their life cycles. Daylength plays an important role in tuber formation, as long days favour vine growth while short days favour tuber formation and growth. Low light intensities favour yam growth and productivity and the crop is therefore shade-loving. Growing yams with stakes yields more than without stakes, due to increased mutual leaf shading.

Yams prefer fertile soils and thrive best in a loose, deep well-drained soil. The crop cannot tolerate waterlogging. In West Africa, the northern limit of the yam zone is about 10°N, beyond which the dry season is too long. The southern limit is set by the dry coastal areas of the low-lying lagoons.

The choice of variety depends on taste, market preference and ecological conditions as well as various socio-economic factors, including availability of labour.

Methods of propagation

The yam plant can be propagated by tubers, bulbils, seed, vine cuttings or tissue culture. By far the most common and commercially viable method of cultivation is by tubers.

Propagation by tubers

The tuber pieces used for planting may be small whole tubers or tuber pieces derived from a large tuber subdivided into smaller pieces. When such divisions are made the setts may be derived from the head, middle or tail regions of the tuber. The small whole tuber is the best planting material since it has a head region, enabling it to sprout readily, and has no cut surfaces to make it prone to rotting. The setts from the head region produce the next best planting material followed by those from the tail and middle regions.

It has been established that the greater the weight of the yam sett the greater the resulting tuber. In commercial yam production, setts weighing 150–300 g should be used for planting. Even though larger setts may produce bigger yields per stand, the yield per unit weight of planting material tends to decrease as the weight of setts increases. For this reason, setts heavier than 300 g should not normally be used.

Propagating yams by tuber means that as much as one-fifth of the yams produced in each season must be saved as planting material. This is a major limiting factor in the availability of planting material, the cost of which is rapidly becoming prohibitive. In order to overcome this problem, researchers have recently focussed attention on the following techniques of rapidly producing seed material in abundance:

1. The minisett technique involving the planting of pre-germinated tuber pieces, ranging in weight from 45–90 g, directly into the field for seed yam production.
2. The microsett technique involving 3–5 g pieces, treated with phytohormones and germinated before planting.

Propagation by true seed

The sexual process which results in seed production has become degenerate in many yam species. Of the major species, only *D. dumetorum*, *D. bubifera* and *D. trifida* flower and produce seed regularly. The three commercially important species have irregular flowering; *D. cayenensis* produces only male flowers and, therefore, cannot produce seeds, *D. alata* rarely produces flowers and seeds and *D. rotundata* has different flowering and seed production behaviour depending on variety. Even in flowering species or varieties of yam, flower production is not assured and seed production is even less reliable. In addition to this relative scarcity of well-formed seeds, propagation of yam by seed is complicated by difficulties in germination. Recent

work at the International Institute of Tropical Agriculture in Ibadan, Nigeria, has revealed that true yam seeds have a dormancy period of three to four months after which viable seeds will readily germinate.

Even though the commercial propagation of yams by true seed is yet to be undertaken the technique of seed production is of extreme value to breeders for improving the crop. The reproductive process which produces seed can also be used for hybridisation, which results in the production of new genotypes for crop improvement.

Propagation by vine cuttings

The method of propagating edible yam plants by vine cuttings involves taking a piece of the vine and allowing it to form roots in water or a moist substratum, before growing the established plant to maturity. Occasionally, successful rooting has been obtained with cuttings planted directly in the field.

To ensure the greatest rapidity of rooting the cuttings should be chosen carefully as follows:
1. They should contain at least one node.
2. They should bear only a few leaves.
3. They should be taken from young and vigorously growing plants, not more than three months old.

The propagation of yams by vine cuttings, while useful for rapid multiplication of desirable clonal material, is not yet a feasible technique for commercial production.

Propagation by bulbils

For those species bearing bulbils, yam cultivation using the bulbils is similar to the method described for tubers. A dormancy period is required after harvesting, the best growth is from the head section of the bulbil and the yield is influenced by the size of bulbil planting material. Yams grown for and from bulbils invariably produce reduced tuber yields. This method is used quite widely because of its convenience and because it leads to good plant establishment.

Propagation by tissue culture

The technique of producing yams by tissue culture

Figure 10.4 Yams are usually planted on mounds or ridges, and staked.

is relatively recent. It involves the rapid multiplication of undifferentiated yam tissue in a nutrient and hormone rich media to stimulate root and shoot differentiation. The procedure offers promise when rapid multiplication of a single desirable plant is required by breeders.

Planting

Yams require a loose soil in which the tubers can grow easily. When a suitable site has been selected and the necessary land preparation made, mounds, holes, banks or ridges are prepared. In loose, deep, sandy loam soils, yams may be planted on the flat. Plantings may be either in the dry season or at the onset of the rainy season. For dry season planting, land preparation begins just before the rainy season ends and before the soil becomes dry and difficult to work. The fresh setts are planted, and they pass through a period of dormancy in the soil before they sprout. Therefore, in the northern parts of Nigeria for example, dry season yams planted in

October or November will normally not emerge until the beginning of the next rainy season in April or May (Figure 10.4).

For rainy season planting, land preparation may be done at any time during the dry season. However, the setts are not planted until the rainy season begins. Such setts, because they have been stored through the dry season, will sprout readily, usually within a month. In the southern Guinea savanna regions of West Africa, February or early March plantings are recommended even though the rains do not become regular until late March or April. Once the rainy season starts, the later the plantings the lower the yields obtained.

Seed setts are cut into the appropriate sizes one to two days before the envisaged date of planting. The time interval allows the cut surface to heal over before the sett is placed in the ground. After the setts have been cut, they should be handled carefully to avoid damage and infection. Traditional farmers sometimes smear the cut surfaces with wood ash which appears to give some protection against rotting. Dipping in fungicidal solutions of Benlate, captan, limewash or Bordeaux mixture also offers some protection. Dusting the setts with 2.5% aldrin or gammalin-A dust protects them and the subsequent tubers against initial insect attacks.

Yams are generally planted at about 1 m spacing, in rows that are 1 m apart. Wider or narrower spacings may be used depending on the sett sizes. The bigger the sett, the wider the spacing. Wider spacings should be used in heavy soils if staking is not intended. Spacing in traditional yam production is variable, depending on the extent of intercropping.

Fertiliser application

The response of yams to fertiliser applications depends on the fertility of the soil and the species and cultivar of yam being grown. In general, yams respond well to nitrogen and potassium fertilisation, but show only a slight response to phosphorus. This is partly because phosphorous is removed very efficiently from the soil by yams and is seldom a limiting factor, partly because many soils used for yam production are naturally rich in phosphorous. For these reasons, many fertiliser recommendations tend to exclude phosphorus completely, recommending such fertiliser combinations as 20:0:20 or ammonium sulphate only.

On some soils yams should be fertilised with a complete NPK fertiliser. In the southern parts of Nigeria, for example, the application of 10:10:20 is advocated, whilst in eastern Nigeria 12:12:18 is used in quantities ranging from 128−250 kg/ha. The use of organic manures and composts is also recommended where they are available.

Weed control

Yams are particularly sensitive to competition from weeds during the early part of their growth. After the first two to three months weeding of yams is unnecessary. Weeds are controlled by hoeing, which is done carefully to avoid any disturbance to the growing tubers.

Chemical weed control by means of herbicides is now becoming more accepted, especially on large farms. Since yams seem most prone to weed competition early in the season, much attention has been directed at pre-emergence herbicides with prolonged activity. In general, the weed control recommendations for major yam growing areas in Nigeria are for hoe-weeding at 3, 8 and 16 weeks after planting and only light weeding thereafter. Where herbicides are used, the application of atrazine plus metolachlor applied pre-emergence and supplemented with a hoe weeding at 12−16 weeks after planting has been found to be most effective.

Pests and diseases

Insect pests are a serious problem in the production of yams. The yam beetle, *Heteroligus* spp., is a major pest. It lays its eggs in moist ground during the early part of the dry season (November or December in West Africa) and the beetles that develop do considerable damage by feeding on the tubers. The holes they make in the tubers not only make the tubers unsightly and unpalatable, but also predispose them to rotting during storage. The control measure is to apply insecticidal dust to the setts before planting. In this regard aldrin, gamma-HCH(BHC) or Gammalin have been found to be effective insecticides.

Termites, grasshoppers and aphids occasionally attack yam plants, but these can be controlled by insecticidal sprays and good crop sanitation. Various species of nematode, including the yam nematode (*Scutellonema bradys*), root knot nematode (*Meloidogyne* sp.) and the root lesion nematode (*Pratylenchus* sp.) attack growing yam tubers. They also produce secondary effects on the tubers by predisposing them to fungal infections, as well as their own direct deleterious effects. The major methods of control are to avoid nematode-infested lands and to avoid planting nematode-infested setts. Crop rotation, to reduce the build-up of nematodes, is also useful. In special cases, fumigation with nematicides may be used.

Various fungal rots attack yams at all stages, including during storage. Control measures include the treatment of setts with fungicides or alkaline materials such as wood ash and lime wash before planting. Careful harvesting, so as not to wound the tubers, and frequent inspection of stored tubers to effect isolation of rotting material are two methods of fungal control.

Mammals, especially rodents, can be serious pests of yams. They eat the shoots and tubers in the field before harvest or consume them during storage. The best control measures for these mammals are fences to keep them away, traps to catch them or poisoned baits to kill them.

Harvesting and storage

Two general practices exist with respect to harvesting. In the first, each plant in the field is harvested twice. This practice is termed double harvesting. The second approach, single harvesting, calls for the harvesting of each plant once only when the season is over. In the practice of double harvesting the farmer carefully digs all the soil around the tuber before the end of the season and removes the tuber, leaving the roots intact. The roots are then returned to the ground and covered with soil. When the plant has senesced at the end of the season, the second harvest is taken. The double and single methods of harvesting both produce about the same yield. However, the former has the advantage that the produce of the first harvesting allows yam tubers to be on the market early, whilst the second harvest produces excellent planting material of good keeping quality.

There are four main post-harvest storage methods for yams. These are barn, platform, underground and cold storage. Barn storage is the commonest method of storing yams in West Africa (Figure 10.5). A yam barn is essentially a framework of vertically arranged wooden poles, each one about 3 m high. It is made from the midribs of palm fronds, arranged about 50 cm apart and held together by means of rigid horizontal wooden sticks. The yam tubers are tied to the vertical poles of the barn, each tuber being placed in its horizontal axis. Yams are sometimes stored on raised platforms constructed in the field. Tubers are placed either vertically or horizontally in a single tier on the platform. The platform may be built outdoors during the dry season and transferred indoors when the rainy season starts. The storage of yams in ditches is also practised. The ditch is dug and the tubers are placed in it and covered with soil.

In the above three methods, there are certain requirements that must be met to reduce spoilage. These include the need for the tubers to be well

Figure 10.5 Yam storage in a yam barn in West Africa.

aerated to avoid conditions of high humidity and the increased risk of rotting. In addition, ample shade must be provided so that the tubers are kept relatively cool. It is important that the stored tubers are frequently inspected so that any rotting ones are detected early and removed.

The fourth storage method calls for cold storage in rooms maintained at about 15°C. The use of air conditioners in such rooms has been found to be effective, since they reduce both temperature and humidity. This storage method is, however, expensive and its use in many yam growing areas in the tropics is limited.

Sweet potato

Sweet potato, *Ipomoea batatas*, is a dicotyledonous plant belonging to the family Convolvulaceae. There are over 40 genera in this family, but only the species *I. batatas* is of economic importance. Sweet potato is indigenous to South America from where it has spread to most parts of the tropics and subtropics. In West Africa, the sweet potato is popular in some of the northern ethnic groups; many West Africans of the forest and southern savanna regions regard such sweet commodities as being suitable only for consumption by children. Nevertheless, most sweet potatoes produced in tropical countries are used as human food and are boiled, fried or roasted. Sweet potatoes are canned, dehydrated, made into flour and used as sources of starch, glucose, syrup and alcohol. They are also fed to livestock. The tender tops and young leaves are eaten as a green vegetable and the vines are used as fodder for livestock (Figure 10.6).

Types and ecological adaptation

Although it is grown in the tropical, subtropical and warm temperate regions of the world, sweet potato is essentially a warm season crop. It likes high temperatures and sunshine and requires considerable amounts of rain during the growing period. Optimal temperatures are in the range of 24−28°C. Sweet potato does best in regions with 750−1000 mm of rainfall per annum, with 500 mm falling within the growing season. Although the crop can withstand drought conditons, yields are considerably reduced if the drought

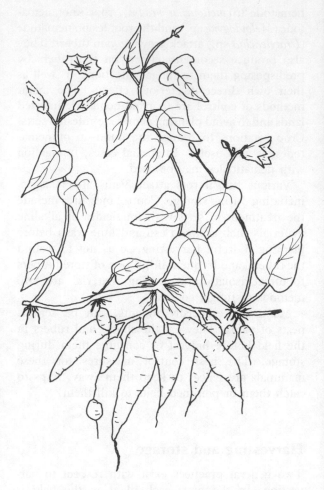

Figure 10.6 A sweet potato plant, *Ipomoea batatas*, showing the edible tuberous roots and leaves.

occurs within the first six weeks after planting or at the time of tuber initiation.

The crop grows best in well-drained sandy loams or loamy soils and does poorly on clay soils, especially when they are waterlogged. It is sensitive to saline and alkaline soils and requires a pH between 5.6 and 6.6.

Sweet potato cultivars are differentiated on the basis of the texture of the tuber after cooking, the colour of the tuber skin, the shape of the tuber, the shape of leaves, the depth of rooting, the time of maturity, the resistance to disease and several other vegetative characteristics. The decision as to which variety to plant has to be taken by the producer. In addition to using the local varieties which suit the preference of the farmers, several improved cultivars are recommended or have shown promise for various parts of the world. These are TIS 2498, 9265, 146/3092, 2534, 2421, 2352 BIS 23, Dokobo and Anioma in Nigeria; Tucumana Morada and Brasilers Blanca in Argentina; Onokeo in Hawaii; la Catemaco and Dulco in Venezuela, and T.25 and A.26/7 in the West Indies.

Planting

Sweet potatoes are propagated by means of tubers or vine cuttings. If the crop is to be grown by tuber propagation the setts should be derived from robust and healthy tubers. Setts should be small and should weigh between 20−50 g. However, because of the very low yield obtained using setts as planting material, the use of vine cuttings is generally recommended as the best method of propagating sweet potatoes.

Sweet potato generally requires a loose soil in which the tubers can grow with little hindrance. Planting is done either on ridges, on mounds or on the flat. Planting on ridges is best because it facilitates root penetration and development and minimises soil erosion. Research results indicate that the higher the ridge the greater the yield, up to a ridge height of 3.6 m. Naturally, the optimal height of a ridge must depend on the soil type and the cultivar being grown.

Flat planting results in a low tuber yield, whilst cultivation on mounds involves more manual labour than cultivation on ridges. In swamps or locations where the water table is high, planting on high mounds is advantageous.

Sweet potatoes should be planted when the rains are established. In areas where the rainy season is very long, planting should be timed such that the crop matures as the rainfall begins to decline. Where irrigation is available, the crop can be planted at any time. Sweet potato does not show a photoperiodic response.

At planting, the vine is inserted into the soil at an angle, with one half to two-thirds of its length beneath the soil and exposing at least two nodes. The vines are normally planted 25−30 cm apart on 60−100 cm ridges. Sweet potato is generally able to compensate for variations in planting density in that as the plant population per unit area increases, the number of tubers per plant, the mean weight per tuber and the tuber yield per plant all decrease.

Fertiliser application

Sweet potato responds well to fertilisers. The exact type and dosage of fertiliser used depends on the soil type, the environment and the cultivar grown. A 10:10:20 complete (NPK) fertiliser at 700 kg/ha is often suggested. Alternatively, nitrogen at the rate of 34−45 kgN/ha, phosphate at 50−110 $k_gP_2O_5$/ha and potash at 84−169 k_gK_2O/ha may be used.

Extreme care should be taken to avoid applying excessive nitrogen, since this causes delayed tuber formation and much vine growth at the expense of tuber growth. The K_2O:N ratio should be maintained at a relatively high level to maintain a high rate of tuber growth.

Weed control

Weeds are a problem in sweet potato cultivation only during the first two months of growth. After this period vigorous growth of the vine causes rapid and effective coverage of the ground, smothering weeds. One hand weeding is recommended at three to four weeks after planting.

Diphenamid at 2.7−4.4 kg/ha, fluometuron at 2.0−2.5 kg/ha and Primextra at 2.0−2.5 kg/ha have all been used successfully for control of weeds in sweet potato plots.

Pests and diseases

Sweet potato is attacked by a number of pests, including the sweet potato weevil (*Cylas* sp.) which has caused extensive damage to crops in the West Indies, West Africa, East Africa, Brazil, Venezuela, Central America, Malaysia and several other tropical countries around the world. The larvae feed on the tubers and burrow through them. Activities of the larvae also promote the spread of fungal infections. The use of insecticides, for example 0.1% carbaryl or fenthion applied three times at three weekly intervals, starting from the time of tuber initiation or the first sign of affected plants, controls the sweet potato weevil. Other methods of control include crop rotation, the use of resistant cultivars and prompt harvesting. It has been reported that the deeper the tubers lie within the soil, the less likely they are to be infested by the weevil. Furthermore, earthing-up around the tubers results in a lower level of infestation.

Other important pests of sweet potato are the beetle *Euscepes* sp., the vine borer (which is the larval stage of the moth *Omphisa anastomosalis*), and the hawk moth (*Herse convolvuli*). Also of importance in some localities are nematodes, damage being caused by the sting nematode (*Belonolaimus gracilis*), the root lesion nematode (*Pratylenchus* sp.) and the root knot nematode (*Meloidogyne* sp.).

Mosaic is a virus disease of sweet potato which has become increasingly important in Africa. It is caused by a strain of the tobacco mosaic virus. Infected plants have small, mottled, malformed leaves and they yield little or no tuber. The disease can be controlled by burning all affected plants. A complex of viruses (the internal cork virus, the leaf-spot virus and the white fly-transmitted yellow dwarf virus) cause the sweet potato virus disease complex, often referred to as the 'feathery mottle complex'. When present together, the three viruses cause severe symptoms which none can cause individually. The disease is characterised by dwarfing of the plants, yellowing of the veins in young leaves and yellow spotting in older leaves. The internodes are also short and the tubers are small. Control is achieved by removing affected plants and by the use of insecticides to control the white flies (*Bemisia* spp. and *Trisleurodes* spp.).

Fungal diseases of sweet potato reported in various parts of the tropics include:

1. Black rot, caused by *Ceratocystis fimbriata*, which can attack the plants in the field or during storage of tubers.
2. Scurf, caused by *Monilochaetes infuscans*, which causes brownish blotches on the roots, tubers and other subterranean parts of the plant, and invades the outer tissues of these organs.
3. Fusarium wilt (caused by *Fusarium oxysporum* f. sp. *batatas*) which enters the plant mostly through wounds and attacks the vascular tissues, especially the xylem.

Harvesting and storage

Sweet potato is normally ready for harvesting three to six months after planting, depending on variety and the environmental conditions in which it is grown. Maturity is indicated by yellowing of the leaves, drying of vines and sap exudation from mature tubers.

Yields vary according to the cultivar used, the location and the production practices employed. While unimproved varieties produce an average of 6–14 tonnes/ha, yields of up to 21–41 tonnes/ha are possible using improved varieties and better technology. The harvested tubers are subjected to curing to promote rapid healing of the wounds inflicted during harvesting. Tubers are left in the sun to cure naturally for four to six hours before storage. The recommended storage methods include:

Pit storage. A pit 0.5m square is dug and the bottom is lined with grass. The cured tubers are treated with wood ash and are stacked to half the pit depth, then covered with dry grass and soil.

Storage in sawdust. A heap of sawdust is placed on the ground and the cured and ash-treated tubers are placed in the heap and covered.

Potato

The potato, *Solanum tuberosum*, which is often referred to as the 'Irish' potato, is not Irish in origin. It is a member of the Solanaceae, a family which includes other important economic crops such as peppers, tomatoes, egg-plants and tobacco. There are at least 100 edible, tuber-bearing, solanaceous species, all natives of North and South America, many coming from the Andean regions of Peru and Bolivia. However, because of its popularity in Ireland, the potato crop became widely known as the 'Irish' potato (Figure 10.7).

The plant is an annual, attaining a height of 0.5 m and producing swollen underground stem tubers when mature. The Irish potato produces flowers, sets fruit and produces seed which is used for breeding purposes. The tubers are cooked in various ways as a staple food or vegetable. They are also used for stock feed and starch manufacture as well as the preparation of spirits and industrial alcohol.

Types and ecological adaptation

Although the origin of the potato is tropical, the natural habitat is at such a high altitude that the crop has adapted readily to the ecological conditions of temperate countries, where it is grown far more widely than in the tropics. Nevertheless, the production of Irish potatoes in the tropics has increased being more important than cool day temperatures. For tuber formation, the optimum soil temperature is in the range of 15−24°C.

The potato can be grown in all types of soil, but the crop does best in deep and well drained clay, silt or sandy loam soils with a pH range of 4.8−6.5. Where scab (*Streptomyces scabies*) is a problem the pH should be below 5.2. Excessive rainfall encourages the spread of various leaf diseases, notably blight caused by *Phytophthora infestans*. Hence dry season cultivation gives the most satisfactory yields.

Many early cultivars imported from temperate areas are now adapted for use in the tropics. Only certified seed potatoes should be used for planting. These must be reasonably free from the diseases transmitted in the tuber, notably blight and virus diseases.

Planting

Potato 'seeds' used commercially are actually tubers or pieces of tubers. Each tuber is usually cut into several sections, each containing an 'eye'. Botanically, the tubers are stems and the eyes are buds.

Land preparation for potatoes involves working the soil to a fine tilth. Organic manure and fertilisers may be applied and worked into the soil at the time of land preparation. Planting may be done on ridges, raised beds or on the flat. On ridges, planting does not have to be too deep, since later in the season the process of earthing up further builds up the ridges, thus covering the developing tubers with an adequate layer of soil.

At elevations above 1000 m potatoes can be grown almost all the year round with good results. At lower elevations, maximum yields are obtained when potatoes are planted during the cool season, usually from November to February in the northern hemisphere. The rainy season crop should be planted as soon as the rains are established.

After harvest, potato seed stays dormant for a length of time before sprouting. If the time between harvest and planting is too short, the dormancy has to be broken. Chemicals are used for this purpose, the concentrations and time of application of these varying with the variety of potato used, the stage of dormancy, the method of application and the prevalent temperature during treatment. The cutting of tubers may, in some cases, be

Figure 10.7 The potato plant is an annual which is widely grown as a cool-weather crop.

greatly in recent years. Much of this increase has been in the highland areas, especially in Kenya and India. In Nigeria, production of the crop is concentrated on the Jos plateau. Potatoes are a cool-weather crop. In temperate regions, potatoes produce tubers when the daylength increases (spring) but in tropical conditions, tubers are produced during short days.

In the tropics, the potato is usually only grown in areas where the elevation is 1000 m above sea level or higher. Some cultivars can, however, be grown satisfactorily at lower elevations during cooler seasons. In general, yields are much lower when the potato is grown at low elevations. For maximum yields, the average temperature should be in the range of 15−18°C with cool night temperatures sufficient to break dormancy. Chemicals used include calcium disulphide, thiourea, chlorhydrin and gibberellic acid. Most of these chemicals are quite poisonous and should be applied in gas chambers. Seed cutting is practised in some countries, especially if the seed size is large. This usually has the disadvantage of reducing the percentage emergence, due to seed piece decay, and the transmission of several kinds of disease with the knife, such as viruses, ring rot, black leg and bacterial wilt. These can be prevented by disinfecting the knife. Measures taken to promote wound-healing of the cut surfaces help to prevent seed pieces from rotting. Large-sized seed potatoes tend to give a higher yield than smaller sized ones, but the intitial seed rate is higher.

Potatoes can be grown at row distances of 50–100 cm, although 75–90 cm row distances are preferred. The distance to allow between plants within a row varies with the growing conditions and whether seed potatoes or potatoes for consumption are being grown. For the latter the plants are at a greater spacing than for seed potatoes.

Fertiliser application

Potato yields are decidedly increased by heavy applications of nitrogen fertilisers. In most studies, potatoes have not responded well to phosphorus fertilisers. A crop yielding 30 tonnes of tubers/ha removes from the soil an estimated 150 kg N, 60 kg P_2O_5 and 350 kg K_2O. The application of fertilisers is thus based on these ratios, although it is advisable to carry out tests to give a correct indication of fertiliser requirements. In Nigeria, for instance, the practice is to apply 50 kg N, 45 kg P_2O_5 and 40 kg K_2O/ha. While phosphorus and potassium fertilisers are applied in old furrows before splitting the ridges, the nitrogen is placed about 8 cm from the seed at planting.

Weed control

Weeds can be controlled either manually, mechanically or by the use of herbicides. In mechanical weeding, the ridges are built up gradually, as soon as fresh weeds start growing.

Contact herbicides, such as dinoseb and paraquat, can be used to kill the growing weeds before the potatoes emerge, after which mechanical weeding should be done. Herbicides, such as linuron and metobromuron, which have a residual effect but do not kill weeds that have already emerged, can also be used. When the final ridge is made the chemical is then applied before the weeds germinate.

Pests and diseases

The common pests of potato include aphids, cutworms and tuber worms. Aphids are important in transmitting viral diseases. A severe attack can cause leaves to curl and plants to be stunted. Aphids can be controlled by spraying with malathion, diazinon or Cygon. Cutworms damage potatoes by eating young plants at ground level and making holes in exposed tubers. Toxaphene, sprayed before planting, can be a good method of control. The larvae of the potato tuber moth tunnel into the tuber as well as the stem, thus rendering the tubers unsaleable. Malathion or Cygon sprays can offer good control.

Common diseases include late blight, *Phytophthora infestans*. This is one of the most important diseases in potato growing areas. The symptoms are the appearance of small brown-black patches on the leaves which, in wet conditions, increase in size along the water-soaked leaf margin producing white spores on the underside of the leaves. Frequently the stems and other plant parts may be killed completely and tubers can also be affected. Chemical control by means of regular spraying with difolatan is reported to be effective.

Early blight, *Alternaria solani*, causes external damage. The symptoms include the presence of dead spots on the leaves with distinct concentric rings. Zineb and maneb sprays give effective chemical control.

Rhizoctonia disease is common in most tropical soils and affects potato stems and tubers at or below ground level, causing wilting by girdling and eventually the death of plants. Control measures involve the use of clean planting material and also treating the soil and seed pieces with Terraclor.

Scab, *Streptomyces scabies*, is a soil-borne organism which is manifested on potato tubers as raised areas, which are actually the brown patches of corky tissue produced by the tubers in an attempt to exclude the organism. Some appear superficially, while others are deeply pitted. The use of resistant varieties as well as clean and treated seed is known to check scab. Its prevalence is also increased by abrasive soils and high pH levels.

Other diseases are caused by various bacteria and viruses. Bacteria cause many soft rots in the field and in storage. Control measures include careful sanitation and field rotation to break the disease cycle, and gentle tuber handling during harvest.

Harvesting and storage

Potato tubers are considered to be ready for harvest when the skin of the tuber does not peel off easily. This can be tested by rubbing the skin off the tubers with the thumb. Harvesting is usually done by hand-digging the tubers. Mechanical harvesters are now being used in many potato growing areas of the world. With good management, yields of 12–15 tonnes/ha have been obtained. Much higher yields are possible.

Cocoyam

The monocotyledonous family Araceae (the aroids) contains several cultivated root crops used for food in various parts of the tropics. Some of the genera, notably *Alocasia*, *Cyrtosperma* and *Amorphophallus* are limited in cultivation and are important only in certain communities in India, Southeast Asia and the Pacific Islands. By far the most important genera of the aroid root crops are *Colocasia* spp. and *Xanthosoma* spp.

Both genera of cocoyam are staple foods, especially in the tropical rainforests of the Pacific Islands and West Africa. The corm and cormels are the major economic parts although occasionally the leaves and petioles are also used for food. The corms and cormels can be consumed after being boiled, baked, roasted or fried in oil. In parts of West Africa, the boiled corms and cormels may be pounded into a paste called fufu, similar to pounded yam. It is also used to make flour. The use of corms, cormels and leaves as animal feed occurs, although it is presently not widespread.

Types and ecological adaptation

In West Africa, the name 'cocoyam' is used for both *Colocasia* and *Xanthosoma*. In other parts of the tropics where the crop is grown, there are

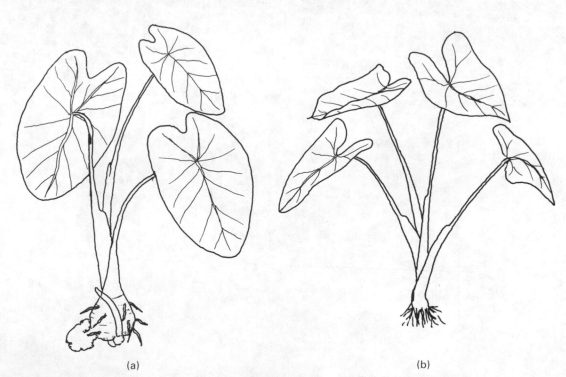

(a)   (b)

Figure 10.8 (a) *Colocasia* cocoyam is a lowland plant while (b) *Xanthosoma* cocoyam is grown mostly as an upland crop.

various common names used which sometimes cause confusion. *Colocasia esculenta*, commonly known as taro, dasheen, old cocoyam or eddoe, is a variable species with many forms and varieties being grown throughout the world. Its peltate leaves distinguish it from *Xanthosoma* which has hastate leaves. Variously referred to as tannia (tanier), yautia or new yam, the leaves of *Xanthosoma* tend to be arrow-shaped, with sharp tips and deep, wide basal lobes and prominent marginal veins (Figure 10.8).

Cocoyams, whether *Colocasia* or *Xanthosoma*, generally show a high demand for moisture. They thrive best where the rainfall is above 2000 mm per annum. However, unlike *Xanthosoma*, which is grown mostly as an upland crop and is unable to withstand continuous flooding, *Colocasia* grows best in lowland conditions. In many areas it is cultivated along streams, rivers and 'fadamas' in submerged or flooded culture. During the dry season, irrigation is usually required.

Cocoyams are warm-weather crops which require average daily temperatures above 21°C. Although they have been successfully cultivated at altitudes up to 2000 m, yields tend to be poor, probably as a result of the relatively cool temperatures encountered at such heights. Unlike *Colocasia*, *Xanthosoma* can tolerate moderate shade, but may die out completely under heavy shade (Figure 10.9).

The soil condition can also be a growth limiting factor. *Colocasia* prefers soils that are either heavy, with a high moisture-holding capacity or well-drained and loamy, with a high water table. In contrast, *Xanthosoma* generally does best on deep, well-drained soils. For all cocoyams, the soil must be of good fertility and have a pH of 5.5–6.5.

Many local cocoyam varieties are available and can be classified on the basis of the colour of their corms and leaves. Any of the varieties may be planted, depending on availability and taste preferences. Whichever variety is grown must be well suited to the ecological conditions of the location.

Planting

The parts of cocoyam used as planting material include:

Figure 10.9 *Xanthosoma* cocoyam prefers a deep well-drained soil and can only tolerate moderate shade.

1. The stem, cut to consist of the whole corm or its apical portion with a small section (15−25 cm) of the leaf stalk attached.
2. Whole corms or portions (setts) cut from larger corms, with each sett having at least one 'eye' (bud) and weighing about 100−120 g.
3. Whole cormels (suckers) or setts cut from large cormels.

The best yields of cocoyam are obtained from stem cuttings, probably because these produce more roots and leaves than the other two kinds of planting material. When using stem cuttings, they may be put in the sun to dry for some hours as a means of curing them and thus reducing the risk of fungal infection. Where corms and cormels are used, a form of pre-germination may be employed whereby the setts are stored in a shady or moist area for a few weeks. At planting only those that have sprouted are used. This method is reported to ensure good plant establishment. Cocoyam setts may be treated with Dithane M-45 to prevent fungal attack. Setts may also be cured to reduce the risk of rotting.

The best type of site for *Colocasia* has an abundant water supply throughout the growing period. However, for good yields the soil should be dry while the crop is finally maturing.

Because of its relatively high nutrient demand, *Xanthosoma* requires a site which is rich in fertility. *Xanthosoma*, with its preference for well-drained and deep friable soils, is often the first crop to be planted on land which has been cleared of forest or bush fallow.

Cocoyams may be planted on flat beds, ridges or mounds. Spacing varies from 30×30 cm to 1×1 m. Generally, close spacing increases the corm and shoot yields per unit area, but decreases the corm yield per plant, the contribution of sucker corms to the yield and the leaf area per plant. Therefore wide spacing is suitable for the production of a few large-sized cormels in contrast to the many small-sized suckers produced by close spacing. A spacing of 60×60 cm is often preferred for optimum yield.

The major determinant of the time of planting for cocoyam is the availability of moisture. In areas with distinct rainy and dry seasons, the crop is planted at the beginning of the rains. With irrigation, planting can be done throughout the year. Planting may also be done during the dry season around 'fadamas' where plants make use of residual moisture.

Fertiliser application

In traditional cocoyam cultivation in Africa and parts of the Pacific Islands, little or no fertiliser is used. This is particularly true where the crop is grown on newly cleared land. However, cocoyams have been reported to respond well to fertilisers. Yields of cormels are increased significantly with the applications of up to 150 kg N, 100 kg P_2O_5 and 200 kg K_2O/ha for *Xanthosoma* and 150 kg N, 60 kg P_2O_5 and 180 kg K_2O/ha for *Colocasia*.

Weed control

Cocoyam fields are particulary prone to weed growth during the first three to four months after planting, when the leaf canopy is sparse and again later in the season during the period of starch accumulation and maturation. Manual hoeing and earthing up the ridges are normally done to control weeds and to enhance corm formation. Two hoe weedings prior to cormel initiation are usually sufficient for good yields. Thereafter, the leaf canopy closes up to suppress weed growth.

Various herbicides have been used to control weeds in cocoyam, including atrazine, simazine, linuron, monolinuron, prometryne, dalapon and diuron.

Pests and diseases

Common insect pests of *Colocasia* include taro leaf-hopper (*Tarophagus proserpina*), cut-worm (*Prodenia litura*) and white fly (*Bemisia tabaci*). All three insects attack the leaves and can easily be controlled by spraying with malathion. The taro leaf-hopper has been biologically controlled using *Cytorhinus fulvus* in Hawaii. In addition to the insects already referred to, the larvae of the gabi moth (*Hippotion celerio*) can also cause serious defoliation in *Colocasia*. The root knot nematode (*Meloidogyne* sp.) is also reported to attack the crop, especially as a ratoon.

Among the more important diseases of *Colocasia* are corm rot (caused by *Phytophthora cinnamoni* and *P. nicotianae* var. *parastica*), taro leaf blight (caused by *Phytophthora colocasiae*) and brown leaf spot (caused by *Cladosporium colocasiicola*). An attack of corm rot results in wilted plants with complete rotting of the corm. The disease can be controlled by using resistant *Colocasia* varieties, especially in poorly drained soils. Taro leaf blight is characterised by circular water-soaked necrotic spots on the leaves, followed by the subsequent collapse of the plant. With brown leaf-spot, reddish brown blotches measuring 2.5–5 cm in diameter are formed on both sides of the *Colocasia* leaf. The use of Dithane M-45 is reported to be effective in controlling the disease.

An important pest of *Xanthosoma* is the dynastid beetle (*Ligyrus ebenus*) which damages growing tubers, but can be controlled with malathion spray. The most common disease is root-rot (caused by *Pythium myriotylum*) which takes the form of wet rot, wilting of leaves and the inability of the plant to form tubers. Severe cases cause death and decay of the entire plant. Root-rot is normally controlled by rotation as well as the practice of removing all infested plants. Research is being undertaken in several centres to introduce genetic variability by floral induction for the breeding of resistant cultivars against the root-rot disease.

Harvesting and storage

The time from planting to harvesting is 5–15 months, depending on the cultivar and the method of cultivation. Usually no serious deterioration occurs if cocoyams are left in the ground for a few weeks after maturity. For upland taro or tannia, harvesting should be timed to coincide with the dry season or should be done during dry weather.

Cocoyams are mostly harvested by hand or using hand tools. In upland culture, pulling of the withered aerial portions of the plant is sufficient to lift the corm and cormels, although this often leaves many of the cormels in the ground. An alternative method is to lift the crop with a hand-tool (hoe, fork, or trowel), followed by manual picking of the corms and cormels.

The two approaches to harvesting are whole and milking. In whole harvesting of cocoyams, the entire plant is uprooted and the corms and cormels are removed. In milking, the soil around the base of the corm is removed and the mature cormels are excised in situ, thereafter the soil is replaced and the parent plant is left to continue growing.

The average yield of cocoyams on a world-wide basis is 5–6 tonnes/ha for *Colocasia* and 12–20 tonnes/ha for the *Xanthosoma* varieties. With good management, yields of 25–37 tonnes/ha can be achieved.

When produced under conditions of subsistence agriculture, cocoyams are not normally stored for any length of time, but are harvested as and when required. Storage of cormels for up to six weeks in ambient conditions is possible, provided they have not been injured during harvesting. Pieces of cocoyam may also be dried and stored as chips or ground into flour.

Carrot

The carrot (*Daucus carota*) is a member of the family Umbelliferae. Another economically important member of the family is parsnip (*Pastinaca sativa*). Both carrot and parsnip are grown for their swollen taproot and hypocotyl (Figure 10.10).

The carrot is known to have been cultivated for more than 2000 years. Although probably native to Europe, the carrot is found in all parts of the world where it is grown as a vegetable crop or occasionally as a forage crop. In the tropics carrots are mostly grown for human consumption. The roots are used as a vegetable and for putting in soups, stews, curries and other dishes, and the grated root is used in salads. Tender roots may also be pickled. The tops are often fed to cattle and other livestock. The seeds contain an essential oil which is used as a flavouring and also in perfumery.

Types and ecological adaptation

The crop is relatively sensitive to temperature. High soil temperatures appear to encourage the production of short roots. Colour is also affected with pale yellow roots being produced as a result of high soil temperatures. For commercial production in the tropics altitudes of over 500 metres are necessary. Although some cultivars may grow at lower elevations yields are generally poor. The crop is tolerant of a wide range of soil types, although a soil pH of 5.5–6.5 is required.

Most carrot varieties produce seeds under tropical conditions if they are left to grow a second year. Varieties which are currently adapted to tropical environments include Amsterdam Forcing, Early Nantes, Danver's Half Long, Oxheart Touchon, Early Gem and Favourites.

Figure 10.10 The edible root of the carrot develops from the swollen base of the taproot.

Planting

Seeds of carrot may lose their viability rapidly. It is therefore necessary to keep the seeds in a desiccator or in cool storage. Seeds should be well rubbed to remove their fine hairs before sowing. A site with a deep sandy loam or silty loam soil should be chosen. Other soil types may be used provided they are deep, contain sufficient nutrients and do not form a crust. The soil in which the crop is grown must be kept constantly moist.

Planting may be done at any time during the year, although dry season planting is reported to produce better quality carrots. Seeds are sown in drills at the end of the wet season on raised beds. Drills are made 0.3−0.4 m apart and the seeds sown thinly. Seedlings are then thinned to 5−7.5 cm apart when only a few centimetres high; thereafter, the plants are thinned to 10−15 cm apart.

Fertiliser application

Carrots remove substantial quantities of nutrients, particularly potash, from the soil. Wood ash may therefore be applied where inorganic fertilisers are not available. An application of 21 g of single superphosphate per square metre, worked into the soil at land preparation, is also beneficial. A top dressing of 7 g of calcium ammonium nitrate per square metre may be applied to the soil at two weeks after germination. This dose of nitrogen should be repeated four weeks later.

The use of fresh manure should be avoided for carrots as it may cause branched or misshappen roots. It is therefore better to grow carrots on a soil which received heavy manuring the previous year.

Weed control

Because carrots grow very slowly for the first few weeks, the plants cannot compete successfully with weeds. Light cultivation is, therefore, done to remove weeds. Avoid deep cultivation which will damage the roots as many roots are found in the top few centimetres of the soil surface. All exposed roots should be covered in order to protect them from excessive heat. Such exposed roots may also turn green and become unmarketable.

The herbicide MCPA applied at the rate of 8−10 kg/ha as a pre-emergence spray has been used to successfully control weeds in carrots.

Table 10.2 Other edible root and tuber crops

Botanical name	Common name	Areas of importance	Botanical name	Common name	Areas of importance
Solenostemon rotundifolius or *Coleus rotundifolius*	Hausa potato, Coleus potato, Country potato	Southeast Asia, Ceylon, Malaysia, Indonesia, tropical Africa	*Maranta arundinacea*	Arrowroot	Tropical America, West Indies, Brazil, India, Sri-Lanka, Indonesia, Philippines
Helianthus tuberosus	Jerusalem artichoke, Canada potato, Root artichoke, Sunroot	Temperate in origin, but also cultivated in subtropical and tropical areas	*Sphenostylis stenocarpa*	African yam bean, Wild yam bean, Girigiri	Tropical Africa, especially Ethiopia, Togo, Ivory Coast, Nigeria
Raphanus sativus	Radish, Radis, Rettich, Ripani	Temperate in origin, but also widely grown in the tropics and subtropics	*Metroxylon sagus*	Sago, sago palm	Southeast Asia, Pacific Islands, Indonesia, Malaysia, Papua New Guinea
Beta vulgaris	Beetroot	Primarily European, but also cultivated in tropical areas	*Dioscoreophyllum cumminsii*	Guinea potato	Forest zone of West Africa to Zaire
Canna edulis	Edible canna, Queensland arrowroot	Australia, Pacific Islands, West Indies, South America	*Icacina senegalensis*	Icacina	Savanna areas of West Africa, particularly Ghana

Pests and diseases

The carrot crop is relatively free of pests and diseases. However, nematodes can be a problem and the control measures include the avoidance of areas known to be nematode infested.

Blight (*Alternaria dauci*) is a fungus which can cause loss of foliage in the wet season. The disease is air-borne and its incidence may be minimised if the carrot is grown in small patches in the middle of taller crops such as cotton and sorghum. Spraying with a copper fungicide is effective against the carrot blight. Bacterial soft rot (*Erwinia carotovora*) can occur in carrots grown in the wet season. It may be prevented by ensuring adequate soil drainage and by growing the crop in the dry season.

Harvesting and storage

Most carrot cultivars mature in 10–12 weeks from sowing, at which time the roots are about 2.5 cm in diameter at the crown. Smaller roots tend to

shrivel rapidly after harvest. Carrots store well in the soil and may be harvested as needed. Roots may be kept in cool storage for several weeks without deterioration.

Other edible root and tuber crops

The various other root and tuber crops of minor importance cultivated in different parts of the tropics are listed in Table 10.2.

CHAPTER 11

Fruit, vegetable and spice crops

In this chapter, fruits and vegetables are defined to include fruits, nuts and vegetable crops. Also included for convenience is the locust bean, a tree crop which is not normally considered as a horticultural crop. Floricultural plants have been excluded from the entire book.

Coverage of all fruit and vegetable crops grown in the tropics is not within the scope of this book, particularly as each crop varies in importance, depending on the ecology of the area and the local community. In addition to the horticultural crops of world-wide economic importance discussed in some detail in this chapter, other edible fruits, nuts and vegetable crops, grown in tropical and subtropical areas but of lesser importance, are presented in Table 11.10.

Tomato

The cultivated tomato, *Lycopersicon lycopersicum*, belongs to the family Solanaceae. The crop originated in Central and South America, but has developed into a great number of cultivated types suited to different environments, many producing large round fruits, which differ markedly from the wild types and the original cultivated plants. Tomatoes are the second most important vegetable in many regions of the world, ranking second in importance to potatoes (Figure 11.1).

Tomatoes can be eaten raw or cooked. Large quantities are used to produce soups, juices, sauces, ketchups, purees and pastes. They are also used in the canning industry and green tomatoes are used for pickles and preserves. The seeds, which are extracted from the pulp and residues of the canning industry, contain 24% oil. The oil is used as a salad dressing and in the manufacture of margarine and soap. The residual press cake is used as stock feed and as a fertiliser.

Types and ecological adaptation

There are five recognised varieties of *Lycopersicon lycopersicum*:

(a) **Cherry tomato,** var. *cerasiforme*, has leaves which are generally smaller, thinner and less acuminate and flowers in longer clusters than the common types. The fruits are globular and regular, with a red or yellow colour. This variety occurs wild and is used mainly in breeding work.

(b) **Poor tomato,** var. *pyriforme*, has pear shaped

Figure 11.1 Tomatoes are the second most important vegetable crop in many regions of the world.

207

fruits. The leaves are of medium size and the fruits are red or yellow in colour. The variety is mostly used as a rootstock.

(c) **Potato-leafed tomato,** var. *grandifolium*, is similar to potato in size and structure. The leaves are large, the primary leaflets are few with entire margins and the secondary leaflets are few or absent. This variety of tomato is mainly used in breeding work. Because of its large leaves, it provides adequate photosythates and shade during adverse temperatures.

(d) **Upright or tree tomato,** var. *validum*, has plants which are short and erect. The plant is compact with crowded and curled leaves. It grows erect and, therefore, rarely needs to be staked. It is propagated either by seed or cuttings and begins to bear in its second year. The fruits are like an egg, reddish to yellow or in some varieties purple when ripe.

(e) **Common tomato,** var. *commune*, has plants which have a looping habit. Fruits are large, many celled and varied in shape, but mostly globular.

Tomatoes are tolerant of a wide range of climatic conditions. Although naturally a warm season crop, they can be grown in the open in any environment where there are three months of frost-free weather. Night temperatures between 13 and 17°C and day temperatures of around 23°C are ideal for optimum growth and yields. In environments where there are distinct dry and rainy seasons, the tomato crop invariably performs best under irrigation during the dry season when insect pests and disease incidences are at a minimum.

The ideal soils for the crop should be light, free-draining and fertile. However, different varieties can be grown in many soil types, with a pH range of 5−7. The soil must be deep and rich in humus. Flowering is not affected by photoperiod and therefore the crop can grow and produce fruits throughout the year.

In recent times, a great many tomato cultivars have been selected or bred to suit different environmental requirements and different uses. In northern Nigeria, for example, types have been selected for both the dry and wet seasons for the different ecological zones represented in the West African region. Types found suitable for processing in Nigeria include Cirio 56, Harvester, Marzanino, Piacenza 0164 and Ronita. Cultivars La Donita, Ife 1 and Enterpriser are fresh market types. In Ghana, Roma VF, Red Top and MH/VF are grown commercially for processing while Molokai (M44A), Anahu (M90) and Improved Zuarungu are reported to show promise as fresh market fruits.

Planting

The land on which tomatoes are to be grown should be free-draining, fertile and a sandy loam. It should not have been planted to tomatoes or other solanaceous crops in the previous year. The land should be cleared and deeply cultivated well ahead of time. Where available farmyard manure at the rate of 25 tonnes/ha could be incorporated into the soil. Otherwise, a compound NPK fertiliser could be used.

Tomato seedlings can be raised in seedbeds on the ground or in seed boxes (pots or trays). When raising seedlings in seed boxes, it is necessary to ensure that the proper soil mixture is used. In a ground nursery, the seedbed should be raised during the rainy season to improve drainage. During the dry season the seedbed should be slightly lowered, for water retention after irrigation. The width of the bed should be 1−1.5 m and may be as long as necessary. It should be prepared to a fine tilth. Before sowing, a seed drill is used to draw planting holes at 7.5−10 cm apart and 10 mm deep. The seeds are sown thinly at 3−4 seeds per 2.5 cm to avoid overcrowding and then covered lightly with grass or other shade material. Seeds germinate five to eight days after sowing, depending on their freshness and the ambient temperature. Once the seedlings have emerged the shade (grass) is removed.

In seedbeds, pest and disease control should be undertaken as early as in the nursery. If the growth of the seedlings is observed to be slow, they should be side-dressed with a solution of 30 g calcium ammonium nitrate in 10 litres of water at the rate of 1 litre/m^2. The seedlings are ready for transplanting when they are about 10−15 cm high and have developed three true leaves.

About three hours before transplanting, the seedbed should be soaked well. Transplanting is done in the late afternoon and the seedlings are lifted with a hand-fork or trowel. Transplanting positions should be marked out with a spacing on the ridges of 60 cm between plants, and on beds 60 × 60 cm. The seedlings should be watered immediately after transplanting.

Fertiliser application

Where available, farmyard manure or compost should be applied during land preparation and allowed to decompose thoroughly before planting the seedlings. In the absence of these organic sources, an NPK compound fertiliser should be used. Once transplanting is completed nitrogen, preferably as calcium ammonium nitrate, is applied as a top dressing at a rate of 250 kg/ha, in two equal doses at two and five weeks after transplanting.

Staking

Tomatoes require staking when the variety has stems which are weak and therefore unable to carry the weight of heavy fruits. Staking also gives support to the plant during strong winds. Individual stakes can be made of bamboos, canes or locally available wood. Raphia or other local twining fibre materials are suitable for tying the plants on to the supporting stakes. The rope should be carefully twisted once around the stake before being loosely looped and tied round the stem of the plant (Figure 11.2).

Fruits of tomato produced from staked plants often suffer less from fungal infections and are cleaner. The positive effects of staking on both the quality and yield of fruits are seen in Table 11.1.

Table 11.1. Effect of staking on the marketable fruit yield (tonnes/ha) of two tomato cultivars grown during the wet season at Samaru, Nigeria

| | Yield (tonnes/ha) | | | |
| | Ronita | | Piacenza 0164 | |
Year	Staked	Unstaked	Staked	Unstaked
1971	32.12	16.77	35.87	8.20
1972	52.33	25.35	49.03	16.94
1973	26.96	6.16	27.70	7.65
Mean	37.14	16.09	37.53	10.93

Source: Quin, J.G. (1975). A further assesment of mulching and staking a rain-fed tomato crop in the Northern Nigerian states. Samaru Research Bulletin, No.275.

Weed control

The critical period for weed control in tomato lies between six and nine weeks after transplanting. Unchecked weed growth results in as much as a 50–70% reduction in tomato fruit yield. Therefore, weeds must be removed as often as possible during the early stages of growth. Hoe-weeding may be done twice before fruiting and once thereafter, making sure that plants that have spread out in the row are not damaged.

A number of herbicides have been found suitable for the control of weeds in tomato fields. In Nigeria, for instance, diphenamid at 1.5 kg/ha followed by a supplementary hoe weeding, diphenamid at 3.9 kg/ha, metribuzin at 0.5 kg/ha or metolachlor plus metobromuron at 1.0 + 1.0 kg/ha followed

Figure 11.2 Tomatoes are often irrigated during the dry season and those varieties with weak stems are staked to enable them to carry a heavy crop.

by two or three hoe weedings, have been found to control weeds and result in increased yields of tomatoes.

Pests and diseases

The common insect pests which attack tomatoes are white flies, aphids and fruit worms. The major diseases include wilts, virus diseases, leaf and stem spots and blossom and storage rots. Nematodes can also be a problem.

Aphids, fruit worms and white fly (which is the vector of the leaf curl virus) may all be controlled by applications of contact insecticides such as carbaryl. Synthetic pyrethroids have also been found to be effective in the control of fruit worms while a mixture of malathion and dimethoate have been successfully used for the control of white flies.

Basamid (dazomet) has been used successfully to control nematodes. Tomato leaf diseases, especially in the vulnerable wet season crop, have been successfully controlled by several prophylactic fungicides such as captafol (Ortho Difolatan) and mancozeb (Dithane M-45). Where continuous rainfall makes fungicidal protection impracticable, the use of suitable synthetic fungicides may partially solve the problem.

Damping-off of seedlings and foot-rot of growing plants are caused by the soil fungi, *Phytophthora* spp. Infection takes place in the potting soil. The use of undecomposed farmyard manure may also contribute to this kind of infection. Sanitation of nursery beds and crop rotation may reduce the incidence of infection.

Harvesting and storage

Fruits of tomato reach maturity 10–14 weeks after seedlings are transplanted. At physiological maturity the crop may be harvested at two stages of fruit ripening: vine-ripe or medium ripe (yellowish green). Vine-ripe fruits are meant for sale within one or two days in close-by markets, while medium ripe harvesting allows for a longer shelf life and a greater distance of transportation. While vine-ripe tomato fruits may be kept for two or three days before spoilage, medium-ripe fruits may have a shelf life lasting seven days. Storage for any length of time beyond seven days may have to be done at a low temperature (10–15°C) and low relative humidity (70–75%). The fruits may be sprayed with Benlate (benomyl) at 1000 ppm to control possible storage decay.

Eggplant

The family Solanaceae contains some of the most important economic flowering plants. The family consists of about 75 genera and over 2000 species. The family contains many valuable crop plants such as the potato, eggplants, tomatoes, capsicums and tobacco. The genus *Solanum* is one of the largest genera in the family, containing more than 1000 species, among which is the eggplant, *Solanum melongena*, which is also referred to as the garden egg, aubergine, melongene and brinjal. A native of India and the East Indies, *S. melongena* is cultivated for its immature fruits which are eaten as vegetables and used to season other foods. Eggplants supplement starchy foods in addition to being a cheap source of protein, minerals and vitamins. The leaves of some species are also used as a vegetable. The alkaloid 'solanin' is extracted from the roots, leaves and fruits of eggplants for therapeutic purposes. Many colour forms and shapes of the fruit have been selected, including white, yellow, green, purplish-red, magenta-purple and black. Shapes include short and long, ovoid and sausage-shaped, tapering or rounded at the tip. The species is found throughout the tropics.

Types and ecological adaptation

The eggplant is a thermophilic plant grown in tropical and subtropical regions wherever the temperature is high enough. In addition, the crop thrives well in temperate areas which have plenty of sunshine. It requires an optimum temperature range of 25–35°C for growth. Garden eggs do well in most soils, but particularly in rich, well-drained loams, and in areas with moderate rainfall.

Garden eggs are usually annual plants, although some varieties may live into a second season. Local cultivars are more commonly used since they are invariably better adapted to prevailing ecological conditions. Such cultivars as Black Beauty, Florida Market and Black Magic are new to the West African region and are reported to be high-yielding.

Planting

Eggplants are usually grown in well-prepared nursery beds and the seedlings are planted out later. No special treatment of seed is required. Seed extracted from mature fruits which have been properly stored to maintain seed viability is preferred to poorly stored seed from immature pods. Nursery seedlings are raised in boxes or on the ground and transplanted when they are four to six weeks old. Seeds are sown at the beginning of the rains. For dry season plantings, seedlings are raised in the nursery towards the end of the rains and are transplanted at the start of the dry weather, preferably before the onset of the cold Harmattan wind in the West African region.

Seedbed preparation for transplanted seedlings varies with the season. In the wet season, planting is done on ridges or raised beds to avoid possible damage due to flooding. In the dry season, when moisture conservation is important, eggplants are transplanted on the flat or in sunken beds.

Although cultivated as an annual crop, the eggplant is naturally a much branched perennial with a strong taproot and an extensively branched root system, which is not widely spreading but penetrates deeply into the soil. Thus, fields for cultivation of the crop should be relatively level and free-draining. It is best if the soil is rich in organic matter and capable of good moisture retention. Loamy and sandy loam soils are preferred.

Transplanting of eggplant seedlings may be done at anytime once the rains have become established. When cultivated during the dry season, transplanting is done after the rains have ceased. In general, wet season cultivation of the crop is preferred, mostly because the higher temperatures give better vegetative growth than in the dry season. This is particularly true in the West African region.

The spacing adopted depends on variety. Varieties like Black Beauty and Florida Market, which generally cover a large area, require a wide spacing of about 90×75cm. For local types such as Samaru-striped (*Solanum gilo*), which has a smaller plant stature, a spacing of 75×50 cm is adequate.

Fertiliser application

Eggplants respond to both nitrogenous and phosphate fertilisers. Applications of about 75 kg N and 60 kg P_2O_5/ha result in substantial increases in yield. The nitrogen fertiliser is applied in split doses at transplanting and just before flowering. The phosphorus is applied as a single dose at transplanting.

Weed control

Weed control is important during the early growth of eggplants when the vegetative cover is still scanty. Two to three hoe weedings are advised during the growth of the crop. Investigations into the use of herbicides are in progress and there is a possibility that this method of weed control may eventually replace or supplement hoe weeding. The most promising herbicides identified under conditions in the savanna zones of Nigeria are metribuzin and diphenamid.

Pests and diseases

Garden eggs are hardy plants which are generally attacked by few pests. However, young leaves can be attacked by leaf-eating caterpillars which sometimes defoliate the plants completely. Control of such caterpillars may be effected by weekly spraying with a contact insecticide such as malathion, carbaryl or a synthetic pyrethroid.

Fungal attack may also occur, especially during the rainy season. They may be controlled by spraying with protectant fungicides such as captafol (Ortho Difolatan) and mancozeb (Dithane M-45). Fungicides may also be used successfully to control leaf-spot.

Harvesting and storage

Eggplants should be harvested as regularly as possible, usually twice a week during the peak harvesting period to avoid over-maturing and to encourage development of new fruits. Continued availability of moisture encourages prolonged fruit-setting and therefore harvesting. The early cessation of the rains invariably causes early plant death.

In the absence of cool storage, fruits harvested green may be kept at room temperature for up to six days before turning red. Ripe fruits may be sun-dried before being stored for subsequent use as a seasoning.

Onion

A member of the family Alliaceae, *Allium cepa* is the most important of the seven species in the genus *Allium*. The onion probably originated in the Mediterranean area of western Asia, but is now cultivated in all parts of the tropics, forming one of the major vegetable crops of the drier regions, where it grows well under irrigation. A herbaceous biennial plant, onions produce a shallow fibrous rooting system from the base of the short stem (Figure 11.3).

The demand for onions is world-wide, the crop being grown and consumed by people of all nationalities. Onions are important items of world trade. The immature and mature bulbs are eaten raw or they may be cooked and eaten as a vegetable. They are used in soups and sauces and for seasoning many other foods, or they may be fried. Onion oil is produced by steaming and is used to a limited extent for flavouring foods. Onion and onion waste can be fed to livestock, but care is necessary to prevent tainting milk and other products. Several parts of the plant have a place in the traditional medicines of many countries.

Types and ecological adaptation

There are three groups of cultivars of *Allium cepa* which are given the rank of botanical varieties as follows:

(a) **Common onion,** var. *cepa*. This is the group of *A. cepa* that contains most of the commercially important bulb forming onions. The group exhibits great variation in the shape and colour of their bulbs, their response to photoperiod and temperature, their storage quality, their pungency and other characteristics. The bulbs are large and usually single, without bulbils in the inflorescence. They are almost always propagated by true seed.

(b) **Aggregatum onion,** var. *aggregatum*. In this group, also called *multiplicans*, the bulbs produce many laterals or shoots which multiply freely and are used for propagation. The group contains three forms, namely potato or multiplier onion, ever-ready onion and shallot. All these produce several bulbs which are used for their propagation instead of seed.

(c) **Proliferum onion,** var. *proliferum*. This is sometimes known as the tree onion. The inflorescence produces a cluster of bulbils or small bulbs instead of seed. Sometimes both bulbils and flowers are borne in the same inflorescence. Propagation is by the inflorescence bulbils.

Figure 11.3 Onions are a major vegetable crop in the drier regions of the tropics where they grow well under irrigation.

Onions grow in a wide range of climatic conditions, but they grow best in a mild climate without excessive rainfall or great extremes of heat and cold. Cool conditions with adequte moisture supply are most suitable for early growth, followed by warm drier conditions for maturation, harvesting and curing.

Although onions can be grown on a variety of soils, it is important that the soil retains water well and is non-packing, stone-free and friable, a good fertile loam being the best. The optimum pH is 6–7. Their need for a high level of available water and a rich nutrient supply is caused by their sparse root formation and consequent low extraction potential.

The crop is photosensitive and bulb formation takes place faster at warm temperatures. Low temperatures and high soil fertility induce bolting. The crop produces flowers during the short-day period. Most exotic cultivars fail to produce bulbs in the short-day conditions of West Africa. Some American hybrids are, however, adapted to these conditions and yield reasonably well. These include Early Texas Grano, Yellow Granex, Red Star, Dissex, White Granex, New Mexico and White Grano. Many local cultivars have been developed which are used for seed production. In northern Nigeria, Wuyar Bijimi and Maiduguri Improved are among the commonly used local varieties. The three main types have white, red and purple skins. The red varieties are most common and their flavour is generally mildest.

Planting

Onions are generally propagated by seed, although bulbs, sliced across the top, may also be used. The seeds may be raised in nursery boxes or on the ground. The seedbed on the ground is made 1–5 m wide and the area is worked to a fine tilth by breaking up clods, removing debris and making the surface level. The ground bed may be as long as desired but should be raised in the rainy season to improve drainage and slightly lowered in the dry season. It should also be slightly lowered for water retention during irrigation. Poultry droppings and well rotted manure should be incorporated into the seedbed and superphosphate applied as a basal dressing at the rate of 21 g/m². Trays may also be used for sowing onion seeds, the size used depending upon convenience.

Before seeds are sown, seed drills are drawn 10–15 cm apart. Drills should not exceed 13 mm in depth. Seeds should be sown thinly and covered with soil and then mulched with dry grass to prevent them from being washed away during rains or watering. The seedbed should be watered when necessary. Germination of seeds takes 7–10 days, after which the mulch should be removed. If growth of the seedlings is slow, nitrogen fertiliser should be side-dressed. Seedlings are ready to be transplanted after 6–8 weeks when they are about 15–25 cm high.

Rainfed onions are generally transplanted on ridges or raised beds, while the dry season crop is transplanted on flat beds at a spacing of 1 × 2 m. In both croppings, the soil should be well prepared and well rotted farmyard manure, compost or a compound NPK fertiliser should be incorporated.

With the use of irrigation water, particularly in the dry parts of the tropics, it is possible to grow three crops of onion in a year, two in the rainy season and one in the dry season. The essence of producing good quality onions involves planting so that the crop matures in the dry season. When planted on ridges, seedlings are spaced every 10 cm within the row, while on the flat the spacing is 15–20 cm within rows.

Fertiliser application

Farmyard manure or compost and 250 kg/ha of single superphosphate should be applied at land preparation. Thereafter, calcium ammonium nitrate is applied at 250 kg/ha in two equal doses, two to three weeks after transplanting and five weeks later.

Weed control

Weeds are particularly serious in onions produced from seed, primarily because the crop germinates and emerges relatively slowly. Because of the scanty foliage, onions compete poorly with weeds. Cultivation for weed control should be light in order not to compact the soil, and to prevent excessive water loss by evaporation in rainfed situations. Chemical weed control which causes minimum injury to bulbs appears to show promise. A pretransplant application of chlorthal-dimethyl at 12.0 kg/ha, fluorodifen at 3.0 kg/ha or oxadiazon at 2.0 kg/ha are reported to effect good selective control of weeds, giving bulb yields that are comparable with hoe weeding.

Pests and diseases

Crickets are particularly important pests in onions at the nursery stage and immediately after transplanting to the field. To control crickets, 0.5 kg Agrocide, 0.65–10 HCH(BHC), is mixed with 45.5 kg of bran and scattered around the seedbed area for use as a bait. Thrips also feed on the leaves. These can be controlled by spraying with Vetox (carbaryl) at the rate of 1.0 kg in 158 litres of water/ha.

The common diseases of onion are purple blotch, *Fusarium* rot and twister. Purple blotch is a fungal disease caused by *Alternaria porri*, which attacks onions during the rainy season. Infected plants develop small white flecks which almost immediately turn brown and velvety in high humidity. As spots enlarge, they form lesions which are distinctly purple in colour. Infected leaves weaken, turn yellow and fall over.

Fusarium rot is caused by *Fusarium oxysporum*, which attacks onion plants in the field and may continue after harvest as storage rot. The leaves die back rapidly from the tip and roots. At the base of the bulb, a whitish mouldy growth appears on the surface of decayed portions.

Onion twister disease is also caused by *Fusarium* spp. predominantly *F. oxysporum* and *F. solani* and less often by *Colletotrichum* spp. The disease occurs during the rains. It causes distortion of the pseudostems and leaves and slender bulb formation. Yield losses can exceed 50% and affected bulbs rot rapidly in storage.

Several protectant fungicides, including Dithane M-45, are recommended for use, especially during the wet season. Crop rotation, disposal of crop residues, planting in well-drained soils and avoidance of fresh manure are reported to be helpful in reducing the incidence of the attack.

Harvesting and storage

Onions are harvested when 50–75% of the leaves have died back. If the bulbs are not harvested at this stage they are used to regenerate new growth, and then set flowers and seed in the second year. In general, onions take about four months to mature. Harvested bulbs should be left in the sun to dry thoroughly so that they will not rot in storage. In the rainy season the crop may be dried under cover, protected from rains, while in the dry season drying is done in the field. It is important to ensure that only whole, sound, dry bulbs are stored and bruised bulbs are avoided. The storage area must be well ventilated and the bulbs arranged in single layers. Properly done, onions can be kept in store for up to six months.

Okra

Okra, *Abelmoschus esculentus*, belongs to the family Malvaceae. Other important members of the family include cotton, kenaf and roselle. the generic name of Okra was previously *Hibiscus*, but this was changed to *Abelmoschus*, while the species name remains *esculentus*.

The okra plant is African in origin but is now cultivated widely in the tropics for its fruit wich is used as a vegetable, both in the green and dried state. The fruit has a high mucilage content and is used in soups and gravies. The ripe seed contains about 20% edible oil. Like the other *Abelmoschus* species, *A. esculentus* also produces stem fibres which find many local uses, but which are not extracted on a commercial scale (Figure 11.4 opposite).

Types and ecological adaptation

Okra is a hot weather tropical crop, susceptible to drought and low night temperatures. It grows well in day temperatures of 25–40°C, with night temperatures over 22°C. Most tropical cultivars are short-day plants. The crop is produced all year round as a rainfed crop in the wet season and an irrigated crop in the dry season. An average annual rainfall of 750 mm is adequate for okra growth. Okra is a deep-rooted plant and irrigation once or twice a week is adequate, provided enough water is given at each application to saturate the soil to field capacity.

The okra plant grows erect to a height of 1.2–1.8 m, although there are some short, spreading varieties which are either slow or quick maturing. Fruits may be long or short in size, broad or cylindrical in shape, smooth or ribbed in

Figure 11.4 The okra plant is cultivated widely in the tropics for its mucilaginous fruit which is used as a vegetable.

texture and green, reddish-green or pale green to yellow in colour. In Nigeria, varieties such as V35, TAE-38, White Velvet and NHAE-36 produce fruits between 45 and 50 days after sowing. The variety Lady's Fingers is widely grown in the wetter parts of the country.

Planting

Okra thrives in moist soils, provided the ground is not waterlogged. Good crops can be raised even on dry coastal plains, on both sandy and clayey land where the rather salty soil is unsuited to the growth of most crops. Okra is propagated from seeds which are sown directly in the field, without having to raise seedlings first.

Okra is usually grown during the main rains, although good crops can also be raised in the dry season under irrigation. For the wet season (main) crop, sowing usually commences when the rains are fully established. Dry season okra can be sown at any time, except where growth conditions are known to be unfavourable. For example, in West Africa the crop is not grown between December and February, to avoid the cold Harmattan weather.

Okra is often interplanted with other vegetables such as tomato and chillies, or with root crops such as yams and cassava.

In order to achieve good germination and establishment, seeds may be floated in water and only those seeds that sink to the bottom used for planting, seeds that float are invariably immature or not viable. Seed dressing chemicals such as Aldrex T or Fernasan D may be used before sowing to prevent attack by soil-borne diseases and pests.

If ridges are used, they should be 75 cm apart. The intra-row spacing may be 60, 45 or 30 cm, depending on whether the variety is a tall or dwarf type. For dry season production on ridges, the seeds should be planted at the side of the ridge (instead of in the middle) so that the roots can easily reach irrigation water in the furrows.

Fertiliser application

Okra responds to fertiliser applications, especially nitrogen and phosphorus fertilisers. Nitrogen at 75 kg N/ha should be split-applied in two equal doses, at three and six weeks after sowing. Phosphorus, at the rate of 30 kg P_2O_5/ha should be applied during land preparation or immediately after seedling emergence.

Weed control

Okra competes poorly with weeds, especially during the early stages of crop growth. Weed control may be effected by three hoe weedings: twice before the plant flowers and once during the early stages of fruit development.

Chemicals are being increasingly used for weed control. In Nigeria, metolachlor plus terbutryne (Igran combi) at the rate of 2.0 + 1.0 kg/ha, effectively controls weeds for up to seven weeks after sowing. A supplementary hoe weeding or application of a post-emergence herbicide is, however, necessary in order to give season-long weed control.

Pests and diseases

The major pests of okra include the okra flea beetle and the okra fruit worm. The flea beetle feeds on the crop by eating out leaf material and leaving numerous characteristically round holes in young leaves. These leaves eventually look like coarse sieves when they expand with further growth. Heavy defoliation of young plants results in stunted growth or death. In addition to this direct damage, the beetles also transmit two viral diseases; the okra mosaic and okra leaf curl viruses. Flea beetles can be controlled by applications of Cymbush (cypermethrin), 2.5 kg/ha at weekly intervals, commencing as early as one week after germination.

The larvae of the okra fruit worm, *Heliothis armigera*, can cause severe damage to okra fruits. The caterpillars cut neat round holes in the fruits as they feed. Both the holes made by the larvae and the distortion they cause render the fruit unsightly and unmarketable. The use of contact insecticides such as Cymbush, sprayed initially when 50% of the plants have flowered, achieves effective control of okra fruit worm. A cultural method of control is to ensure that maize and cowpeas are not grown near fields in which okra is being cultivated, since plots of maize or cowpeas could constitute a source of *Heliothis*.

A major disease of okra is leaf-spot. Protectant fungicides such as captafol (Ortho Difolatan) and mancozeb (Dithane M-45) may be used with success to control leaf-spot. Dazomet (Pasmid) may also be used to control nematodes.

Harvesting and storage

Okra starts to flower seven to nine weeks after sowing, and continues flowering throughout the rainy season, or as long as water is available for growth. In order to have high-quality and non-fibrous fruits which are highly mucilaginous, it is best to harvest four to six days after fertilisation of

the flowers. However, in general the first set of fruits is harvested two to three months after sowing and thereafter the harvest is continued over the period of ripening. Where okra is grown for seed, the pods are allowed to ripen before being harvested and dried. When it is not easy to detach pods by lightly twisting the fruit with the hand, they can be carefully cut using a knife.

In the absence of cold storage, fresh okra fruits may keep at ambient temperatures for only a few days, after which they may begin to rot. The traditional method of storage is to line the inside of containers, usually baskets, with moist jute sacks before putting in the fruits. The packed containers are then placed in a cool environment. Fresh okra fruits can also be sliced and dried in the sun before storage.

Mature pods harvested for seed are hung in small bunches to dry. If seeds are extracted from dried fruits before being stored they must be kept in airtight containers until the next planting season.

Chillies

Chilli peppers, *Capsicum* spp., belong to the family Solanaceae. There are four species of *Capsicum*, but only two, *Capsicum annuum*, the sweet pepper and *Capsicum frutescens*, the red pepper or chilli pepper, are widely cultivated. Chillies originated in the West Indies, Peru and Mexico and were spread over the tropics and subtropics mainly by birds. West Africa has now become an important producer of chillies for international trade.

Sweet peppers, with their mild flavour, are eaten raw in salads and cooked in various ways. Chillies are used for culinary purposes and for seasonings. They provide the 'hot' ingredient in curry powder, which is made by grinding roasted dried chillies with turmeric, coriander and other spices. Chilli peppers also have medicinal uses, internally as a powerful stimulant and carminative and externally as a counter-irritant.

Types and ecological adaptation

In *Capsicum annuum*, the flowers are borne singly in the leaf axils. When ripe, the fruits are red, yellow or brown, but immature fruits of the large mild kinds are often picked while still green for use in salads. This species is generally large-

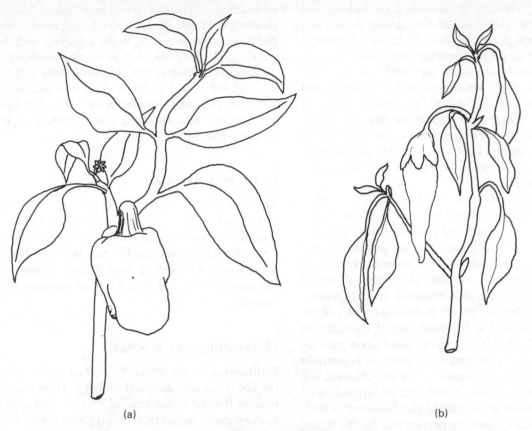

Figure 11.5   (a) Sweet peppers, *Capsicum annuum*, are milder than (b) chilli peppers, *Capsicum frutescens*.

fruited. In contrast, the flowers of *C. frutescens* are borne in clusters of two or more in the leaf axils. The fruits are, in general, much smaller than those of *C. annuum*, bright red in colour and with a wide range of shapes and sizes. Pungency is again variable, but in general it is greater than in *C. annuum* (Figure 11.5).

Chillies are cultivated in a wide range of environmental conditions, in both tropical and subtropical areas. When grown as a rainfed crop, chillies can be successful in areas with an annual rainfall of between 600–2500 mm. The crop can also be raised entirely under irrigation. Both the moisture content of the soil and the prevailing temperatures have important effects on the growth and yield of chillies. In general, maximum growth occurs in the temperature range of 21–27°C. Early blossoming and maturing of fruit are favoured by high temperatures, which also tend to reduce the setting of fruit. At later stages of blossoming, a

lowering of temperature therefore results in enhanced fruit setting.

Chillies require a well-drained, fertile soil of a heavy loam type. Poorly drained soils can cause the plants to shed their leaves and to be susceptible to diseases, resulting in poor yields.

There are many varieties of chillies of temperate origin which have become adapted to the tropics and produce well. Examples grown in parts of Africa are Yolo Wonder and California Wonder. However, there are many good local varieties of both the mild and hot types which may be grown successfully.

Planting

No special seed treatment is required other than that the seeds should be extracted from mature fruits, and should be well cleaned and properly kept before sowing. The seedbed for raising seedlings is made 120–150 cm wide and as long as necessary. The soil is pulverised by forking and breaking up the clods and removing stones and straw. A compound fertiliser 15:15:15 is applied at the rate of 22 g/m². Seeds are then sown thinly in drills about 10 cm apart and with one seed per centimetre. After sowing, the bed is mulched, and heavily but carefully watered so as not to wash out the seeds. When the seedlings emerge, the mulch is removed. In the rainy season, the beds should be raised about 15 cm above the surrounding ground and lowered during the dry season for water retention.

A well-drained fertile clay loam should be selected. The pepper plant is not very sensitive to soil acidity, but strongly acid soils should be avoided. For chillies grown during the dry season, the site must be near a good source of water for irrigation, although peppers are able to withstand longer periods of drought than tomatoes and eggplants.

Most chillies are grown as rainfed crops and are therefore planted at the beginning of the rains, in June in West Africa. An irrigated crop may be planted towards the end of the rains, in September in West Africa, to make use of the residual soil moisture at that time, which can be supplemented with irrigation water during the flowering period.

Ridge sizes for growing peppers may be 75–90 cm, with an interplant spacing of 60 cm. When planted on the flat, a closer spacing of 60 × 60 cm and 60 × 30 cm may be adopted for sweet and chilli peppers respectively.

Fertiliser application

About a week before transplanting the soil should be cultivated deeply and a generous amount of farmyard manure should be applied. In addition, 250–500 kg/ha of a compound fertiliser, 15:15:15, should be added to the soil before ridging. A top-dressing of 65 kg N/ha may be given in two doses, at two weeks after transplanting and at the first set of fruits. Too much nitrogen may cause a poor fruit set.

Weed control

Weeds should be removed as often as necessary. Mulching materials may be applied to suppress weeds and conserve moisture. Two or three hoe weedings may be done, but cultivation should not

be too deep, to avoid damaging the root system. A number of herbicides have been found to consistently combine effective weed control with high yields, comparable with those achieved with hoe weeding. In Nigeria, for example, mixtures of alachlor with linuron and chlorbromuron at 1.0 + 0.5 and 1.0 + 1.0 kg/ha respectively, and pendimethalin plus metobromuron at 2.0 + 1.0 kg/ha are effective in controlling weeds with a supplementary hoe weeding.

Pests and diseases

The major insect pests of pepper are white flies and aphids which are also vectors of the most common disease of the crop, the pepper mosaic virus. White flies and aphids may be controlled by applications of contact insecticides. Control of the mosaic virus lies in the use of resistant chilli varieties.

Harvesting and storage

Fruit peppers are generally picked green whilst the hot types are harvested red-ripe. However, the time of harvest is determined by use. The skin of mature green peppers is shiny, waxy and fairly crisp. Hot peppers may be dried in the sun and stored to be sold later as dried chillies.

Pepper

Pepper, *Piper nigrum*, is a climbing vine. It is a member of the family Piperaceae. Members of this family are characterised by their pendulous, dense spikes of small flowers and by the absence of a perianth, but the presence of a subtending floral bract. Pepper is grown throughout the tropics in Asia, Africa and South America. Today, Malaysia, India and Indonesia produce about 90% of the world's pepper (Figure 11.6).

Pepper is one of the oldest and most important of all spices and is still used extensively as a food flavouring. It is used as a condiment and flavouring for many savoury dishes, in preserving and pickling and in the manufacture of sauces and ketchups. The two forms of pepper grown commercially are 'black' and 'white' pepper. Black pepper consists of the dried ground whole fruits while white pepper comes from the dried ground seeds.

Figure 11.6 A branch of a pepper plant, *Piper nigrum*, showing the dense spikes of peppercorns.

Types and ecological adaptation

Pepper is grown in high rainfall areas and requires some kind of shade for good growth, especially in the early stages. The crop requires a dry period for maturing, ripening and harvesting of the peppercorns. It needs a soil which is free-draining, slightly acidic and rich in organic matter.

There are many local varieties which have been successfully cultivated in various countries. Some of the best yielding varieties from Asia include Balamcotta, Killuvali and Cheriakodi.

Planting

Pepper is usually propagated from cuttings rather than from seed, although the seeds can be made to germinate. When selecting material for cuttings, young vines of about one year old which have

hardened sufficiently are chosen. Pieces of 30 cm in length are cut, trimmed of all leaves except one and planted in a 1:1 forest soil : sand mixture in a nursery box. For rapid rooting, it is necessary to dip the cut end of the vine in a rooting hormone such as IBA (indolyl butyric acid). The cuttings are planted at a spacing of 15 cm apart and watered regularly. After about six weeks, adventitious roots should form, indicated by the formation of new buds. The cuttings are then ready for planting in the field.

Land for pepper production does not need thorough cultivation. The land is only sufficiently cutlassed to allow easy passage. Tree felling is done selectively, leaving some trees to provide shade and support where this is necessary. Shade is essential for the proper growth of pepper plants. In the absence of shade, flowering is heavy, but fruit drop is also very high. When preparing land therefore, a certain amount of natural shade should be allowed to remain, to cut the sunlight by a third to one half. Trees left to provide such shade should be tough and able to resist wind damage. Burning is not advisable as some of the trees left to provide shade may be affected by fire. After these operations, the field is marked out and pegged at the required spacing. The recommended spacing is 2.5 × 2.5 m with supports in alternate positions to allow for easy access during training and pruning.

In its young stages, pepper grows like a saprophyte and needs an organic medium for its nutrition. Holes for planting must be filled with a compost material, prepared from sawdust or decomposing vegetable matter.

Planting should be done at the beginning of the rains. Early planting is necessary so that the plants receive maximum rainfall and are fully established before the dry season commences. At planting time the cuttings should be growing vigorously. The leaves should be removed, leaving two or three per cutting. This is necessary to prevent transpiration shock. The cutting is then placed in the prepared hole and the soil is pressed firmly around it.

Fertiliser application

Pepper is normally grown on newly cleared land and therefore benefits from the high levels of available nutrients in the forest soil and from the leaf litter provided by shade plants. However, it is advisable to put rock phosphate at the bottom of the planting holes. About 0.5–0.75 kg of calcium ammonium nitrate should be applied to each stand after every harvest.

Weed control

If weeds such as climbing leguminous plants and grasses are allowed to establish amongst young pepper plants they become difficult to remove later, especially when they are entwined with the crop. Weeding within a row is done cleanly and weeds near the plant are removed using a hoe. The strips between the rows are slashed with little or no disturbance to the soil.

Pests and diseases

No serious pests attack the pepper crop. However, occasionally young leaves may be eaten by caterpillars. These may be controlled by spraying regularly with insecticides. Nematodes may also attack older seedlings. This attack may be controlled by the application of a nematicide, although it is advisable not to plant pepper in land infested with nematodes.

Phytophthora root rot may attack pepper plants at the nursery stage, especially in conditions of high moisture. Soil sterilisation for nursery plants is recommended as a means of control. Speckled necrosis has been observed on pepper leaves. The affected vines have a reduced root system and the plants may die off later. The application of wood ash is reported to control this disease. During the rainy season, leaves may produce a large number of small patches on both sides, which can be controlled by spraying with Dithane M-45.

Harvesting and storage

When to harvest the peppercorns depends on which type of pepper (black or white) is being produced. For black pepper, the berries are picked when they are just mature, but before they ripen. The harvested berries are heaped together for a day or two to enable fermentation to take place. The fruits are then removed from the spikes and steamed in boiling water for five to ten minutes and then spread out in the sun for about seven days. After sun drying, they are put in the oven at a temperature of about 60°C for four to five days.

For processing into white pepper, the berries are harvested when fully ripe. They are then put

into a sack and placed in running water for seven days to decompose the pericarp. Then they are put into punched polythene bags and washed in running water for three days. This removes all the pericarp and makes the seeds white. The white pepper seeds are spread out on mats to dry in the sun for three days and then further oven-dried at 60°C for four to five days. Pepper prepared in these ways may be ground and stored in bottles as black or white pepper.

Plantains and bananas

Plantains and bananas, *Musa paradisiaca* and *Musa sapientum*, belong to the family Musaceae. They are derived from two wild diploid species, *Musa acuminata* (A genome) and *Musa balbisiana* (B genome). There are many cultivars believed to have been formed as a result of hybridisation between these species. There is strictly no botanical difference between plantains and bananas. Plantains are, however, more starchy and need to be cooked to be palatable. The fruits of plantains are also generally larger in size than those of bananas (Figure 11.7).

Both bananas and plantains originated in Asia. Bananas are grown in both the tropics and subtropics, with the West Indies and Central America producing most of the crop for export. In West Africa, the Cameroons and the Ivory Coast are important banana exporting countries. In other places, they are produced for local consumption as a dessert fruit which is eaten in the fresh ripe form.

Plantains have become important as a staple food crop in several countries in West and East Africa. They are usually eaten in the unripe form by roasting as thin slices in oil or by boiling and then pounding, either alone or with cassava, to make 'fufu'. When ripe, they are pealed and roasted on charcoal fires or sliced and fried in oil. A flour can also be made from both bananas and plantains.

Types and ecological adaptation

The major characteristics of the main cultivars of plantains are summarised in Table 11.2. Plantains and bananas can be cultivated in a wide range of tropical and subtropical climates. The optimum conditions are monthly means of about 100 mm

Figure 11.7 A large bunch of Dwarf Cavendish bananas.

rainfall and 27°C temperature. Serious checks to growth occur when monthly means are below 50 mm rainfall or 21°C. Thus, their cultivation is largely within the 15°C winter isotherm, corresponding to latitudes of approximately 30° north and south. The cultivar Dwarf Cavendish is considerably more tolerant of temperature fluctuations than any other, and is the predominant cultivar grown near the limits of the temperature range. Tolerance to drought is achieved by the use of the hybrid cultivars AAB and ABB, as those of AAA and AA are less hardy. Hybrid cultivars, therefore, dominate in areas with a long dry season.

The essential requirement for soil is good drainage. This depends on soil structure and depth and a very wide range of soils can be used. A soil pH between 6 and 7.5 is best. Plantains and bananas should not be grown in windy regions, since they fall over readily in strong winds. Intercropping with tall trees such as coconuts can help protect the banana crop.

In Nigeria, the commercially important plantain varieties include the French and Horn types. With bananas, Gros Michel and Dwarf Cavendish are the most widely grown. There has been a recent introduction of Poyo, which is a Cavendish group from the Cameroons. In general the choice of variety to grow depends on availability of planting material, market preferences and ecological factors.

Planting

Where and how planting materials are obtained to establish a new plantation is often a problem. While planting materials might be obtained from producing plants, the quantities are necessarily limited, particularly since allowing many suckers to grow interferes with the pruning system. Thus, special nurseries need to be established on clean land, using clean planting materials. All plants with borer or nematode symptoms must be thoroughly trimmed and disinfected. In the nursery, a density of about 4000 plants/ha is maintained at a spacing of 1.6×1.6 m.

As most suckers are produced during the fruiting stage, the plants should be left alone for at least six months before suckers are taken. These suckers are then set out in another part of the nursery and treated as described previously. The system permits a ten-fold increase in suckers in a year.

The site selected for growing plantains and bananas should be free-draining, yet capable of retaining a reasonable amount of water. The nutrient status of the soil should be high, especially in potassium. The area should be sheltered from strong winds or provision should be made for windbreaks.

Both bananas and plantains should be planted at a time when the young plants are able to make optimum use of available water in the soil. Planting should also be timed so that the bulk of the crop is harvested to coincide with periods of high market prices. In West Africa this is best achieved by planting in June or July. If the soil is heavy, then planting should be earlier. In elevated areas where rainfall is adequate most of the year, planting should

Table 11.2. Characteristics of main cultivars of plantains and bananas

Group	Cultivar	Characteristics
AA	Sucrier	Very sweet, thin-skinned fruits, often highly preferred. Short fingers and low yielding
AAA	Gros Michel	Large fruits with good bunch shape. Sweet and of good eating quality
	Highgate	A dwarf mutant of Gros Michel, used as alternate parent in breeding
	Cavendish	The group includes Dwarf Cavendish, Giant Cavendish, Robusta and Lacatan which are strong flavoured, sweet and of good eating quality. They are the basis of all major export trades in bananas
AAAA	IC2	Tetraploid *M. acuminata* bananas have only been produced by breeding. IC2 is a cross between Gros Michel (AAA) and *M. acuminata* subspecies *malaccensis* (AA)

Group	Cultivar	Characteristics
	Bodles Altafort	Bodles Altafort is a cross between Gros Michel and Pisang lilin
AB	Ney Poovan (Lady's Fingers)	A small group of diploid hybrids between *M. acuminata* (AA) and *M. balbisiana* (BB). A widely distributed but unimportant clone. Plants are rather slender, lacking in vigour, with sweet acid, white-fleshed fruits
AAB	Mysore	Vigorous, high-yielding with a sweet-acid flavour
	Silk	Less vigorous, with white flesh and very attractive sweet-acid flavour
	Fome	Plants vigorous, hardy and not very prolific. Used as a dessert banana
	Pisang raja	Plants vigorous, sweet fruits eaten as dessert
	Maia madi	Plants with compact bunches of large blunt-ended fruits

be undertaken during the wettest times of the year.

The planting materials used are sword suckers, maidens, peppers, water suckers, bullheads or bullhead sections (bits). Water suckers are poor planting materials. All planting materials must be prepared by paring all reddish brown and dark tissues from the corm, including old leaf sheaths, roots, borer tunnels and nematode casts. This operation should be carried out in one section of the field. Planting materials of different types should be planted in separate sections of the field. This means all sword suckers should be planted together, bullheads together, and so on. Before planting, the suckers should be dipped in a 0.1% dieldrin solution to protect them against borers.

The recommended spacing for pure-stand crops is 3.6–4.5 m apart for tall varieties, such as Gros Michel bananas and French plantain types, on fertile soils, giving a population of about 480–750 plants/ha. On less fertile soils and particularly in drier areas, the spacing recommended is 3.6 × 2.7 m, giving about 890 plants/ha. Dwarf varieties of banana such as Dwarf Cavendish, as well as the short types of True Horn plantains, are grown at 2.4 × 2.4 m, thus giving 1680 plants/ha.

Fertiliser application

Although plantains and bananas are commonly grown on fertile soils, manuring is very beneficial except in exceptional cases. On fertile soils plantains may not need additional fertilisation at the initial growth stages, but subsequent ratoon crops must be fertilised.

A steady supply of nitrogen is essential for optimum yields. About 110 g of nitrochalk (calcium ammonium nitrate) per plant (mat) has been shown

Group	Cultivar	Characteristics
AAB	Plantain	There are two types: (a) French plantain type which is vigorous and high-yielding with heavy bunches of numerous medium-sized fingers. Cooking is necessary (b) Horn plantain type, which may further be divided into True Horn and False Horn plantains. They are both hardy but less high-yielding than the French type. Bunches consist of few, very large fingers in an open and often irregular configuration. Cooking is necessary in both. In the True Horn, the male axis is absent while in the False Horn the male axis is present or degenerates early

Group	Cultivar	Characteristics
ABB		Due to two genomes from *M. balbisiana* members of this group are very vigorous and drought resistant. Suited to dry climates and exposed situations
	Bluggoe	High yielding with straight blunt-ended fingers. Cooking banana
	Pisang awak	High yielding dessert type with indifferent flavour. Fruits are somewhat seedy when cross-pollinated with edible or wild diploids
ABBB		Only one natural tetraploid banana is known, but ABBB types can be produced artificially by crossing ABB clones, such as 'Bluggoe', with *M. balbisiana*
	Klue teparod	Used for sweatmeats, as the fresh flesh has a disagreeably fibrous spongy texture. Plants vigorous. Bunches have no male axis. Fruits massive, blunt and grey

to increase plantain and banana yields. It should be broadcast during the rainy season around each plant about 0.6 m away from the base. About 200 g of single superphosphate per plant is needed, but this is best applied into the planting hole before planting the sucker. A higher amount of potassium is required, an application of 320 g K_2O per mat usually being considered adequate. If and when available, farmyard manure and mulch may be applied to increase the amount of organic matter in the soil.

Weed control

If the land is properly prepared before planting and the correct spacing is adopted, weed control in plantains and bananas should require little effort. It can be achieved by cutlassing and by the use of herbicides. It is necessary to ensure effective weed control especially around the mats in young and newly planted fields. When cutlassing, the weeds must be cut close to the ground especially immediately surrounding each mat. Any form of cultivation should be shallow to minimise damage to the roots, which are only superficial.

Both pre- and post-emergence herbicides are effective in weed control. The application of $2,4-D$ and $2,4,5-T$ in a diesel or kerosene solution has been found to effectively control broad leaved weeds, while dalapon, diuron, MCPA and chlorbromuron applied singly or in various combinations are effective in controlling perennial grasses.

Pests and diseases

Larvae of the banana borer, *Cosmopolites sordidus*, feed in tunnels in the corm, which is reduced to a blackened mass of rotten tissue. The leaves turn yellow, wither and die prematurely. Other pathogens can then attack the plant which eventually dies or is blown over. Control of the borer is achieved by using clean planting materials. The incidence of attack is less when suckers are treated with insecticides before planting. If attacks are severe, then field treatment with aldrin, dieldrin or chlordane should be carried out.

The burrowing nematode, *Radopholus similis*, is a major pest of the roots. All cultivars are susceptible and corms must be pared and treated with warm water before planting. If field attack is observed, plants should be treated with 10% granules of Furadan at the rate of 20 g per mat at two and six months after planting, and every subsequent four months.

Thrips, *Hercinothrips bicinctus*, discolour or crack the skins of the fruit and may harm the flesh. Insecticides such as HCH (BHC) and dieldrin give effective control.

Panama disease, caused by *Fusarium oxysporium* var. *cubense*, is a serious disease of bananas and plantains. The first symptom is yellowing of the lower leaves which subsequently hang downwards. The vascular tissue inside the stems and rhizomes then turns purple. Some plantain varieties and all the Cavendish bananas are resistant, but Gros Michel is very susceptible. The only way to prevent the disease is to use resistant varieties.

Cigar-end rot, caused by *Verticillium theobromae*, makes the ends of the fruit look like cigar ash. The damage occurs when the crop is grown in areas which are not well suited to banana cultivation. The disease can be controlled if the withered inflorescence beyond the developing fruits is removed or by early spraying with Dithane M-45.

Other management practices

There are a number of measures specific to plantain and banana cultivation. These include:

Bunch clearing

As soon as a hand or two become exposed, the young bunch should be cleared of any leaves or trash that touches or might touch it. Green leaves should be bent away from the bunch and all dead leaves cut off. The spade leaf and the first bract should be removed or bent back. The bunch should be checked weekly until it is harvested. If leaves are allowed to touch or rub against the bunches, they cause unsightly scars and reduce the market value. This problem is not, however, as important in plantains, where skin bruises do not cause serious damage to the pulp.

Deflowering

Deflowering consists of the removal of the withered style and perianth. It is necessary in Gros Michel banana and the French plantains as well as the Cavendish group. The dried flowers may damage the backs of the next hand and may also cause cigar-end rot.

Sleeving

Sleeving (or bagging) of bunches protects the fruit against sunburn, dust, spray residue, insects and birds. The bag is left open on the underside. Bagging also increases the average temperature and advances the time of harvest. Generally, blue plastic bags are preferred (Figure 11.8).

Propping

This practice serves to protect bearing plants from falling over and from wind damage. It is necessary because the weight of the bunch can pull the plant over. Two props, forming a triangle, are better than one. Forked branches or bamboo poles are placed against the stem on the side where it is leaning over.

Removal of male bud

The removal of male buds is said to promote fruit development. Bunch weight is reported to increase by 2−5% when male buds are removed. The practice also reduces the frequency with which birds visit the flowers, therefore preventing other possible damage.

Earthing up

Earthing up protects the plants against wind damage and also helps the suckers to develop properly.

Harvesting and storage

Under optimal conditions, it takes nine months from planting to the harvest of plantain and banana

Figure 11.8 Sleeving of banana bunches protects the fruit against sunburn, dust, spray residue, insects and birds.

crops. This period may extend to 18 months, depending on climate, cultural practices and cultivar. From flowering to harvest takes 80–90 days under ideal conditions and up to 120 days under sub-optimal conditions. Whatever the circumstances, plantains and bananas are always harvested green.

For home consumption, the bunch is left hanging until the fingers are full (rounded). However, for market the bunches are cut at a less mature stage in order to lengthen the shelf life. Harvesting is done by removing the props and then nicking the pseudostem with a cutlass in such a way that it comes down gently and the bunch can be removed from the plant. The pseudostem is then cut back about 1 m above the ground. Fruits may be transported as intact bunches, which can be stored for relatively long periods, or cut into fingers (Figure 11.9).

Figure 11.9 Transferring banana bunches from a lorry to railway trucks in Cameroon.

In tropical conditions, plantains and bananas are non-seasonal crops and are produced all year round. Unless they are stored in optimal storage temperatures of 12–14°C, they ripen quickly after harvest, their shelf life being only a few days. Like cassava, plantains can be dried by cutting the peeled pulp into pieces and sun-drying the pieces as chips. They can be stored in this form or after grinding into flour, for relatively long periods. Bananas are always eaten as ripe fruit or as fresh green cooking bananas.

Avocado pear

The avocado pear, *Persea americana*, belongs to the family Lauraceae. The avocado originated in Central America and has been mainly produced in the southern USA, Brazil, Hawaii, Israel and tropical Australia. It was introduced into West Africa years ago and is now grown in many countries. The fresh, smooth, buttery pulp of the avocado pear is eaten as a vegetable. Avocado oil is highly priced and is widely used in cosmetics.

Types and ecological adaptation

Three races of cultivated avocado are generally recognised. These are the Mexican, the West Indian and the Guatemalan types. The main characteristics of the different races are compared in Table 11.3. In general, the best pears have small nuts and more flesh (Figure 11.10).

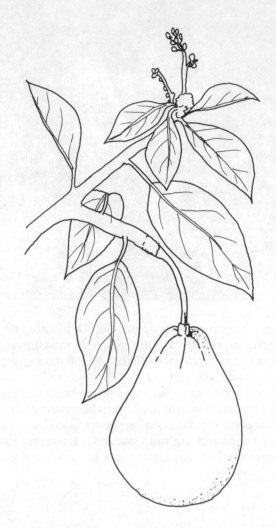

Figure 11.10 The avocado tree bears pear-shaped fruits with a high oil content.

Table 11.3. Comparative characteristics of the different races of avocado pear

Characteristic	Mexican	Guatemalan	West Indian
Leaf scent	Anise	None	None
Fruit size	Small	Variable	Variable
Fruit skin	Thin	Warty	Leathery
Months to ripen	Six	Nine	Six
Oil content	High	Medium	Low
Seed size	Large	Small	Large
Seed cavity	Loose	Tight	Loose
Tolerance of cold	Yes	Moderate	No
Tolerance of salt	No	No	Yes

Avocado pears grow well in closed forest country and in moist hilly districts, especially at altitudes of 300–900 m. They will grow at altitudes below 300 m, but do not thrive in open and grassy country or too near the sea. The influence of climate on avocado trees is related to the ecological races. The average temperature requirement varies from 12–28°C, while the rainfall required ranges from 650–1500 mm. The dry season must be well marked. In contrast to mangoes and cashews, the blossom in avocado pears is not harmed by rain, unless it carries on for a month or more. Wind is an important factor as avocado branches are brittle and break easily. The provision of windbreaks in plantations is therefore vital.

The avocado tree does best on a rich, well-drained soil whereas stiff and clayey soils are unsuitable. The optimum pH is between 5–7.

There are a great many varieties of avocado pear. Many are hybrids between the races. For example Fuerte, the most popular avocado cultivar in the world, is a Mexican x Guatemalan hybrid. It has a pear-shaped fruit with a high oil content (up to 26%), horizontal branches and a tendency to alternate bearing. It reacts well to girdling. The cultivar Hass belongs to the Guatemalan group. It is self-fertile, with medium sized, roundish fruits that turn purple on ripening. Pollock, Simmonds, Booth 8, Booth 7, Lula and Choquatte are all Florida cultivars which are well suited to the tropics. In the Ivory Coast where the production is at sea level, the cultivars Gottfried and Pernod (Mexican), Benik (Guatemalan) and Black Prince, Simmonds and Waldin (West Indian) are grown and all produce well.

Planting

The avocado pear is propagated either by seed or vegetatively by budding or cleft-grafting. Trees grown from seed are inclined to be variable in yield and in the size and number of fruits produced. Indeed, some seedling trees produce no fruit at all. Because of the unreliability of seedling trees, avocado pears are often propagated vegetatively.

Where propagation is by seed, the seeds are sown in nursery beds at 30×60 cm and the young plants are transplanted when six months to a year old. Seeds quickly lose their ability to germinate, unless they are stored in dry peat or sand at 5°C. Removing the seed coat and making thin slices at the top and bottom of the seed may speed up germination.

As a crop native to the rainforest, avocado trees must usually be grown in high rainfall areas, unless irrigation can be provided, allowing them to be grown in arid environments, as is done in Israel and elsewhere. Also, because of the brittle nature of the stems, the trees should be grown in areas with light winds or they should be provided with windbreaks. Avocado is a sun-loving crop which should be grown without shade. Because of problems with *Phytophthora*, the site selected should be free-draining and not exposed to waterlogging.

The best time to transplant avocado budlings (or transplant seedling trees) is at the beginning of the rains. Planting early enables the plants to establish before the dry season sets in. If irrigation is used, it is best to use sprinkling or trickling, as flooding may promote root rot.

Spacing varies with climate, soil fertility and cultivar, from 6–12 m on the square. The modern practice is to plant at 5×5 m (400 trees/ha) and gradually thin to an ultimate spacing of 7×10 m (140 trees/ha). As most growers tend to be reluctant to cut down bearing trees, it is better to plant wider so as to avoid having to thin the trees later. This permits intercropping with vegetables between the rows, except tomato and eggplant which carry *Verticillium*, a fungus to which avocado is particularly sensitive. Mulching is generally advantageous.

As the tree grows, dead or spindly branches and any branches which tend to fill up the centre should be removed and the tree should be pruned to a good shape. It is important to remember that avocado pear fruits on young wood.

Fertiliser application

The roots of the avocado pear are near the surface and the tree can be manured by leaf mulching or by digging in green manure, wood ashes or animal manure. If inorganic fertilisers are available, the crop should receive N, P_2O_5 and K_2O in the proportions 1:1:1 for young trees and 2:1:2 for older trees.

Weed control

Avocado trees have shallow root systems and therefore hoeing to control weeds must be light and carefully executed. Annual weeds are better controlled by herbicides such as monuron and simazine. Against perennial weeds, oil and paraquat can be safely used, but dalapon and bromacil may only be applied outside the root zone. Cover crops are recommended only in humid areas where competition for moisture would not have serious effects.

Pests and diseases

The pests of avocado include scale insects, weevils, mealybugs, leaf-cutting ants and mites. Spraying with insecticides is an effective means of control. The most serious disease of avocado is root rot caused by *Phytophthora cinnamoni*. It attacks small roots, thereby causing leaf fall, die-back of branches and tree collapse. The attack may be checked by spraying with copper-based fungicides. Prevention of the disease is by the use of suitable root-stocks as well as proper field sanitation. Other fungal diseases are caused by scab (*Sphacelona*), anthracnose

(*Colletotrichum*) and *Cercospora*. Minor attacks of viruses and parasitic nematodes are also occasionally observed.

Harvesting

Avocado pear trees come into bearing in 4–5 years and full maturity is attained 7–8 years later. The fruit is harvested at the beginning of the rains when it is almost ripe. Mature fruits ripen within a week at 27°C and in a month at 5°C Avocados ripen best at temperatures of 15–24°C and should be stored or transported at 13°C. If pears are picked before they are mature, they are unlikely to ripen and the fat content will be lower. Mature fruits should be picked and must not be allowed to fall to the ground or get bruised.

Ginger

Ginger, *Zingiber officinale*, is a member of the family Zingiberaceae. The plant is a perennial, and even though its aerial parts die out annually, it can perennate by means of underground portions. The underground rhizomes are much branched, giving rise to primaries, secondaries and tertiaries, the last formed ones having young buds at their tips. The root system is superficial and fibrous. The above-ground stem is unbranched and thin, formed by the sheathing petioles of the leaves. The leaf blades are narrow and lanceolate and are at right angles to the stem. Flowers, which are seldom produced, are borne on a stalk which varies between 15–30 cm in length and arises directly from the rhizome (Figure 11.11).

Ginger originated in Asia, probably in India, where it has been cultivated since ancient times. It is presently also grown in parts of West Africa and the West Indies, with Jamaica producing the finest quality ginger in world trade. The crop is used extensively as a flavouring in foodstuffs and beverages. Large quantities are used in the manufacture of ginger-beer, ginger-ale, gingerbread, ginger-oil, ginger-essence and ginger chocolates, pastries and biscuits. It is also used to some extent

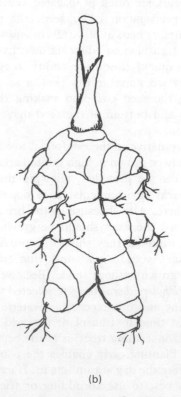

(a)

(b)

Figure 11.11 The ginger plant is a perennial which seldom flowers, but is grown for the production of ginger from its rhizomes. (a) Flowering ginger plant and (b) ginger rhizome, from which dried ginger is made.

228

in pharmaceuticals and perfumes. Ginger oleoresin is the substance which gives ginger its pungent properties. It is obtained from the rhizome by extraction.

Types and ecological adaptation

Ginger is essentially a tropical crop which requires warm weather during its growing season, with optimum temperatures being between 28–35°C. Dry weather is required during the harvesting period to facilitate curing. It is grown mainly as a rainfed crop, although it may be successfully cultivated using irrigation water. For successful cultivation, a well distributed rainfall of between 1500–3000 mm is necessary. However, the crop does not tolerate waterlogging, which causes rotting of the rhizomes. In regions with a light rainfall of 750–1000 mm, it is best to use irrigation. The crop thrives in closed forest regions under shade and especially where the soil is well-drained and rich in organic matter. It thrives up to an altitude of 1200 m.

In the Guinea savanna region of Nigeria the ginger variety Taffin Giwa is considered to be the best from the commercial point of view.

Planting

Ginger intended as seed for the following season is left unharvested in the ground. The plot containing the seed ginger (pieces of rhizome used for propagation) should be heavily mulched with grass at the end of the rains, when all the leaves have dried up. If the ginger which is intended for planting is lifted well ahead of planting time, it should be stored in a well ventilated shady place on a layer of sand, and then covered with grass.

The rhizome is the part of the plant from which ginger is propagated. Planting may be done directly into the field without raising plants in a nursery. For optimal development of the rhizomes, the soil should be well cultivated to an appreciable depth. During this operation the soil must be broken down to a fine tilth.

Planting material should be obtained from fingers (rhizomes) specially stored for the purpose. Good, healthy fingers are selected and broken into small pieces (seed), each containing two or three 'eyes'. The use of whole hands as planting material results in small sets in which the fingers are bunched together.

Ginger is normally planted as soon as the rains are established. In general, the closer the spacing the greater the total yield per unit area of land, even though the yield per plant is decreased. The best yields are obtained at spacings of between 20 × 20 cm and 20 × 30 cm.

Fertiliser application

Ginger benefits from a liberal application of manure. Where farmyard manure or compost is available, it should be applied at the rate of 7.5 tonnes/ha before cultivation. In the absence of farmyard manure, nitrogenous fertilisers at the rate of 50 kg N/ha should be supplied in two doses, when the plants have established and six weeks after.

Weed control

The control of weeds is important, especially in the early stages of crop growth. The degree of weeding required will depend on the amount of weeds present and the cultural practices undertaken. A well cultivated and properly mulched field should require less weeding. Weeding is normally done by hand-pulling, since the use of a hoe would be harmful to the rhizomes, which are located near to the surface of the soil. Little work has so far been undertaken on the use of chemicals in weed control in ginger. Considering the risk to the rhizomes of hoe weeding and the expense of hand weeding, the eventual use of herbicides should greatly increase the economic value of the crop.

Pests and diseases

The two important pests of ginger are shoot borers and root knot nematodes. Shoot borers bore into the shoot and damage it. Control is effected by periodic spraying with 0.2% endrin at monthly intervals. The root knot nematodes which infect ginger include *Meloidogyne incognita* and *M. javanica*. Both nematodes can be controlled by soil fumigation with ethylene dibromide and Basamid.

Ginger is known to suffer from many diseases, both in the field and in storage. Among the major diseases are leaf spot, caused by *Colletotrichum zingiberis*, soft rot, caused by *Pythium* sp., bacterial wilt caused by *Pseudomonas solanacearum*, rhizome rot in storage, and bacterial rhizome rot. Most of these diseases can be effectively controlled by spraying with appropriate fungicides. In addition, soil drenching, where possible, is beneficial. Bacterial rhizome rot may be controlled by dipping the rhizomes in a 0.2% solution of Dithane M-45. For disease control in storage the rhizomes should be dipped in 0.1% mercuric chloride.

Harvesting and storage

Harvesting may be carried out from nine to ten months after planting, when the leaves begin to turn yellow and the shoots begin to lodge. In West Africa the harvest period is normally from November to the end of April for planted ginger. For 'ratoon' ginger, which usually matures earlier, harvesting is between seven and nine months after the start of new growth. At harvesting the ginger is carefully dug and the soil is removed from the rhizome. During harvesting, the minimum amount of damage should be done to the hands, which should not be broken up.

Storage of replanting material is an important aspect of ginger cultivation. Since harvesting is completed several months before the following planting season, it is vital that rhizomes should be stored in the correct conditions of temperature and humidity. Rhizomes remain viable for over six months if stored at 13°C and 65% relative humidity.

The best way to store ginger is in the form of the cured product. Curing is a skilled operation. Dry weather during curing is essential, as wet conditions cause mould formation, which gives the ginger a musty odour and bad flavour. Immediately after lifting, the rhizomes are cleaned of any earth and roots, then immersed in a tank of water and thoroughly washed. The water is constantly changed and finally the ginger is soaked in fresh water to facilitate peeling of the skin. Peeling is a delicate operation, and special peeling knives are used for this purpose. Peeling involves scrapping off the outer skin without damaging the cells below, which contain much of the essential oil on which the aroma and flavour of good quality ginger depends. Peeling and washing of rhizomes should be carried out simultaneously. The more efficient the washing the whiter the product. The peeled ginger is then allowed to remain in clean water overnight, and is washed again the next day in lime water. The treated ginger is then sun-dried for five to six days and bagged. In this dried condition the ginger can be kept for over a year.

Melon

Melons belong to the family Cucurbitaceae. There are about 90 genera in this family, the three most important being *Cucurbita*, including squashes, pumpkins, gourds and vegetable marrows, *Cucumis*, including cucumber and sweet melon, and *Citrullus*, including watermelons. Several species in these genera provide edible fruits used mainly as vegetables, but sometimes eaten as dessert fruits. Not only are the fruits and seeds edible, but the hard rind of the fruits of certain species, provide a variety of drinking vessels and utensils which find many uses in all parts of the tropics. The seeds contain a high-quality edible oil.

Figure 11.12 A watermelon fruit raised on a grass pad to help prevent fungal infections and insect damage.

Types and ecological adaptation

In general members of the genus *Citrullus* can be distinguished from *Curcurbita* spp. and *Cucumis* spp. by their deeply pinnately lobed leaves, branched tendrils and corolla tube, which is divided almost to the base.

The cultivated species of melons are of special importance in the drier regions of the tropics since they thrive well on sandy soils under irrigation. In the West African region the main areas of melon production include the forest and southern Guinea savanna ecological zones. Melon is usually sown as the first crop in a mixture with such other crops as maize, sorghum and yam at the onset of the rains. The crop tolerates high temperatures and low humidities. It prefers a sandy loam soil which is rich in organic matter. Acid soils are not suitable for the production of melons.

Many different varieties of melon are adapted to local environments. The choice of variety to grow should be dictated by taste as well as local environmental conditions. Melons do not do well when transplanted, so are grown directly from seed sown in the field.

The water melon, *Citrullus lanatus*, is widely cultivated in many areas of the tropics, especially in Africa.

Planting

Melon seeds are generally sown at the beginning of the rains or towards the end of the rainy season. They can be sown at any time if sufficient water is available. Crops grown under irrigation during the dry season generally perform better than those grown under rainfed conditions. Planting may be done on ridges or on flat beds. Two or three seeds are sown per hole at a spacing of 1.5×1.5 m to 2×2 m. Seedlings are subsequently thinned to one plant per stand (Figure 11.12).

Fertiliser application

In general, melons have not shown much response to the application of fertilisers, especially when grown on relatively fertile and well-drained soils. However, for maximum yields, applications of 50 kg N, 30 kg each of P_2O_5 and K_2O/ha may be made before planting. Subsequently, an additional 20 kg N/ha may be side-dressed at flowering, 15 cm away from the seedlings.

Weed control

Effective weed control may be achieved with two hoe weedings before the crop forms a good cover. Pre-emergence herbicides have been evaluated for weed control in 'egusi' melon crops (*Cucumeropsis edulis*). Mixtures of alachlor with either prometryne, chloroxuron or chloramben at $1.0 + 1.5$ kg/ha are reported to combine acceptable weed control with high melon seed yields.

Pests and diseases

Pests and diseases are generally not a problem in melon cultivation. The most damaging insect is a fruit-fly which lays its eggs in developing fruits of melons. Vetox (carbaryl) used at the rate of 1.7 kg in 170 litres of water/ha may control the pest. Control of fungal diseases may be achieved by spraying with Benlate.

Harvesting and storage

'Egusi' melons should be harvested when the fruit stalk attached to the vine (neck) is dry. The seeds may be extracted by breaking the fruits and heaping them together to ferment, allowing the pulp to rot. The seeds are then dried and stored. Melons which are cultivated for their fruits are harvested just before they are fully ripe.

Pawpaw

The family Caricaceae has only four genera. The genus *Carica* is an important genus with over 22 species. *Carica papaya*, the pawpaw, is of tropical American origin, but is now cultivated throughout the tropics. Cultivation of pawpaw is, however, hardly ever extensive, and the crop is usually grown in small gardens adjacent to homesteads. Hawaii is one of the only places in the world where it is produced in significant quantities for export, as well as for local consumption. As well as being eaten as ripe fresh fruits, pawpaws are also used for making soft drinks, jam, icecream flavouring, crystallised fruit and canned fruit in syrup. Unripe fruits are cooked as a substitute for marrow or apple sauce. Papain, prepared from the dried latex of immature fruits is a proteolytic enzyme similar in action to pepsin and is used in meat tenderising preparations, in the manufacture of children's food, chewing gum and cosmetics, as a drug for digestive ailments and in the tanning industry for bathing hides. Papain finds its main use in the brewing industry where its presence prevents the cloudiness in beer caused by precipitation of protein during chilling. The process of applying shrink-resistance to wool and silk textiles also involves the use of papain.

Types and ecological adaptation

The pawpaw is a fair-sized tree, 3–6 m in height and with prominent leaf scars on unbranched, mostly hollow stems, with leaves at the top only. The leaves are large and deeply lobed with long hollow petioles. The plant is usually dioecious, producing male and female reproductive structures on separate plants. However, hermaphrodite (bisexual) trees also occur. All parts of the plant contain small latex vessels throughout the tissue, which produce latex freely when cut. The latex contains the proteolytic enzyme papain.

Pawpaw thrives in a variety of soils and is not particularly demanding in its environmental requirements, which are quite similar to those of banana. The crop prefers fairly high temperatures, although it can withstand low temperatures for short periods. The minimum temperature for growth is 15°C whereas 22–26°C is considered the optimum range. The crop requires adequate rainfall, although it does not need very high amounts. Day-length has no effect on pawpaw. The best quality fruit, determined largely by sugar content, develops in full sunlight, with its final four to five days to full ripeness on the tree. With a shallow root system, a delicate tree like pawpaw does not tolerate strong winds and must be protected from them.

Pawpaw fruits vary in shape and size, but are generally elongated or globular, sometimes reaching 50 cm in length. The seeds of pawpaw are small, rounded, dark green or brown and about the size of a pea. Dried seeds retain their viability for a considerable time. However, the best germination is obtained from fresh seeds which may even begin to germinate in the ripe fruit. The colour of the edible flesh varies with the variety, some being yellow-orange while others have pink pulp. Local varieties are invariably well adapted to local conditions and those with the preferred fruit quality are grown. However, such varieties as Solo (developed in Hawaii), Hortus Cold (South Africa), Improved Peterson (Australia) and Betty (United States of America) are reported to be high-yielding and of good fruit quality (Figure 11.13).

Planting

Pawpaws are generally propagated by seed, either directly in the field or in the nursery prior to transplanting to their permanent positions. The

Figure 11.13 Pawpaw fruits are generally elongated or globular in shape and are grown to ripeness on the tree.

crop may also be raised by cuttings taken from trees with side branches.

The commonest method of propagating pawpaw is by raising seedlings in the nursery before planting out. Seedbeds are made to a suitable size in the ground or in boxes and the seeds are planted thinly. Germination takes three to four weeks and thereafter the plants grow rapidly. The small seedlings are transplanted into deep polythene bags and are set out in their permanent positions in the field six to eight weeks after sowing in the nursery, when they are about 20 cm in height.

The best sites for growing pawpaws are rich in nutrients and are well-drained. The plant cannot withstand waterlogging. A pH range of $6-7$ is preferred. Where possible the site should be shaded from strong winds which might cause the uprooting of plants.

Whether propagated by seedlings raised in a nursery or seedbed, or from cuttings, pawpaws are spaced at 3×3 m in the permanent site. Before planting out, holes measuring $1 \times 1 \times 0.5$ m deep are dug and filled with a mixture of topsoil and well-rotted manure. Seedlings are then planted in the holes to the same height as they were growing in the nursery.

If cuttings are used, side branches should be used and cut at the point where they join the main stem. Branches should be no longer than $0.6-1.2$ m, and the cuttings should be planted in wet weather after their main leaves have been removed. Although the cuttings may die back at first, they quickly make new growth.

In general, it is advisable to plant several male plants among a predominantly female population so that pollination may take place and good fruit and seed can be formed. An average of one male tree to $20-25$ females should suffice.

Fertiliser application

Pawpaw responds well to fertilisers, particularly nitrogen and phosphorus. Mulching is reported to be advantageous. Four weeks after planting out in a permanent site, each tree should be supplied with 225 kg/ha of a complete fertiliser compounded to give N, P_2O_5 and K_2O in the ratio of 1:3:1.

Weed control

Pawpaw may be kept free of weeds by regularly hoeing lightly around the base of the plants and slashing between the rows. The control of weeds by chemicals may also be adopted. The use of either paraquat or diuron at $0.5-1.1$ kg/ha has been reported to effectively control weeds in pawpaw plantations. It should be noted, however, that pawpaw is sensitive to hormonal herbicides such as 2, $4-D$.

Diseases and pests

The pawpaw is attacked by a number of insects, fungi and other organisms. Among the disease-causing fungi is *Colletotrichum* which causes fruit spotting. The most serious disease, mosaic, is caused by a virus transmitted by aphids. Another important disease, bunchy-top, which is caused by a mycoplasma, is transmitted by a leaf hopper. Other insects, mites, birds and rodents can be pests of pawpaw. In general, control is effected by insecticidal and fungicidal sprays, by crop sanitation and by soil drainage.

Root knot nematodes and soil-borne fungi may also cause problems.

Harvesting and storage

The best fruits are produced by trees which have been prevented from forming lateral branches. Although more fruits may be produced by a branching pawpaw, they are invariably smaller and of poorer quality than those produced on trees that have not branched.

Fruits turn yellow when fully ripe. For local consumption, fruits may be harvested at full yellow, but when intended for market the fruits should be allowed to retain streaks of green. Harvesting should be done with a sharp knife, with the fruit stalk remaining attached to the fruit.

Fruits harvested at the correct time will store for a few days at a temperature of 21°C during which time a good flavour will develop. When being transported over a long distance, the fruits should be packed in containers which are padded with a soft material.

In many tropical countries latex is collected from the fruit and used in preparing papain. Immature fruits are tapped and the latex is dried and preserved until the papain is extracted later. The outer skin of the green pawpaw fruit is cut with a sharp knife and the latex is collected and strained through a brass wire mesh and heated on trays at 38°C for 12−14 hours. The dried product is creamy white and crumbly. Yields of papain are highest in the first year of tapping and are reduced in subsequent years.

Grape

Grape, *Vitis vinifera*, also referred to as grapevine, is a member of the family Vitaceae. There are four species within the family, but only two, *Vitis vinifera* or European grape and *V. rotundifolia* or muscadine, both of which tolerate hot climates, are of economic importance. Of the two, *V. vinifera* is the most widely cultivated species world-wide. Originating in Russia, the grape is one of the most delicious, refreshing and nourishing fruits in the world. It is raised as a table fruit and for making into wine. Grapes occupy more land than any other single fruit crop, and account for nearly half of the total world production of all fruits.

Types and ecological adaptation

Grapevine types are generally classified according to their intended use as follows:

Table grapes. These have medium to large-sized sweet berries, black, purple or greenish in colour with small or no seeds.

Raisin grapes. These types have large, medium or small-sized berries. The fruits may or may not have seeds, but must contain a high percentage of sugar.

Unfermented juice grapes. These types have either black or greenish fruits and a strong characteristic aroma.

Wine grapes. These types also have black or greenish fruits. The fruits have a strong characteristic aroma and are more acidic than other types.

Grapes are grown in temperate and subtropical regions where the seasons are clearly defined into winter and summer. The winter season is necessary for breaking bud dormancy while the summer favours optimum development and ripening of fruits. The vines shed their leaves and rest in winter, make new shoots in spring and mature in summer. In the tropics, the grapevine is evergreen and yields poorly, unless special techniques of pruning are employed. Humid tropical conditions encourage pests and diseases and are therefore unsuitable for growing grapes. Rains during flowering and ripening result in poor fruit-set and berry splitting. The ideal temperature range for grape cultivation is from 10−20°C, while some cultivars can grow in temperatures up to 32°C. Grapes can be grown on a variety of soils, but light sandy loams which are free-draining are ideal.

The grape industry in West Africa is of special interest in the sense that it has arisen out of the activities of amateur growers. Today, grapevine cultivation has been taken up seriously, particularly in Ghana and Nigeria.

Specific varieties are generally favoured for their intended uses, but can be used for other purposes as well. Varieties which have proved successful in several tropical areas and which are in current use in the Northern Guinea and Sudan savanna regions of Nigeria include Aneb-e-Shahi (table type), Thompson Seedless (raisin type), Black Hamburg (unfermented juice type) and Muscat of Alexandria (wine type).

Planting

The site for grapevines must have soil which is deep, loamy and with good structure. It must contain organic matter and be well-drained and aerated. A pH of 6 is preferred and the soil must be practically salt-free. Humid tropical regions are unsuitable for grapes as these favour the growth of fungi. A period of dry weather during ripening is necessary for flavour improvement.

Although grapevines are planted in pits, it is preferred to plough and harrow the land at the beginning of the rains. Pits 75 × 75 × 75 cm in size are dug some 4−6 weeks before planting and filled with equal amounts of well-rotted farmyard manure and topsoil. About 2−3 weeks before planting 2−4 kg of superphosphate are applied to each planting pit. The spacing adopted between plants depends on the variety, the method of training and the soil type, with the range of spacing varying from 2.5 × 2.5 m to 5 × 5 m. In a commercial vineyard the distance between rows should allow work with a tractor (Figure 11.14 opposite).

Vigorous and healthy-rooted cuttings 6−13 months old are used. They are planted at the centre of the pits and the soil is pressed around them gently. Thereafter the plants are carefully irrigated. Planting out is done as soon as the rains are established, although where there are irrigation

Figure 11.14 Grapevines grown as cordons.

facilities, especially in the savanna region of West Africa, planting should be done in January when the weather is dry and the risk of fungal attack is reduced.

Training

Grapevines are trained to many shapes and forms which relate to the different types of support used. Many modifications of grapevine training have been adopted all over the world. The basic systems of training are as follows:

Single stake system. A live plant stake, for example *Erythrina indica*, is used to support the vine (Figure 11.15a). The vines are tied to the stake and cut to 2.5 m. Four main branches are allowed to develop, which are further divided into sub-branches. The small branches are pruned for fruiting every year. To avoid shade, the stake plant is often pruned and used as mulch or green manure.

Arbour system. The Arbour system is also often referred to as the 'overhead horizontal bower system'. The bower is made of galvanised pipes or wooden posts (Figure 11.15b). If wooden posts are used, these should be treated with tar or used engine oil to prevent possible termite attack or early decay. These structures can be erected after the grapevines have been planted. On the horizontal arbour, thin wire (4 mm guage) is tied at 50 cm height to form a horizontal frame to support the bearing vine.

Trellis system. Also referred to as the Kniffen system, the trellis system involves stretching three wires on posts (Figure 11.15c). The vines are allowed to develop on either side and are trained along the wires. This system is not suitable for high-yielding and vigorous varieties.

Overhead trellis system. The main stem is allowed to reach the overhead trellis and is trained on three to five wires (Figure 11.15d). This is like a small overhead arbour system.

Pruning

Pruning is the most important cultural operation for successful production of grapes, especially

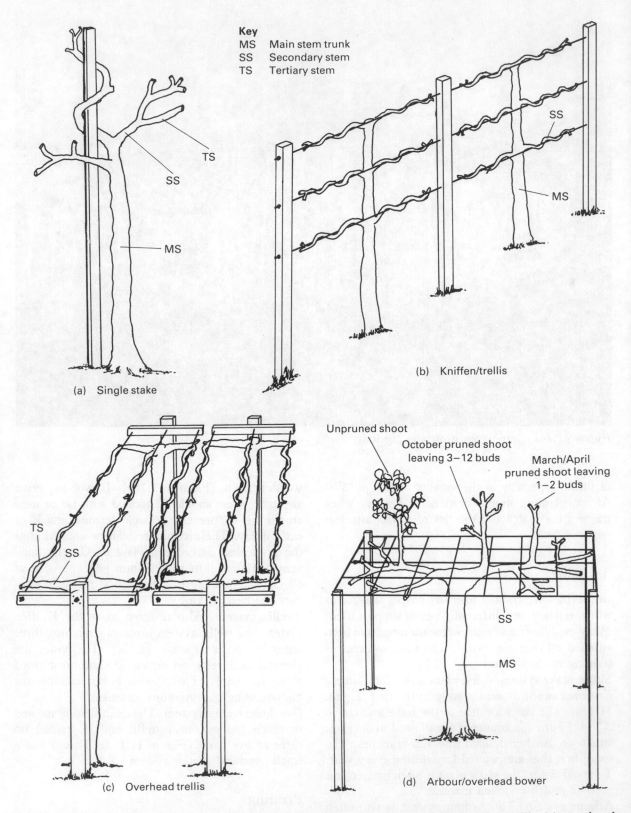

Key
MS Main stem trunk
SS Secondary stem
TS Tertiary stem

(a) Single stake

(b) Kniffen/trellis

(c) Overhead trellis

Unpruned shoot

October pruned shoot
leaving 3–12 buds

March/April
pruned shoot leaving
1–2 buds

(d) Arbour/overhead bower

Figure 11.15 Methods of training grapevines showing (a) single stake (b) kniffen/trellis (c) overhead trellis and (d) arbour/overhead bower.

under tropical conditons. An unpruned and un-trained (wild) vine grows as a robust climber and, if it fruits at all, produces small clusters of poorly flavoured fruits.

The pruning operation consists of the removal of excessive growth, either by leaving a single bud or a shoot with 3–12 buds intended to bear fruits. The main purpose of pruning is to get a uniform distribution of fruiting buds on the vine in order to get a good crop and maintain the productive-ness of the vine for a long period. For this purpose, the grower should know the growth and fruiting habit of the vine. Most of the buds are potential flower buds and would produce fruits if conditions at the time of differentiation were favourable. The inflorescence primordia are formed in the buds in summer during the growth of shoots. The floral parts are, however, not formed until they are forced to grow after a September or October pruning.

Not all shoots on vines have equal potential to bear clusters. Buds on older wood (one to three basal buds) form fewer inflorescence primodia than those on the median portion (four to 12 buds), and the buds at the top of the shoots produce tendrils instead of flower clusters. The vine usually dif-ferentiates more fruit buds than it is capable of bearing each year. Therefore, the fruit buds are reduced to the right number by pruning.

In temperate and subtropical regions, the vine remains dormant during winter and starts growing in spring. In the tropics the vine does not become dormant due to favourable temperatures for growth. It is therefore cut to a single bud after harvest in the dry season. The vines are forced to rest artificially by removing all vegetative growth during the hot period of the year. The shoots grow until September or October. These are pruned again, leaving 3–12 buds in different varieties as shown in Table 11.4.

Table 11.4. Pruning of grapevines for different varieties

Variety	Fruit type	Number of buds
Aneb-e-Shahi	Table	5–7
Black Hamburg	Unfermented juice	3–5
Muscat of Alexandria	Wine	4–6
Thompson Seedless	Raisin	9–12

The fruiting spurs produce fruits which mature in 4–5 months. Care after pruning consists of removal of all tendril-bearing shoots, disease af-fected berries and shoots. The berries and bunches are thinned for production of quality grapes.

Fertiliser application

The grapevine needs large quantities of manure and fertilisers. Experience has shown that bulky organic manures like farmyard manure, compost or green manure should be applied to the crop at 25–30 tonnes/ha. In addition, 300–400 kg of single superphosphate should be applied at the time of summer pruning, after the crop is har-vested, and 200–300 kg of calcium ammonium nitrate with 100–150 kg of muriate of potash should be applied per hectare in two equal doses at the time of each pruning.

Weed control

Special attention should be given to keeping the vineyard free from weeds by occasional shallow cultivation. Most of the grapevines' feeding roots are in the top 60–80 cm of soil so the depth of cultivation should be minimal. In the rainy season, if the crop continues to make extensive growth, it is advisable to spread green manure or sow a cover crop to avoid excessive vine and weed growth. Young grapevines are very sensitive to herbicides. Therefore, no herbicides should be used for the first three to four years, particulary 2,4–D and 2, 4,5–T, both of which can kill the vines.

Pests and diseases

Various insects and mites attack grapes. The common pests of grapes in the tropics include:

Flea beetles (*Scelodonta strigicollis*). The grub of this beetle feeds on the roots while the adults feed on sprouting buds. Control is achieved by dusting the vines with lindane or toxaphene.

Thrips (*Rhipiphorcthrips cruentatus*). Damage is caused by lacerating the leaf and fruit surfaces and then feeding on the juice. The infested leaf turns brown and dries up. Control is by spraying with malathion or any systemic insecticide.

Vine girdle beetle (*Sthenias grisator*). This beetle causes damage to grapes by girdling the stems and eventually causing them to dry up. Control is by spraying with malathion or carbaryl.

Termites (*Odontotermes obsesus*). Termites feed on the roots and stems and the vine eventually dies. Application of chlordane as a drenching spray around the plant base has given some control.

Rodents and birds. These large pests can cause considerable damage by feeding directly on the berries. Subsequently, large yield losses are recorded. Fencing round the plants and the use of mechanical scarers may partially solve the problem.

Because the grapevine fruits during dry weather in West Africa, diseases are not usually a problem. However, some of the important diseases are:

Powdery mildew (*Uncinula necator*). This is seen as a white growth on berries and upper leaf surfaces. Young leaves become curled and older ones become brown. Flowers turn black and affected berries remain small and cracked. The appplication of either dust or wettable powder of sulphur at weekly intervals is effective.

Downy mildew (*Plasmopora viticola*) affects twigs, buds, flowers and fruits. A white growth appears on the lower surface of the leaves which later turn brown. Badly affected parts should be pruned and destroyed. Spraying with copper compounds also gives some control.

Anthracnose or Bird's eye disease (*Gloeosporium amoelophagum*) is characterised by the appearance of dark brown oval spots on the leaves, twigs and fruits. This disease is most serious during the rainy season. Badly affected parts must be pruned and destroyed. Spraying with copper compounds also helps.

Harvesting and storage

Grapevine fruits do not ripen or improve after harvest. The bunches should, therefore, be picked when they are fully ripe. Ripeness in grapes is judged by a combination of indications, including the waxy bloom on the fruit, characteristic colour changes, light thickening of the juice, easy detachment of the berries and their sweetness. Yields vary from 10−25 tonnes/ha, depending upon variety and the method of training and pruning as well as other management practices.

Storage at 5−10°C keeps berries for over one month without loss of quality. The berries may, however, be processed in various forms for wine.

Pineapple

The pineapple, *Ananas comosus*, belongs to the family Bromeliaceae. One feature of this family is the production of compound fruits, formed by fusion of the parthenocarpic fruitlets with the bract and the central axis of the inflorescence. There may be a hundred or more individual fruitlets arranged spirally around the thick central axis, and the whole forms a broad, almost cylindrical multiple fruit, tapering at the top and surmounted by a rosette of short stiff spirally arranged leaves referred to as the 'crown' (Figure 11.16).

The pineapple originated in tropical America

Figure 11.16 The pineapple, *Ananas comosus*, is widely grown throughout the tropics, mostly for local consumption and for canning and juicing.

but it is now widely cultivated in other parts of the tropics and subtropics. There is a small export trade in fresh fruits from West Africa, especially the Ivory Coast and Senegal, but most pineapples grown commercially are canned or made into juice. The fruit is also made into jam and sugar syrup, with the latter being obtained from the milled juice. Alcohol and citric acid are also manufactured from pineapples. The fruit residues after milling the juice can be fed to cattle as bran. The leaves yield 2–3% of a strong white silky fibre. Usage of pineapples in local medicine has been recorded.

Types and ecological adaptation

The many cultivars of pineapple may be divided into four groups: Cayenne (also called Smooth Cayenne), Queen, Spanish and Abacaxi. The characteristics of these cultivars are summarised in Table 11.5.

Cayenne is the most commonly grown pineapple in the world. It combines spineless leaves with high fruit production, high fruit quality and resistance to gummosis. The fruit is cylindrical in shape and ideal for canning.

Red Spanish has long narrow leaves with spines. The fruit is roundish with few but large flat eyes. This cultivar is good for fresh fruit export, but only fair for canning.

Abacaxi is widely grown especially in Brazil. Plants are erect with rigid fruits, which are pyramid shaped. It produces fairly well with average fruit size. It is poor for canning and is therefore eaten fresh. However, even though it is very juicy, its pale colour makes it unattractive for this purpose.

Queen is an old cultivar grown mainly in Australia and South Africa. The fruit is conical in shape and the number of fruits produced is fair, but fruit size is very small. It is preferred for fresh

fruit. A related cultivar of Queen is 'MacGregor' which is the main commercial fresh fruit cultivar in Australia.

The pineapple plant grows well within the tropical and subtropical region up to 35° north and south of the equator. They are generally hardy plants and can withstand considerable drought. The crop is grown without irrigation, rainfall of as low as 600 mm being adequate. However, in such low rainfall regions, dew which forms on the leaves and percolates down to the roots is essential for growth and yield. Pineapples have an ability to store water for later use. Temperatures of 21–27°C are optimal. There are varietal differences in temperature requirements and tolerance. The cultivar Queen, for example, is suited to regions of low temperature; the cultivar 'Cayenne' is suitable for growing in the subtropics or at elevations up to 100 m in the tropics. Cultivars 'Spanish' and 'Abacaxi' are suited to the lowland tropics. If they are grown in cooler locations the resulting fruit becomes acidic with a lowered sugar content. Their minimal temperature range is 10–16°C.

Pineapples are grown in a wide range of soil types, but the most essential feature should be good drainage. As most acidic soils are highly leached, they necessarily require heavy fertilisation for optimal fruit yields.

Planting

Pineapples are propagated vegetatively either from the crown of leaves surmounting the fruit, from small shoots known as 'slips' which develop on the peduncle just below the fruit, or from suckers produced in the leaf axils lower down the stem. However, the different types of propagules grow at different rates although in general large shoots

Table 11.5. Major characteristics of different groups of pineapple

Characteristic	Cayenne	Queen	Spanish	Abacaxi
Spines	Nil	Yes	Yes	Yes
Fruit form	Cylindrical	Conical	Globular	Pyramidal
Productivity	High	Fair	Fair	Fair
Average fruit wt (kg)	2.5	1.0	1.5	1.5
Flesh colour	Yellow	Golden	White	Pale
Slips	Few	Few	Many	Many
Canning	Very good	No	Fair	No

Source: Samson, 1980.

grow faster and mature sooner than small ones. This implies that planting materials should be graded and only one type planted in any one field.

The stem (or stump) section method of propagation entials forcing into growth dormant axillary buds on the stem, crown and shoots. Mature plant stems are stripped of their leaves and roots, and cut longitudinally into quarter pieces. The inner part of each piece is pared away, resulting in a truncated wedge. These slices are then pressed into the soil in a nursery bed with the outer surface of the piece upwards and lightly covered with soil. Within a few weeks small plantlets develop from the dormant axillary buds. These are then planted in a nursery bed until large enough to be placed in a field. This method is reported to produce up to 25 plantlets from a single plant stem.

Pineapple is a sun-loving crop and should therefore be cultivated in an area with adequate sunshine. There should be no shade plants nearby to cast shadows on the growing crop. The soil should be free-draining and preferably a sandy loam type, rich in organic matter.

Planting of pineapple from suckers is usually done at the beginning of the rains. However, the tolerance of planting material to drought depends on the type. Suckers, slips and crowns may be planted late in the season, but other planting materials such as shoots and stem sections need to be planted when the rains are established.

Seed treatment depends on the type of planting material to be used. Slips and shoots may be dried for one or two weeks and may be kept in storage for several months before planting. If crowns are used, the material should be dried for a few days and should be planted immediately without storage. As already explained, stem sections may also be used to achieve rapid propagation. These should

Figure 11.17 A double row spacing for pineapple cultivation allows easy access between the rows.

be treated with a mercurial fungicide prior to planting.

Spacing markedly affects yield and fruit size, and close spacing is used to increase yields and produce small-size fruits for canning. For the fresh fruit market, a somewhat larger fruit is often preferred. A spacing of 100×60 cm (39 000 plants/ha) is used for fresh market production, and 100×45 cm (43 800 plants/ha) is used for canning. A double row spacing is more convenient as it allows easy access between the rows. The double row beds are normally 1.5 m apart from centre to centre, and the two rows within a bed are 0.5 m apart with the plants 28 cm aparts within the row. This gives a population of 47 000 plants/ha (Figure 11.17).

Fertiliser application

Pineapples need a lot of nitrogen and potassium, some calcium and magnesium and much less phosphorus. The following formula is used for fresh fruit production: $4:2:11:2$ g/plant of $N:P_2O_5:K_2O:MgO$, which is then extrapolated for various planting densities. For processing pineapples, the quantity of fertiliser is halved. Fertilisers can be applied in a solid form to the soil or in a liquid form to the lower leaf axils, although the latter generally results in better yields. It must be divided into small monthly applications for nitrogen and fewer applications for potassium. Nitrogenous fertilisers must be stopped about two months before flower induction.

Weed control

The techniques of mulching and pre-planting cultivation greatly reduce weed growth. Mulching with black polythene has been found to be more effective for weed control than the use of organic mulches. Mulching is used in combination with periodic hand eradication of weeds.

Pests and diseases

The major and most widespread problem in pineapple is infection by nematodes. These are normally controlled by soil fumigation which may be carried out during ploughing, some weeks before planting. Crop rotation also reduces nematodes.

Mealy bug wilt, which is caused by a virus and spread by mealy bugs, is a serious pineapple disease. The control of the disease is by periodic spraying to eradicate the mealy bugs. Heart and root rots, caused by *Phytophthora cinnamoni* and *P. parasitica*, also affect pineapples, particularly in wet weather, at cooler sites and on poorly drained soils. Improved drainage and the use of raised beds are suggested as means of control.

Harvesting and storage

The pineapple is a fruit which does not continue to ripen after harvest so that it has to be ripe or nearly ripe when harvested. The fruit is usually harvested when the surface colour is between 'colour break' and 'quarter yellow' for transportation. In the tropics and subtropics where pineapples are consumed as fresh fruits, they are harvested when fully ripe in order to get the full flavour which lasts four to five days. For export, half ripe fruits may be held at $7-10°C$ for $10-20$ days during shipment. At room temperature pineapples may be stored for up to seven days.

Citrus

The genus *Citrus* is one of about 150 genera belonging to the family Rutaceae. The family is characterised by leaves which usually possess transparent oil glands and flowers which contain an annular disc. Within the family, seven subfamilies are recognised but only one is of major agricultural interest. The 'true' citrus group consists of six related genera: *Ermocitrus*, *Poncirus* and *Clymenia* (all with one species each), *Fortunella* (with four species), *Microcitrus* (with six species) and *Citrus* (with 16 species). All six genera may be grafted on to each other and can be crossed to produce hybrids.

Although citrus fruits originated in tropical Southeast Asia, they are now cultivated most extensively in the subtropics, in areas with a Mediterranean climate. Those citrus fruits with a sweet taste are eaten mostly fresh or the segments may be canned. The juice may be extracted and used in squashes and cordials or for flavouring. The waste pulp is used for cattle feed. The peel is a source of essential oils used in flavouring and perfumery and also as a source of pectin. Essential oils may be obtained from the flowers and leaves. Citric acid may also be obtained from various *Citrus* species.

Types and ecological adaptation

The eight species of *Citrus* which are cultivated are shown in Table 11.6.

Two other species, *Poncirus trifoliata* (trifoliate orange) and *Fortunella margarita* (kumquat), are not widely cultivated.

In addition, there are intrageneric hybrids (within the genera), for instance:

Tangor   mandarin × sweet orange
Tangelo   mandarin × grapefruit
Lemonime   lemon × lime

There are also intergeneric hybrids (between genera) such as:

Citrange   *Poncirus* × sweet orange
Citrumelo   *Poncirus* × grapefruit
Limequat   Lime × kumquat

Indeed, there are trigeneric hybrids (between three genera) as for example Citrangequat.

Different *Citrus* species require different types of climate. In general, however, they do not stand severe frost and are, therefore, not widely grown in cold regions. Vegetative growth in most *Citrus* species virtually stops at temperatures below 10°C.

In the humid tropics, the orange does not produce good quality fruit. Some varieties of orange grown in the humid tropics deteriorate rapidly with increasing humidity. This is probably due to both too much leaf growth and to a low incidence of sunshine which causes fruits to develop too fast, making them acid and watery. In humid climates, flowering and fruiting of oranges is also poor, since most varieties need a cool or dry spell to flower. In

Figure 11.18 A young orange tree in fruit in Ethiopia.

equatorial regions, oranges generally increase in quality (sweetness and colour) from wetter to drier zones or from sea level to higher elevations when sunshine is not reduced. However, there are varietal differences. For example, Valencia oranges can be grown successfully at sea level while Navel oranges require higher elevations. Mandarins are similar to oranges in their climatic requirements but generally are slightly better adapted to humid climates. Pummelos, grapefruits and limes do well in humid climates so these species of citrus should be grown in the wetter regions of the tropics (Figure 11.18).

For good growth, citrus trees require light, well-

Table 11.6. Cultivated species of *Citrus*

Botanical name	Common name
C. sinensis	Sweet orange
C. aurantium	Sour (or bitter) orange
C. reticulata	Mandarin or Tangerine
C. paradisi	Grapefruit
C. grandis	Shaddock or Pummelo
C. limon	Lemon
C. medica	Citron
C. aurantifolia	Lime

drained, deep soils. On such soils, the trees will grow strongly and will be able to stand periods of moisture stress. Varieties and rootstocks susceptible to foot rot require a well-drained soil and the absence of flooding. Some varieties can stand heavier soils than others, however, and generally, limes, pummelos and mandarins can be grown on heavier soils than oranges. The rough lemon rootstock is better for shallow soils because it produces an abundance of surface roots. Most citrus species do not stand very acid soils, although very alkaline and saline soils are also undesirable.

There are very many citrus cultivars. The choice of variety depends on many factors. A list of popular varieties is given in Table 11.7.

Planting

An area with good shelter from winds is normally selected for the citrus nursery. In areas with heavy rainfall, gently sloping ground is desirable for free drainage. Waterlogging must be avoided.

For the preparation of a good seedbed, deep cultivation and fertiliser applications are necessary. The seedbed should be about 1.2×1.8 m wide and as long as practicable. The beds are raised to about 15 cm above surrounding ground level. Fertilisers are applied as a basal dressing at the following rates: single superphosphate at 68 g/m^2 and potash and nitrochalk at 34 g/m^2 each. The use of organic matter in the ground nursery must be avoided as this may introduce disease organisms.

The seeds of the desired rootstock (usually rough

Table 11.7. Popular citrus varieties

Cultivated species	Varieties
Sweet orange	Valencia, Shamouti, Ruby, Washington Navel, Mediterranean Sweet, Parson Brown, Agege 1, Umudike, Etinan, Bende and Meran
Mandarin	Dancy, Clementine, Ponkan and King
Grapefruit	Marsh Seedless, Duncan, Red Blush, McCarty and Thompson
Lemon	Eureka, Lisbon and Villa Franca
Lime	Mexican, Tahiti, Key, Rangpur Lime and Persian.

lemon, but also Rangpur lime, sour orange and grapefruit can be used) are sown in rows about 20 cm apart, with the seeds 10 cm apart within the rows. The seedbed is then mulched and watered heavily. Four months later, the seedlings are ready to be transplanted onto the budding bed, where they are planted 45 cm apart in the row and also mulched heavily and watered. After eight weeks, 14 g each of calcium ammonium nitrate and muriate of potash are applied per seedling. After seven to 12 months the seedlings are ready to be budded.

Budding may be done at the beginning of the rains or at the end, but not at the height of the rainy season. This is because diseases are prevalent during the rains due to the favourably humid conditions that encourage the growth of disease organisms. The budding may be carried out any time during the dry season, provided irrigation facilities are available.

Budwood is selected from trees that are known to be disease free, high-yielding and have other desirable characteristics. The varieties should be known to be about ten years old, at which stage a tree would have shown its bad and good characteristics. The budwood should be approximately the same diameter as the root-stock and taken from mature wood of the current season's growth, but not the succulent terminal portion. The two should, as far as possible, be round and not angular. Budding should be done at a height of 35–50 cm above the ground. This is to prevent fungal disease attacks as the humid microclimate near the ground surface favours the growth of fungi. In dry areas, budding at a height of 30 cm is satisfactory.

The budded seedlings should be kept well watered and in good growing condition. About 14 days after budding the wrapping material should be removed to see if the bud has taken (in which case it will be green). If the bud is dry then it has not taken and another budding should be done on the seedling. When the green bud piece has sprouted, the stock should be cut back to it, with the cut sloping away from the growing bud. The cut surface may be coated with a suitable waterproof material. Suckers and stock growth below the bud union, and unnecessary branches on the seedling, should be removed from time to time until the seedling is ready for transplanting into the field.

Budded citrus plants have several advantages. They remain true to type and bear early. The plants have few small thorns and a spreading habit so that fruit picking is carried out at low cost. On suitable root-stocks, budded plants are resistant to many root diseases. The major disadvantage is that budded plants may be infected with virus diseases.

The two other ways of propagating citrus are from seed and by marcottage or cuttings (Figure 5.15). The seeds of most citrus varieties have been cross-pollinated and therefore produce variable trees. Marcottage and cuttings are recommended for limes and mandarins. Marcotting is done by tying earth around a suitable branch shoot. Sometimes the bark is also ringed. A strip of polythene is used to tie soil to the branch. These operations encourage roots to grow from the branch into the ball of soil and after a variable period, the branch together with the ball of soil and roots is cut off and planted.

Producing citrus by cuttings is faster than using marcottage. The shoots cut off should be firm, not too green and soft, but not too old and hard. They

may be pruned of leaves, treated with rooting hormone and planted in polythene bags which are about 20 × 30 cm in size, laid flat and filled with well-drained soil. The cuttings should be given partial overhead shade (about 50%) and watered regularly until they have rooted (Figure 11.19).

A rich, deep, loamy soil is ideal for citrus. The crop will, however, survive and do well in soils too shallow and poorly aerated for other crops such as the avocado pear. On the other hand, citrus trees may not thrive in soils that are shallow and wet where mango trees would grow well. The pH should be in the range of 5.5−6.5.

The best time to plant is at the beginning of the rainy season. The nursery soil is well moistened prior to digging out of the young plants to ensure that as many roots as possible remain intact with the plant. In citrus cultivation spacing has traditionally been relatively wide, with spaces between the trees largely for the entry of tractors for field operations and harvesting. An indication of

Figure 11.19 Citrus seedlings in a nursury, growing in polythene bags.

244

Table 11.8. Planting distances used in citrus plantations

Cultivated species	Spacing (m)	Trees/ha
Orange	6.7 × 6.7	222
	7.2 × 7.2	192
	10.0 × 10.0	100
Grapefruit	9.0 × 9.0	120
	12.0 × 12.0	70
Mandarin	6.0 × 6.0	270
	9.0 × 9.0	120
Shaddock	10.5 × 10.5	90
	12.0 × 12.0	70
Lemon	6.7 × 6.7	222
Lime	4.5 × 4.5	496
	6.7 × 6.7	222
Citron	6.0 × 6.0	270

minimum and maximum planting distances used in various *Citrus* species, based primarily on tree size, is given in Table 11.8.

Fertiliser application

Various citrus species require the application of inorganic fertilisers. It is common practice to apply nitrogen, phosphorus and potassium, and magnesium where it has been shown to be deficient. Where trace elements have been shown by foliar analysis to be deficient, they should be supplied by spraying.

Nitrogen is the most important fertiliser during the developmental and bearing stages of the citrus tree. Excessive potassium should be avoided, and in newly cleared and burnt land, potassium fertilisers may be omitted for the first few years. Recommended rates of fertilisers per hectare, are summarised in Table 11.9.

It is possible that yields may be increased by rates up to twice these stated values. In soils below pH 5.0, magnesium limestone at rates from 500 to 1000 kg/ha may be used to replace the magnesium sulphate whilst rock phosphate could replace the single superphosphate.

Copper and zinc deficiencies are common and should be treated by spraying with 5 kg copper sulphate and 5 kg zinc sulphate in 10 kg lime/ha. Iron and manganese deficiencies may also occur in soils with a high pH and may be treated by spraying at rates of 10 kg/ha each of iron or manganese sulphate in lime. Fertilisers should be applied over the six months preceding fruit maturity, or in six equally spaced applications where fruiting is non-seasonal.

Weed control

A number of methods are used to prevent or at least check competition from weeds. These include hand cultivation, disc harrowing, slashing, mulching, the use of herbicides and cover cropping. Clean weeding in a circle around the plants and brushing low in between must be done at least three times a year.

Traditionally cover crops have been grown between citrus trees for all or part of the year, in the latter case the soil being worked by tillage at appropriate times. However, in more recent times the control of inter-tree soil cover and weed growth has been effected largely by minimal cultivation techniques involving herbicides. Under erosion-prone conditions complete weed control may not be desirable, and a system that gives periodic control using contact herbicides is practised. A wide range of herbicides have been found safe for citrus. For example, diuron at 2.5 kg/ha gives good weed control if applied every three to six months on normal soils. Paraquat or dalapon may be used to kill grasses. The use of 2,4-D at about 1 litre/ha controls creeping weeds between the rows, but must not be sprayed on the foliage of the trees.

Table 11.9. Feriliser rates for citrus crops (kg/ha)

Number of years after planting	Urea	Ammonium sulphate (kg/ha)	Potassium chloride (kg/ha)	Magnesium sulphate (kg/ha)	Single superphosphate (kg/ha)
1	100	50	nil	50	50
3	100	50	25	50	50
5	150	100	50	100	100

Pests and diseases

A number of insects attack citrus, including scale insects, aphids, false codling moth, the Sudan millet bug and various beetles. Perhaps the most important of these is the soft brown scale insect *Coccus hesperdium*. The insects occur on the fruit, leaves and branches or trunks and may cause the die-back of twigs and the premature fall of leaves and fruit. Control of scale insects is achieved by spraying with 0.5% dieldrin, followed every three months by trunk banding with the same insecticide. Symptoms of an attack by the citrus mealybug *Pseudococcus citri*, as well as the method for its control, are similar to those for the scale insect. HCH (BHC), at the rate of 0.3 litres of 20% HCH (BHC) per 455 litres of water also control the mealy-bugs. Aphids of the genus *Toxoptera* are frequently found on the foliage during dry spells, but they do not usually survive in wet weather. The false codling moth, *Argyroploce curvipes*, causes serious losses in citrus plantations by piercing the fruits, thereby causing them to rot and fall from the tree. It is most common during the first half of the rainy season. Control is by proper field sanitation and prompt harvesting. The Sudan millet bug, *Agnoscelis versicolor*, may be controlled by hand-picking and by spraying with HCH (BHC).

The major disease of citrus is gummosis caused by the fungus *Phytophthora*. It attacks the tree at ground level, resulting in death of the cortex and exudation of gum (hence the term 'gummosis'). The leaves turn yellow and the plant eventually dies. Control of gummosis is by the use of resistant root-stocks, such as sour orange and trifoliate orange. Another important disease recently described in certain parts of Nigeria is the citrus brown spot caused by *Phaeosariopsis* sp. This fungal disease appears in lesions on leaves and fruits during the wet season. Excellent control of foliar and fruit stages of the disease has been obtained with fungicidal mixtures that include benomyl (Benlate) applied at 7 or 14 day intervals. Citrus scab, caused by the fungus *Elsinoe fawcettii*, is found more often on young seedlings than on old trees. Sour orange and rough lemon trees are most susceptible, with grapefruit rather less susceptible to the disease. In an area where the disease is known to occur, it is best to remove any rough lemon or sour orange seedlings from the vicinity. The disease is difficult to eradicate, but may be avoided by protective sprays of copper-based fungicides such as Bordeaux mixture or Perenox, starting before the first flush of growth at the beginning of the rains.

More than 20 virus and virus-like diseases of citrus have been described. Virtually all these have been spread throughout the tropical world by infected budwood. Some of the virus diseases, particularly tristeza and greening which are the most serious diseases of citrus, are also reported to be spread by insects. Exocortis is spread by budding knives, pruning shears and other tools as well as by the hands of workers. Tristeza or 'quick decline' disease is transmitted by several aphids which spread the disease. The most important aphid which transmits tristeza is the black citrus aphid, *Toxoptera citricidus*. Control is by the use of tolerant root-stocks and control of the aphids. Tolerant root-stocks include Cleopatra, Sampson or Lake Tangelo, trifoliate orange and rough lemon.

Among non-insect transmitted virus diseases of citrus are psorosis, xyloporosis (cachexia) and exocortis. It is often not easy to determine the presence of absence of xyloporosis because of its similarity with tristeza. Psorosis seems to occur in orange, lemon, tangelo and grapefruit scions, regardless of the stock. The disease causes small eruptions in the bark which subsequently splits, forms a gummy substance and dies out. Leaf mottling and concave depressions in the leaves may also occur.

In general, control of non-insect transmitted virus diseases may be achieved by using budwood which is free of virus as well as the possible use of nucellar seedlings. It should be emphasised that orchard sanitation is a good insurance against most pests and diseases.

Harvesting and storage

Most citrus varieties start bearing fruit after about two years and reach full bearing about ten years after planting in the field. At this stage, the average yield of fruit is about 13 t/ha; however, well managed fields are capable of yielding up to 26 t/ha.

The fruits of citrus are invariably harvested manually, therefore substantially increasing the production costs. Research efforts are under way to develop satisfactory abcission-inducing techniques to facilitate mechanical harvesting, as for example, by shaking the tree. Because of the importance of fruit colour particularly in the tropics, ethylene is widely used to induce colour development.

To prolong the storage life of citrus fruit beyond the normal shelf life of a few weeks, various techniques have been effectively employed, including the use of low temperatures, controlled atmosphere (CO_2 biphenyl), waxing, fungicides and treatment with growth regulators. Processed juices may be stored for prolonged periods in cans, bottles or in the frozen form.

Guava

The guava, *Psidium guajava*, is a native of Central America, but it is now grown in all parts of the tropics. It is a member of the family Myrtaceae which contains over 3000 species, 150 of which belong to the genus *Psidium*. The species *Psidium guajava* is by far the most important commercially. The guava tree or shrub grows to a height of 6–9 metres. The fruit is a pear-shaped berry in which numerous seeds are embedded in the flesh.

The ripe juicy sweet guavas are eaten fresh or the fleshy mesocarp is stewed and made into pies.

The fruits, after the removal of the seeds, are made into preserves, jam, jelly, paste, juice and nectar. The greatest commercial use is for the production of guava jelly, the best jelly, being made from the common sour wild guava.

Types and ecological adaptation

Guava trees grow readily in most soil types and tolerate both wet and dry conditions. However, they grow best in rich, well-drained soils and in areas where the rainfall is average and the altitude is medium to high.

There is much variation in the shape, size, colour of flesh and flavour of the fruits produced by the different guava trees. Where propagation is by seed, there is considerable genetic variation from tree to tree. The best guavas have a mild and pleasant flavour and smell. Others produce highly acid and acrid fruits. There are several types cultivated in the tropics, but in West Africa only a few are of importance. These are the purple or Chinese guava (*P. cattleyanum*) which has large, high quality fruits, and the yellow-skinned guava with pink flesh and a sharp flavour. Both purple

Figure 11.20 A guava tree being sprayed to give protection against fruit flies, in Ethiopia.

and yellow-skinned varieties are excellent for eating raw.

Planting

Guavas may be grown from seed or by cuttings, but generally cutivation is from seed. Seeds retain their viability for some time and are easily spread by birds. Seedlings are raised in nursery beds and are ready for transplanting when about 30 cm high.

Guava seedlings are transplanted into their permanent field positions early in the rainy season. The suggested spacing is 4.5–7.5 m, depending on soil, the climate and the cultivar. Often transplanted seedlings tend to die back, but unless the weather is very dry they soon recover.

Fertiliser application

Very little information is available relating to fertiliser use in guava cultivation. When grown in rich soils especially under optimum conditions, guavas are hardly ever fertilised. Where there are poor soil conditions it is advisable to make applications of 30 kg N, 36 kg P_2O_5 and 36 kg K_2O/ha.

Pests

The main pests of guava are sucking insects such as *Homoeocerus pallens* and fruit flies, *Ceratitis anonae*. These attack only the ripe fruits, so their spread can be controlled by harvesting the crop before full maturity, by good crop sanitation, or by spraying with an insecticide (Figure 11.20). The shoots are also attacked by capsid bugs, *Helopeltis bergrothi*.

Harvesting

Guava trees usually begin to bear fruit after 2–3 years. The fruits ripen over a period of time and should be picked just before they attain full maturity, especially if they are to be transported to other locations. Yields vary according to variety and level of management and range from 10–25 tonnes/ha.

Mango

The mango, *Mangifera indica*, belongs to the family Anacardiaceae. This family is characterised by

Figure 11.21 Mango trees are grown widely throughout the tropics and subtropics, mostly for local consumption.

plants with a resinous sap and alternate leaves. Cashews, *Anacardium occidentale*, and pistachios, *Pistacia vera*, are other economically important members of the family. Mangoes originated in the Indo-Burma region, but their production is now widespread throughout the tropics.

The ripe mango fruits are eaten raw as a dessert fruit and are used in the manufacture of juice, squash, jams, jellies and preserves. The juice is also canned. Unripe fruits are used in pickles, chutneys and culinary preparations. Leaves are fed to livestock, but prolonged feeding may result in death (Figure 11.21).

Types and ecological adaptation

Climate and other environmental factors affect the development of mango trees and the quality of their fruits. The best regions for commercial mango production are those with a humid rainy season alternating with a well defined dry season during the flowering period of the plant. The trees readily grow in all tropical and subtropical climates with a mean temperature of 21−27°C. They perform best in areas commonly known as the 'dry tropics', where they can withstand temperatures as high as 48°C. An annual rainfall as low as 200−400 mm is adequate for mango growth. Where the dry season coincides with flowering, a good crop of high quality, tasty fruits is usually obtained.

Mangoes are adapted to a wider range of soils than most tropical fruit trees, provided the soil is sufficiently deep and well-drained. Rich soils should be avoided as they cause the trees to grow profusely without setting a crop. A pH range of 5.5−7.0 is preferred. Although mango trees are found at elevations of up to 1200 m, it is not advisable to plant orchards higher than 600 m. They require a lot of light, but daylength does not have a marked effect.

There are a great number of mango varieties from India and other Asian countries which have been successfully introduced to different tropical environments, including Africa. The common popular varieties cultivated in various parts of the tropics are the following:

Alphonso. In this variety the fruit is medium to large in size. The peel surface is smooth and the pulp is fibreless and sweet. The base is round with a prominent ventral edge and round dorsal edge. The apex has a ventral edge which is broadly beaked.

Tavmour. The fruit is medium to large in size and the peel surface is smooth. The pulp is fibreless and sweet. The base of the fruit is symmetrically round with the dorsal edge slightly depressed. The apex is slightly beaked with a round dorsal edge.

Peter. The fruit is generally large in size and the peel surface is smooth. The pulp is fibreless and sweet. The base of the fruit is concave with a round ventral edge and dorsal edge slightly depressed. The apex is markedly beaked with a round dorsal edge.

Dabsha. The fruit is large in size and the peel surface is rugose. The pulp is fibreless and sweet. The fruit is slightly concave with round symmetrical edges and a beaked apex.

Mabrouka. The fruit is small to medium in size and the peel surface is very smooth. The pulp is fibreless and sweet. The base of the fruit is concave with a prominent vental edge and a round dorsal edge. The apex is markedly beaked.

Peach. The fruit is large in size and the peel surface is smooth. The pulp is fibreless and sweet. The base of the fruit is symmetrically round. The apex is beaked with a round dorsal edge.

Julie. This variety has a medium sized fruit. The peel surface is rugose and the pulp is fibreless and sweet. The base of the fruit is slightly prolonged with a round ventral edge and a dorsal depressed edge. The fruit apex is slightly beaked with the dorsal edge markedly prominent.

Hindi Sinar/Dasari. The fruit is small to medium sized and oblong-oblique in shape. The base is rounded to obliquely rounded and the beak is absent. The apex is rounded. The pulp is fibreless, thick and sweet and the juice is scanty to moderate in quantity.

Planting

Many mango trees are the result of seedlings which have grown from discarded seeds. Mangoes produce two kinds of seeds: polyembryonic and mono-embryonic seeds. In polyembryony, each seed produces a sexual seedling plus 1−5 nucellar seedlings which are genetically identical to the female plant and therefore grow into true clones. Mono-embryonic seeds give rise to only one sexually produced seedling per seed and in this way do not replicate the parent plant. Although a high degree of uniformity can be achieved from nucellar seedlings, uniformity can only be achieved in monoembryonic varieties by vegetative propagation. All the exotic varieties, with the exception of Taymour, are monoembryonic and therefore need to be vegetatively propagated.

Seedling stocks are grown from local seed and the improved varieties are budded onto them. Different types of budding have been used with success. In West Africa seeds are sown in nursery beds in April or May and spaced 45 × 45 cm apart. The seeds must be from ripe fruits. Two or three weeks after germination takes place, the seedlings are transplanted. Local mangoes are polyembryonic and the seedlings should be selected to eliminate the sexual ones among each set of nucellar seedlings. The sexual seedling is usually the smallest and the one that develops directly from the micropylar end of the seed.

When the seedlings have stems about 1.5 cm in diameter (about 1 year from planting), they are suitable for budding. The time of budding is critical as the operation should coincide with a new flush of growth in the mango trees. At this time the bark separates easily from the wood. A new flush of growth is easily recognised as the young leaves are 'wine coloured' (red) at this time. The trees usually have new flushes of growth several times during the rainy season, starting in April or May in West Africa.

Budwood should be selected from towards the end of strong healthy young branches, but not from the last growth. Budwood should be cured for about ten days before budding. This is achieved in situ by removing the leaves or cutting off half of each leaf on the branches selected for budwood. This practice serves to reduce nutrient and water requirements and lessens the shock after budding.

Patch budding is the most commonly used method in bud grafting of mangoes, although the 'T', the inverted 'T' or the shield budding methods may also be used. Patch budding consists of completely removing a rectangular patch of bark from the stock and replacing it with a patch of bark of the same size containing a bud of the desired variety to be propagated. A rectangular incision measuring about 4 × 2.5 cm or more, depending on the size of the bud, is made in the stock at a place where the bark shows brown colouration, indicating maturity. The same patch is removed from the scion, with a dormant bud in the centre.

With the handle of the budding knife or blade, the flap on the stock is removed. The patch of the scion bud is then inserted into the stock and tied with a suitable binding tape. Alternatively, only three of the sides of the rectangular patch on the stock are cut so that the flap of the bark lifts upwards. In this case, the bud patch inserted into the stock and the flap of stock bark are used to cover the inserted bud, and are tied securely.

The buds are examined three to four weeks after budding and if they have taken, the flaps of the stock covering the bud or the binding tapes are cut off. Whenever the bud or graft has sprouted, the top of the stock should be cut off several centimetres above the bud. When this is done the sprout should have grown to a height of 25 cm. The plants should be watered regularly and suckers and undesirable branches should be removed. The beds should be heavily mulched during the dry season.

The site selected for growing mangoes should be deep and well-drained. Although the crop is tolerant of many soil types, a deep alluvial or loamy soil is preferred. Transplanting into the field is done one year after budding. It should be carried out early in the growing season after the rains have become established. Holes measuring 1 × 1 × 1 m should be dug well ahead of the time of transplanting and filled with well rotted farmyard manure or compost, watered and allowed to settle and further decompose. The transplants are then removed from the nursery, each with a big ball of earth on its roots and placed in the middle of the holes. The base of each plant is then pressed in firmly. Some of the leaves should be removed to reduce transpiration and transplanting shock.

The spacing adopted for mango trees varies from 5 × 5 m to 12 × 12 m, depending on the variety. Grafted trees of short varieties like Julie may be spaced between 5 × 5 m to 7.5 × 7.5 m while the large varieties may take spacings between 10 × 10 m and 12 × 12 m.

Fertiliser application

Generally an application of 450 g to 2.2 kg of ammonium sulphate per plant per year may be made, according to tree size. This should be split into two or three applications and the fertiliser should be spread on the surface. It may also be worked into the soil, but due care must be taken to avoid damaging the root system. The fertiliser should be spread in a circular area extending from about 60 cm from the trunk to 60 cm beyond the canopy of the tree. For the first five years the following mixture may be used per tree per year, in addition to the nitrogen: 230 g single superphosphate, 230 g potassium sulphate and 13.5 kg famyard manure.

Weed control

When the trees are young, weeds should be removed as regularly as possible. This is done by under-brushing between the trees and clean-weeding around the individual plants by careful cultivation. As the trees become established, weeds tend not to be a problem because of the shade cast by the trees. Little is known about the use of herbicides for weed control in mango plantations, especially in tropical Africa. However, studies elsewhere have shown that terbuthylazine + paraquat (4.0 + 1.0 kg/ha) applied post-emergence has been found to be effective in controlling many of the weed species in young mango plantations.

Pests and diseases

The mango weevil, *Sternocochetus mangiferae*, can cause considerable damage to the mango crop. The larvae of the weevil feed on mango seeds and usually enter small, green, immature fruits without marking the skin. They cause premature fruit fall and rotting of mature fruits ready for market. Methods of control include good crop sanitation and prompt removal of all fruits that fall to the ground.

Attack by mango hoppers or jassids can also be serious. The nymphs suck the juice from the inflorescence and thus drain away the vital plant sap. Spraying with 0.1% dieldrin gives effective control of the hoppers. The best time to spray is when the flower buds have just opened.

Also of importance on mango trees are scale insects. About 62 different species of scale insects occur on the crop. The larvae of several species of fruit fly may also render the fruit inedible. The use of oil sprays with 0.1% oil emulsion gives effective control of scale insects, while fruit fly attack may be reduced by crop sanitation and prompt harvesting.

Anthracnose, caused by the fungus *Colletotrichum gloeosporioides* is the most widespread disease of mangoes. It is favoured by high relative humidity and rain during flowering. Under such favourable conditions blossom-blight, which is one of the various phases of anthracnose, may entirely destroy the flowers, thus leading to a total loss of harvest. Brown spots appear on leaves which are attacked and such leaves crinkle and die leaving the twig bare. Other diseases of mango include powdery mildew, caused by *Erysiphe cichoracearum* and scab, caused by *Elsinoe mangiferae*. Benomyl at the rate of 0.3 g/l, or Dithane M-45 give effective control of anthracnose and powdery mildew.

Harvesting and storage

The mango is an irregular fruit bearer in the sense that a period of fruitfulness is often followed by a period of unfruitfulness. This irregularity in bearing is known as periodicity in cropping. The year of a heavy or medium crop is taken to be the 'on' year, whereas the year in which the tree produces a poor crop or fails to bear any fruit is called the 'off' year. The 'on' and 'off' years alternate, and this phenomenon is known as alternate bearing or biennial bearing. In addition, the mango is characterised by a natural heavy drop of fruits at all stages. Severe losses through other means occur. Less than one out of every one thousand perfect flowers develops into a mature fruit. Rain, high humidity, attack by pests and diseases, a low C/N ratio, mineral deficiencies, a low percentage of perfect flowers and hormonal imbalances have all been blamed for alternate bearing. As a result of one or more of these causes, not enough vegetative growth takes place in the 'on' year to support flowering in the 'off' year.

The effect of alternate bearing can be lessened by stripping flowers or thinning fruitlets in an 'on' year. This is achieved either by hand or by means of chemical growth regulators such as naphthyl-acetic acid (NAA). Applications of fertilisers may also help, particularly if they contain Mg, Zn and Mn. Otherwise, the tree may be partly girdled. To achieve this, each year a different branch is ringed and will fruit heavily. In this way, the alternation is shifted from within the orchard to within the tree. Efforts should also be made to genetically select regular bearers and discard alternate bearers.

Mango trees begin to bear fruit from the fifth year after planting and give economic returns from about the eighth year. By the time the trees are about 20 years old, each tree should be producing 600–800 fruits per year. Good varieties like Taymour produce as many as 1000 fruits a year.

The fruits are generally picked when they begin to change colour or after a few ripe fruits have dropped from the tree. For the local market, it is advisable to wait until the fruit is slightly soft. However, if the fruits are to be transported by train or lorry to distant markets, they should be picked when still green and firm.

Picking is done by hand. The picker climbs up the tree with a bag and knife, or a special mango picker consisting of a bamboo pole with an attached knife and a cloth bag held open by a ring. Fruits are sized in three classes: small (200–270 g), medium (270–320 g) and large (over 320 g).

Before marketing, unripe fruits may be ripened for six days at 16–21°C. Fruits can, however, be stored or transported at a temperature of 9°C and a relative humidity of 85–90%.

Cashew

The cashew, *Anacardium occidentale*, belongs to the family Anacardiaceae, which also includes the mango, *Mangifera indica*. The cashew is cultivated for its seeds which are the cashew nuts of commerce. The origin of the cashew is tropical America, notably Peru and Brazil and also the West Indies. The crop is now widely cultivated in Southeast Asia, Africa and the West Indies. In Africa, cashews are widely grown in both East and West Africa, and are of particular importance in Tanzania, Nigeria and Mozambique. Annual world output of cashew nuts is about 0.2 million tonnes, a high proportion of this coming from wild and cultivated trees in tropical America, India, East Africa and the Mediterranean area.

In addition to the nut from the cashew the pericarp of its achenes yield cashew shell oil which is extracted with solvents or steam for industrial use. The seeds contain around 50% by weight of an edible non-drying oil, and the pedicel and receptacle of the flower grow to produce a large fleshy 'cashew apple' or 'pear' which is eaten as a fresh fruit, or the juice may be extracted and fermented into

Figure 11.22 The large cashew apple is subtended by the smaller cashew nut.

wine. The sap of the tree bark provides an indelible ink (Figure 11.22).

Types and ecological adaptation

The tree is hardy and drought resistant, but grows best on well-drained soils and in areas with an annual rainfall of at least 900 mm. It can survive with less rain, but the yield decreases unless it is irrigated. More than 900 mm of rainfall is tolerated if drainage is good and if there is a sufficiently long dry period to allow proper flowering and fruit set. Rain during the flowering period causes a high incidence of fungal infection.

Cashew produces well on soils which are too poor and dry for other crops. The drought tolerance of cashew is explained by the extensive horizontal growth of its roots. It prefers a pH of near neutral.

Cashew grows well between sea level and 750 m. It grows poorly at higher altitudes where the temperatures are low. The preferred temperature range is 15−32°C.

No major breeding work has been done on cashew. There is great variation in performance between seedling trees owing to a high proportion of cross-pollination. There is therefore great scope for the selection of high-yielding clones and the adoption of vegetative propagation techniques. For the present, if high yields of superior qualities are desired, the best clone should be selected and propagated vegetatively, as is being done in many countries where cashew production is carried out on a commercial scale.

Planting

The seeds are either planted directly into the field or raised as seedlings in nurseries before being transplanted to a permanent site. In the latter, seedbeds are made and seeds planted at 0.3 × 0.3 m apart at the beginning of the rains. Mulch is then applied and the beds are watered regularly. The seedlings are raised for one year before transplanting out in the field.

When planted at stake, the seeds should be spaced about 3 cm apart in a triangle in each hole. The holes should be about 0.3 m wide and not deeper than 5.0 cm, this being the depth at which the seeds should be placed. Under favourable conditions, germination occurs in 2−4 weeks. Some form of protection for the young seedlings is necessary and can be achieved by shading each plant with two or three coconut or palm leaves. About a year after planting the most vigorous of the three seedlings is selected and the others thinned out.

Cashews will grow in soil that is unsuitable for other crops, including sandy or hilly land and poor lateritic soils. However, they should not be grown in heavy clay and poorly-drained silty soils.

If they are to be planted from seed, then sowing may be done when the rains are fully established. Nursery-raised seedling trees may be planted out as near the beginning of the rains as possible, thus ensuring that they are not transplanted into dry soil.

It is recommended that the seeds for sowing be selected from high-yielding trees. The nuts should be well dried and subjected to a floatation test by immersing them in a solution of 150 g sugar or 100 g salt in one litre of water. Those seeds which float should be rejected as they have been shown to give poor germination rates or even fail to germinate.

Cashews planted at stake at 6 × 6 m are later thinned to 12 × 12 m. Transplanted seedlings are also put out on the permanent site at 12 × 12 m. Before the young trees begin to yield at the age of two to three years, annual food crops can be grown between the rows.

No fertiliser or manure application is routinely practised, partly because good yields are obtained even on the poorest of soils.

Weed control

Due to the close canopy of cashew leaves, weed control is of minor importance. However, vegetation between the rows should be checked by regular slashing. Clean weeding under the trees should be done for weed control purposes and also for easy identification and collection of fallen fruits.

Pests and diseases

Heliopeltis anacardii and *H. schoutedeni* are two species of fruit worm which can cause considerable damage to cashews. They attack the leaves, young shoots and inflorescences causing long black lesions and die back. The lateral buds can make new growth giving the plant a bushy appearance. The damage caused by the fruit worm extends for a considerable distance down the shoot and is seen as a browning of the tissues. Dieldrin sprays or dusting with HCH (BHC) both give good control of *Heliopeltis*.

Occasionally, a fungal complex is observed on the inflorescences and fruits during wet weather. Fungicidal sprays may help control the disease although the damage is usually minor.

Harvesting and storage

Cashew trees begin to produce fruits in their second or third year and are in economic bearing by their

253

fifth year. Average yields of dry nuts on well managed plots range between 1000 and 1200 kg/ha. When at full yield by their tenth year, as much as 2000 kg/ha of dry nuts can be obtained. Cashews also yield the edible receptacle that is often mistaken for the fruit. The plant is not normally grown for the apple-like edible receptacle and so yield figures of this product are not available.

Nuts should only be collected after they have dropped to the ground to ensure that no unripe fruits are harvested. The frequency of harvesting is largely dependent on the weather. In wet weather the collection of nuts should be a daily exercise to avoid deterioration, discolouration and rotting. In dry weather the length of time between collections can be up to seven days.

The cashew 'apples' are separated from the nuts and any fragments of apple flesh still attached to the nuts are cut away. The nuts are then spread out to dry in the sun for a few days, until their moisture content is about 7%. Processing cashews involves removing the kernels (seeds) from their shells. This operation is hampered by the presence in the shell of a toxic substance called CNSL (cashew nut shell liquid). CNSL is composed of 90% anacardic acid, which causes severe skin blistering on contact. CNSL has various medicinal and industrial uses and is extracted before shelling by roasting the cashew nuts in a bath of oil, or more frequently CNSL at 180°C. The nuts are then shelled and the kernels are roasted or packed for export. The shells yield about 50% CNSL and the kernels contain about 45% oil, 26% carbohydrates and 20% protein.

Date palm

The date palm, *Phoenix dactylifera*, is a member of the family Palmae. There are 12 species in this genus, but *P. dactylifera* is the only commercially cultivated species. The probable origin of the date palm is the Persian Gulf and Western India. It is widely grown in the hotter regions of the world, but requires low night temperatures for optimal growth and development.

Not only does date palm supply one of the main articles of food for the inhabitants of the hot dry regions of the world, but also virtually every part of the tree is used for a variety of purposes. The fruit, consumed in its fresh or dry state, contains a high proportion of sugar (60–70%) and other carbohydrates. The leaves and stems are used to construct houses. Door posts, window frames and roofing rafters are made from the trunks. The leaves and strong fibres extracted from the leaves are made into crates, mats, baskets, ropes and containers of all types. Date seeds are fed to donkeys and camels, and an intoxicating drink can be prepared from the sap which exudes from cut flowering spathes.

Types and ecological adaptation

There are hundreds of varieties of dates, many of them being indistinguishable in the vegetative state, However, they can be classified according to their fruit type as follows:

Soft dates. In these varieties the fruits contain a comparatively small proportion of sugar (about 60%) in the form of glucose and fructose. The fruits have a soft fleshy consistency, and they do not dry out easily. The majority of the dates appearing on the world market are of this type.

Semi-soft dates. These have a firmer flesh than soft dates. They do not dry out to the same extent as the dry types, but they retain a much softer consistency. The sugar content is higher than in the soft dates. The fruit is usually harvested before it is completely ripe.

Dry dates. These have a higher sugar content than the other two types, between 65 and 70%, in the form of sucrose. The fruits ripen and dry out to a hard consistency even while still on the palm.

Temperature, rainfall and humidity are the main climatic factors affecting date palm production. For successful growth and good yields, date palms require a long hot growing season and a rainless period immediately before and during the fruit ripening period. Successful culture is thus found in regions with annual average temperatures between 28–33°C. The plants can, however, tolerate temperatures as high as 50°C and can withstand fairly low temperatures, but only for short periods. Low annual rainfall (50–200 mm) characterises the important date growing regions. Indeed, during the critical period of fruit ripening, rainfall must be totally absent or of negligible amount, although the root system must always be supplied with water. High relative humidity is detrimental to the growth of leaves and fruits.

Dates are grown in a wide range of soils. The

maximum water-holding capacity, consistent with good drainage and aeration is desirable. Deep sandy loams appear to be the best types of soil. Dates grow better in soils containing more alkalies or salts than most other plants. When grown under irrigation care must be taken to ensure good drainage and aeration, especially in the early stages of growth.

Among important varieties cultivated in some Middle East countries are Mishring and Deglet Nour, both of which are of the soft type. In the dry regions of West Africa, the cultivars Fika, Dutse and Tripoli are probably the most widely grown.

Planting

Date palms are either grown from seed or propagated from off-shoots found at the base of tree trunks. Propagation by seed has the disadvantage that seedlings resulting from sexual recombination may not resemble the female parent in vegetative and fruit characteristics. Consequently, commercial planting from seedlings may lack uniformity in type. Another disadvantage is that as yet there is no method of distingushing male seedlings from females. However, if propagation by seed is necessary, seedlings may be raised by planting 4−5 seeds in holes 45 cm in diameter and 25 cm apart in a nursery. After the second year, the seedlings are thinned to two per stand. Thinning is again done to a single plant per stand when one of the plants shows its sex. Being a dioecious plant, the principle is to eventually have one male date palm to every 50 females.

It should be noted that pre-germination of seeds hastens germination which, under normal circumstances, takes four to six months. Seeds can be pre-germinated by placing them in polythene bags containing water for three days. Thereafter, the water is drained off and the bags are weighted and kept in the dark for one day. The bags are then kept open for aeration and the seeds are moistened daily with a mist sprayer. Within one week the technique enables the attainment of a germination percentage of 64−95%.

In propagating date palm vegetatively, off-shoots which develop from axillary buds on the trunk are removed from trees 3−5 years old. To promote rooting, the base of the off-shoot should be in contact with moist soil for at least a year prior to

cutting. The size of the off-shoot, when ready for cutting, should be a maximum of 0.3−0.5 m in diameter. Generally, no green leaves should be cut off from an off-shoot until it is removed from the parent palm, as the growth of an off-shoot is in proportion to its leaf area.

Since the date palm is grown in areas with very low rainfall, early planting with the rains is desirable so that the plants can make use of the rain before the dry season sets in. Where irrigation is available, planting may be done at any time. Nursery-raised seedlings or off-shoots are planted in the permanent site at a spacing of 9 × 9 m in deep holes to which rich soil and well rotted manure has been added.

Fertiliser application

Although it is desirable to apply fertilisers to maintain quality production in date palms, little information is available on fertiliser response. However, where available farmyard manure is used at the rate of 5−15 tonnes/ha. Dates are reported to respond to nitrogen and in general each palm benefits from 2−3 kg N on most soils. It is desirable to apply the nitrogenous fertiliser in two equal doses.

Weed control

One metre radius around each palm should be cleanly ring-weeded and the inter-rows should be slashed occasionally. Information is presently lacking on herbicide use in date palm cultivation.

Pests and diseases

The date palm is attacked by a number of pests, among which are the following:

Date mite, *Oligonychus pretepsis*, causes serious drying and scarring of the date surface during growth. The skin of the infected fruit becomes hard and then cracks. Sulphur dusts or sprays, or Tedion have been reported to be effective in controlling the mites.

Fruit beetles, *Carpophilus* spp., may completely destroy dates in commercial growing districts. The beetles damage ripening and curing dates on the tree, in the ground and in store by entering the fruits, usually at the calyx, and then feeding on the pulp. Spraying with malathion emulsion has been effective. Infection in the store may be controlled by fumigation.

Date bug, *Ascarcropas polmarium*, causes irregular brown areas of damaged tissues. Honeydew is also secreted by the insect. The heavy accumulation of honeydew on pinnae and fruit cause them to become discoloured. Nymphs of the insects also feed on the pinnae. Repeated heavy infestation can cause weakening and death of some palms. Spraying with malathion is effective against the bugs. In large plantations the use of aerial sprays using ultra-low volume techniques gives good control.

Fruit wasps, *Polistes* spp., cause serious damage to date fruits. Control is by enclosing the bunches in cloth bags prior to ripening.

In addition to the insects referred to above, many kinds of bird are attracted to dates and cause loss during fruit ripening, especially in localities where there are only a few palms.

Among important diseases that attack the date palm are ophalia root rot, diplodia and black scorch. Ophalia root rot is caused by two related fungi, *Ophalia pigmentata* and *O. tralucida*. Infection by these fungi results in rotting and abortion of the roots, followed by loss of vigour, stunting and eventual failure to fruit. Adequate irrigation and improved cultural treatments are the only control measures.

Diplodia disease is caused by the fungus *Diplodia phoenicum* which sometimes affects leaf stalks and off-shoots of the palm. The leaves develop reddish or yellowish-brown streaks in the midrib and the off-shoots may die. Control is by removing infected leaves and dead tissues and by spraying with copper carbonate. Pruning tools should be disinfected with formalin.

Black scorch of dates, caused by *Thielaviopsis paradoxa*, does most damage to young palm leaves which become stunted, distorted and blackened as though scorched by heat. Inflorescences and fruit stalks are also affected. No treatment for this disease is known at present.

Various fruit rots and other diseases attack date palms in various localities. Proper field sanitation reduces such attacks.

Harvesting and storage

Date palms begin to bear after six years and are in full production at ten years. They have a life expectancy of 60−70 years or more.

Since all the fruits on any one bunch do not ripen at the same time, several pickings are usually required in any one harvest season. Dry dates are often left until all the fruits are fully ripe, then the entire bunches are cut off. Picking may be done with the aid of light weight ladders used to reach tall palms, or using a picking belt and saddle or by a curved blade in the form of a sickle with a long handle.

The fruit contains varying amounts of moisture which may reach 20% or more in the soft dates. Thus only fruits that have been properly dried can be kept for long periods without refrigeration. The higher the moisture content the more perishable the fruit. In general, storage is best in a cool dry place.

Locust bean

The locust bean, *Parkia spp.*, belongs to the family Leguminoseae (Mimosaceae). Among the economically important species in the genus *Parkia* is *P. clappertoniana*, popularly referred to as the West African locust bean. *P. filicoidea* which used to be classified as a distinct species, is now considered to be synonymous with *P. clappertoniana*.

The locust bean tree grows up to 18 m in height and 4 m in girth. The tree branches extensively. Leaves are often defoliated by caterpillars and are invariably scanty when the plant is in flower. The flowers are scarlet and the flower stalk is glabrous measuring up to 30 cm in length. The pods are 15−30 cm in length and about 2.5 cm in width. A yellow mealy substance encloses the dark brown or black seeds.

The locust bean is tropical African in origin and the tree occurs extensively in the savanna ecological zone as well as in deciduous forests. It is an important crop in these areas and virtually every part of the locust bean plant is of value as food or fodder. The seed is processed into a form used for food seasoning. The yellow pulp surrounding the seed is edible in many forms. The immature fruit is good fodder for livestock. The leaves are rich in nitrogen and are used to feed livestock or as manure. The bark of the tree as well as the empty pods are used in tanning and for hardening floors. The wood is easily worked and is used for house posts, mortars, bowls, hoe-handles and also as fuel.

Types and ecological adaptation

In the West Africa region, the species *P. clappertoniana* is found in Nigeria, Benin, Togo and Ghana, where it is well adapted to climatic conditions in the dry savanna. The slightly larger-leafed *P. filicoidea* syn. *P. clappertoniana* occurs mainly in south and east tropical Africa. The species *P. biglobosa* grows largely in southern Mauritania, Senegal, Gambia, Mali, Guinea Bissau, Guinea, Sierra Leone, Liberia, the Ivory Coast and parts of Ghana.

Planting

Although locust bean trees are not presently cultivated, they are owned and protected. Plants are established from seed and the seedlings are transplanted to their permanent site. Scarification or acid treatment of seeds may be desirable to enhance germination. Apart from occasional termite damage in its early growth stages, the locust bean is relatively free of pest and disease attacks.

Processing

The fruits are produced as bunches of pods, each of which comprises about 41% pod-case, 25% seeds and 34% of a yellow mealy pulp. The empty pod cases are traditionally soaked in water for a number of days and the solution produced in this way is used to harden laterite floors and the sides of indigo pits. Examination of the aqueous extract of the pod-case shows that it is rich in tannin, which acts as a binder for the soil.

The yellow pulp surrounding the seed is sweet and edible. It is a valuable carbohydrate food, with 4% protein, and is consumed in many forms, depending on locality.

The seeds (20% semi-liquid oil and 30% protein) are boiled for about 24 hours to soften the seed-coats. Thereafter they are pounded and washed several times to remove their husks. The cooked kernels are then allowed to ferment for two to three days. The strongly smelling ferment produced by this process (referred to as 'daddawa' in Hausa and 'iru' in Yoruba), is mashed into cakes or balls and is used as a food seasoning. Comparatively rich in food value (16% carbohydrate, 29% fat, 37% protein), the ferment tends to take the place of 'maggi' in food seasoning in many localities, particularly in Nigeria. Studies are presently underway to remove the strong smell which is the major cause of the limited use of the locust bean seed cake.

Other edible fruit and vegetable crops

Other edible fruit and vegetable crops of varying importance in different parts of the tropics are listed in Table 11.10.

Table 11.10. Other edible fruit and vegetable crops grown in the tropics

Botanical name	Common name	Areas of primary importance	Usable part
Cucumis sativus	Cucumber, green gourd, yellow gourd	Indonesia, Central, East and West Africa, Philippines	Fruit, leaves, stems
Amaranthus hybridus	African spinach, bush greens, amaranth, spinach greens	West Africa, Malaysia, Indonesia, Central America	Leaves, flowers
Prunus amygdalus	Almond, bandam	Morocco, Portugal, Turkey, Afghanistan, Algeria	Nut
Averrhoa bilimbi	Bilimbi	Malaysia, West Africa, Asia	Fruit

contd.

257

Botanical name	Common name	Areas of primary importance	Usable part
Artocarpus altilis	Breadfruit	West Indies, India, Malay Archipelago	Fruit, milled juice as sealing material, leaves
Averrhoa carambola	Carambola, star fruit, caramba, kamobola	Indonesia	Fruit
Durio zibethinus	Durian, civet fruit, durian champa, Crab's eyes	Malaysia, Singapore, tropical Asia, East Africa	Fruit, seeds
Ficus carica	Fig	Southern Arabia, Mediterranean region	Fruit
Actinidia chinensis	Chinese gooseberry, Kiwi fruit.	China, Australia, Southeast Asia	Fruit
Artocarpus heterophyllus	Jackfruit, Jack nangka	Southern India, Malaysia, Burma, Brazil	Fruit, leaves
Litchi chinensis	Litchi, Ling King, Laichi, Lychee	India, Hawaii, Japan, Australia, New Zealand, Mauritius, West Indies, Brazil, USA	Fruit
Euphorbia longana	Longan	Southeast Asia	Fruit
Eriobotrya japonica	Loquat, Japanese plum	Japan, China, Australia, India	Fruit
Macadamia ternifolia	Macadamia	Australia, Hawaii	Nut (70% oil)
Garcinia mangostana	Mangosteen, Mang-khut, Mangustai	India, Java, Phillipines, Malaysia	Fruit
Olea europaea	Olive	Mediterranean area	Fruit
Passiflora spp.	Passionfruit, purple fruit	Brazil, Kenya, Australia, New Zealand, Hawaii, Indonesia, Western USA	Fruit
Grewia asiatica	Phalsa	India	Fruit
Pistacia vera	Pistachio	Western Asia, Turkey, Iran, Syria, Afghanistan, Italy	Nut
Nephelium lappaceum	Rambutan	Malaysia, Singapore	Fruit
Annona muricata	Soursop, durian europe, durian belanda, durian maki	Tropical America, Malaysia	Fruit
Chrysophyllum cainito	Star apple	West Indies, Central America	Fruit
Talinum triangulare	Water leaf, Lagos bologi, Bologi spinach	West Africa, Indonesia, Malaysia	Leaves
Psophocarpus tetragonolobus	Winged bean, Manila bean, Goa bean, Asparagus bean	Indonesia, Malaysia, Central and West Africa	Young pods, leaves, stems, ripe seeds, tuberous roots
Bertholletia excelsa	Brazil nut	South America, Brazil, Malaysia	Nut

CHAPTER 12

Beverages, stimulants and insecticides

This chapter discusses some of the major non-alcoholic beverages of the world, including coffee, cocoa, tea and kola, which are all cultivated widely in various parts of the tropics. These crops owe their popularity to their refreshing and stimulating properties, which result from the presence of variable amounts of caffeine or related substances. Caffeine generally acts on the central nervous system and, if taken in small amounts, increases mental activity, reduces fatigue, and aids digestion by stimulating the production of digestive juices. It also has a marked diuretic effect, increasing the excretion of uric acid. However, if taken in excess, caffeine has markedly harmful effects. On average, caffeine comprises about 1.5–1.7% of coffee beans and 2–3% of an infusion of tea leaves. The stimulant in cocoa is in the form of the related alkaloid theobromine, which comprises 1–2% of the cocoa bean.

The second group of crops included in this chapter are essentially drug plants which have a stimulating and slightly narcotic effect when smoked or chewed. The narcotic and stimulating properties of tobacco and similar drug plants are due to the presence of variable quantities of the alkaloid nicotine. The processed leaves of tobacco contains 1–3% nicotine.

The third group of crops includes pyrethrum. It has insecticidal properties due to the oleoresin content, which is extractable from the dried flowers.

Coffee

The genus *Coffea* belongs to the family Rubiaceae. Of the many genera in this family only two, *Coffea* and *Cinchona*, are important economically. The genus *Cinchona* produces quinine.

Out of the 50–60 species in the genus *Coffea*, only four are of economic interest: *C. arabica*, *C. robusta*

syn. *canephora*, *C. liberica* and *C. stenophylla* (Figure 12.1).

With the exception of mineral oil, coffee is the most valuable commodity in international trade and is the most important foreign currency earner for the third world. For example in 1980, coffee was responsible for the transfer of over US $13 billion from the developed countries to the developing world, compared with about US $3 billion for cocoa and only US $1 billion for tea for the same period.

The dried beans (seeds) are roasted, ground and brewed to make a stimulating and refreshing beverage (Figure 12.2). The *robusta* and *liberica* species are generally lower in quality and are often used as coffee blends and fillers respectively. Coffee pulp and parchment are used as manures and mulches and may be fed to cattle.

Types and ecological adaptation

The distinguishing characteristics of the important and economic species of *Coffea* may be summarised as follows:

Arabian coffee

Coffea arabica, commonly referred to as Arabian coffee, is the most favoured coffee for its superior quality. Arabian coffee originated in Ethiopia. There are two types: arabica and bourbon. The latter is considered the best type. It grows well at high elevations both in the tropics and subtropics. The humid lowlands of the equatorial regions are unsuitable for its growth. In Nigeria, it grows well in the Mambila highlands where it produces high-quality coffee.

Robusta coffee

Coffea robusta, which is synonymous with *Coffee canephora* and is popularly referred to as robusta

Figure 12.1 A fruiting branch of a coffee bush showing berries ready for harvesting.

coffee, is the second best coffee in international trade. It originated in the Congo basin. Robusta coffee is a lowland plant that does well in areas less than 600 m above sea level. The fruits are large, although the seeds are fairly small. It is the most widely grown coffee in West Africa, particularly in the forest zone.

Liberian coffee

Coffea liberica, known commonly as Liberian coffee, is indigenous to Liberia. It is a lowland type. The plant and berries are large, but the produce is of low quality. It is not widely grown in West Africa or elsewhere in the world.

Upland coffee

Coffea stenophylla is a native of Sierra Leone. It grows well in both upland and lowland conditions.

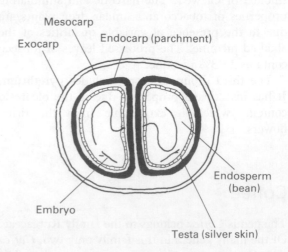

Figure 12.2 Transverse section of a coffee berry showing the two coffee beans surrounded by the outer layers which are removed during processing.

Its yields are usually low, but it provides coffee of good flavour. The ripe fruits are distinguished by their light blue colour.

Each type of coffee is adapted to either lowland or upland conditions. For upland types, growth is best in cool temperatures, while for the lowland types warmer temperatures are preferred. The average temperatures for coffee regions are in the range of 13–28°C, the best being 24°C.

The ideal rainfall regime for coffee is one of even distribution over the major part of the year with a short dry season lasting one or two months to facilitate flower initiation. Moisture supply is most critical during the phase of rapid berry expansion, about 10–17 weeks after flowering. Heavy rainfall during harvesting is not desirable.

Coffee is very sensitive to strong winds and it is an essential estate practice to retain windbreaks of trees in the field. Many coffee varieties are photoperiodically sensitive, being short-day plants. However, in equatorial zones they do not generally show photoperiodic behaviour, since the critical daylength seems to be about 13 hours. However, at higher altitudes flower initiation is often linked to the short-day season.

Coffee grows on a great variety of soils, from deeply weathered tropical soils to coastal swamps. The overall requirement, however, is that the soil must be well-drained and should be reasonably deep (1–2 m) so that the roots can easily penetrate. In general, the soil requirements are closely tied to the climate. For example, a poor or shallow soil is satisfactory in regions of adequate and evenly distributed rainfall, whereas a superior soil is necessary where rainfall is only marginal. The optimum soil pH for coffee is in the range of mid-acid to near neutral (pH 5–6).

The choice of variety depends on the ecological conditions in which it is to be grown. In general, the *robusta* and *liberica* types are adapted to lowland conditions, while *arabica* is an upland type. *C. stenophylla* grows under both upland and lowland conditions.

Planting

Coffee is normally propagated from seed, although vegetative propagation is also possible. For propagation from seed, robust seeds are selected from freshly harvested berries. The seeds are depulped and then washed thoroughly to remove mucilage. They are air-dried by spreading out in the shade and the berries are carefully shelled to ensure that no damage is done. The seeds are then soaked in water for 24 hours and those that float are discarded.

There are two stages in the nursery: a pre-nursery and a proper nursery. In the pre-nursery, the seeds are sown in boxes in a medium of fine river-bed sand, and the box is covered with wet sacking or transparent polythene sheeting to retain moisture. Germination takes 4–6 weeks.

When the seedlings are about 10–15 mm high or when they have developed two true leaves, they are gently transferred to polythene pots containing a rich mixture of topsoil and sand. The taproots are placed in a hole at the centre of the potting medium and the soil is firmed around the roots. The seedlings are kept in the shade and watered regularly until they are ready to be transplanted into the field, after about 9 months in the proper nursery.

A well-drained deep soil, rich in organic matter is ideal for coffee. The crop is generally grown in forest soils. It requires shade, at least during the initial stages of growth of the young plants. Therefore some shade trees should be left on the site during land preparation. These also serve the additional purpose of forming windbreaks against strong winds. Planting out in the field is done when the rains are fully established. It is important that the transplanted seedlings should be well established before the rainy season ends to ensure that they survive the first dry season. Suggested spacing for coffee is 2.7 × 2.7 m. To improve access between the rows, the crop can be spaced at 3.0 × 2.4 m; either spacing gives a population of 1330 plants/ha. However, with a good supply of nutrients and soil moisture, yields are better at the closer spacing of 2660 plants/ha (2.7 × 1.4 m). This is the practice in areas of high rainfall or where the crop is grown under irrigation.

Fertiliser application

Nitrogen is by far the most important nutrient in coffee growing and good yields are never maintained without its regular application. Economic responses to nitrogen fertilisation are almost always recorded, except under heavy shade where the level of response becomes so small that applications are unprofitable. The amount of nitrogen to be applied depends upon the level of yield expected. A small yield of 750–880 kg/ha of beans requires about 90 kg/ha of N, a 1300 kg/ha yield requires 110 kg N/ha, while a heavy crop of 2000 kg dry beans/ha needs nitrogen in the range of 135–155 kg/ha. The nitrogen is applied in three split doses. The first at the beginning of the rains, then one month later and the third during the minor rains.

Response to phosphate is generally low. The use of phosphate fertilisers is therefore limited to inclusion in the planting holes. The importance of potassium seems to be related to obtaining a balance with nitrogen. Deficiencies of micronutrients such as iron, boron and zinc have been reported. Such deficiencies are remedied by foliar sprays.

Pruning

Pruning is one of the operations in coffee management which enhances high yields. The system of single stem pruning results in the formation of bushes with a single stout upright stem. By this approach the crop is borne on lateral branches. The main problem with single stem pruning is that eventually the tree grows too tall and spindly, with an umbrella shape. It is for this reason that multiple stem pruning is adopted. Old stems are periodically removed and the problem of the bush becoming too tall does not arise. The single stem bush may be converted into a multiple stem bush when it becomes too tall.

Weed control

Grass weeds are particularly bad for coffee because of their ability to immobilise nitrogen. However, clean tillage also has a bad effect on both the soil and the plant roots and often causes a decline in yield. While the cultivation of leguminous cover crops between the rows has the advantage of suppressing weed growth, the competition for water often seriously affects production, particularly in dry climates and during the dry season. Effective ground cover plants widely used in West Africa include *Fleminga congesta*, *Leucaena leucocephala*, *Pueraria javanica*, *Mimosa invisa* and *Stylosanthes gracilis*.

Mulching, using heavy dry straw grasses, has a controlling influence on weed growth. The practice of mulching is a superior field treatment, particularly in situations where continuous stress is likely to affect production. Chemical weed control is widely used in coffee estates. Herbicides in current use, singly or in combination, are aminotriazole, dalapon, Gramoxone, simazine, diuron, fluometuron and atrazine.

Pests and diseases

The coffee stem borer, *Bixadus sierricola*, is a common pest in many coffee growing areas. The eggs are laid on the stem and the larvae frequently ring the bark of the tree before boring into the stem. The presence of 'sawdust' round the base of the tree is a sign of borer attack. The pest is prevalent on weedy farms, and good farm sanitation is effective in reducing its incidence. If an outbreak is severe, systemic insecticidal sprays may be used.

Another serious pest of coffee is the berry borer (*Stephanoderes hampei*). The adult female bores through the fruit and lays her eggs. The eggs hatch and develop into adults which may migrate to infest other fruits. In addition to using insecticidal sprays as a means of control, timely harvesting has also been reported to be effective against the spread of the berry borer.

Coffee berry disease, caused by the fungus *Colletotrichum coffeanum* var. *virulans*, is the most important disease of coffee. It attacks the berries, particularly in wet conditions, and causes the young fruits to shrivel and turn black. It also affects the leaves and twigs. Control is by destroying infected material and by spraying regularly with a mixture of copper sulphate and calcium carbide in water.

Coffee leaf rust, caused by the fungus *Hemileia vastatrix*, is another important disease. It is seen as lesions on the lower surface of the leaf, which release orange coloured spores. It can be controlled by the use of copper fungicides. Robusta coffee is relatively resistant to this disease.

Harvesting and processing

Multiple stem coffee comes into bearing about 2 years after planting and full bearing is reached in 4 years. The average bean yield of clean coffee is about 630 kg/ha. With good management, including good disease control, a yield of 1250 kg/ha of dry beans is possible. Indeed, some well managed coffee estates record regular yields of 2.5 tonnes/ha.

Coffee harvesting, usually referred to as picking, is done entirely by hand. Only uniformly ripe berries can produce good quality coffee. The berries are conveyed to a central preparation area for processing. In general, processed coffee is of two types: 'dry' and 'wet' processed coffee. Dry processed coffee gives 'hard' coffee, and is usually done when the fruits have been left to dry on the tree or have been dried after picking.

The dried fruits are repeatedly run through hulling machines which remove the dried pulp and parchment. Thereafter, fruits are further cleaned to remove their silver skins (seed-coat or testa).

Wet processing of coffee is done when a high quality product is desired. In this process, the freshly picked ripe fruits are fed with water into a pumping machine that removes the pulp. The resulting 'parchment beans' are left for 2−4 days in fermentation tanks where the remains of the sticky pulp is broken down by microorganisms. The beans are then dried in the sun for some 2−3 weeks, using raised trays. During this period the beans are turned to enhance uniform drying. Thereafter, the beans are put through hulling and polishing machines to remove the parchment and silver skins. At this stage the beans are green-grey

in colour and are ready for market. Coffee beans processed in this way may be stored for months before being used for various preparations.

Coffee is graded and priced primarily on its unroasted appearance. Roasting is usually done by the purchaser or the consumer. During roasting, the 'green' coffee loses about 16% of its weight. After roasting, the coffee is ground, and can then be vacuum packed.

'Instant' coffee, also referred to as soluble coffee, is brewed into a beverage after roasting and grinding, and then evaporated partially or wholly into a solid form. 'De-cafeinated' coffee is produced by treating the beans with steam or a benzene derivative before roasting. The quality of the final product is a function of the way the coffee is picked, processed, roasted, treated and blended as well as the cultivar used and the environment in which it is grown.

Cocoa

Cocoa, *Theobroma cacao*, belongs to the family Sterculiaceae. This family includes 50 genera and about 700 species of tropical trees and shrubs. The genus *Theobroma* contains 20 species, but only *T. cacao* is of economic importance. The home of

Figure 12.3 Cacoa pods growing on a mature branch of a cacoa tree.

cocoa is the tropical rainforest of South America, but the principal producing area is now West Africa, particularly Ghana, Ivory Coast and Nigeria. Other producing areas include the West Indies, Brazil, Indonesia, Ceylon and Central America (Figure 12.3).

The dried cocoa seeds, known as beans, are used in the manufacture of chocolate. Cocoa butter contains 50–57% of a pale yellow non-drying fat. It is used in the production of confectionery, cosmetics and in some medicinal preparations. The residue from the beans after the cocoa butter has been removed provides, after processing, cocoa powder, which is used for making the beverage cocoa and for flavouring. The husk is a good source of potash and is used as a fertiliser and in the manufacture of soap in rural industries.

Types and ecological adaptation

All the wild, semi-wild and cultivated cocoa varieties are referred to as *Theobroma cacao*. They all intercross readily, although the diversity of types within the species is considerable. The prevalence of cross fertilisation, the length of time the crop has been cultivated in different environments, the ease with which the different types hybridise and the fact that until recently propagation has been entirely by seed, are all factors which have contributed to the present heterogeneity and heterozygosity in the crop.

Although there is a general similarity between all types of cocoa in their vegetative characters, considerable variation exists between them in fruit and seed characters and in hardiness and vigour of growth. Depending largely on differences between the fruits and seeds, cocoa can be conveniently divided into two main groups, the 'criollo' and the 'forastero' types, which are fairly easy to distinguish.

The criollo group is characterised by the presence of a red pigment in the pericarp of their fruits. The mature pods are yellow or red. The pods are characteristically deeply furrowed and tend to be rough and warty with distinctly pointed ends. The seeds of criollo cocoa are large and rounded with white or pale violet cotyledons, and they produce a better quality beverage than the forastero types. The group is more common in Central and South America than in West Africa.

In the forastero group the mature pods are green or yellow, with a much thicker or woodier wall than in the criollo type. The pods are much less deeply furrowed than those of the criollo group and they may even be smooth. The seeds are flat, with dark-red or deep-purple cotyledons. The West African crop is derived mainly from the amelonado type of the Amazonian forastero, whilst West Indian cocoa is from the complex trinitario group. A second type of cocoa cultivated in West Africa is described as the Amazon type. Amazon cocoa normally comes into bearing about two to three years earlier than the amelonado type of Amazonian forastero. In addition to these two types, there are hybrids of the amelonado and the Amazon which are also cultivated. Such hybrid cocoa plants produce more vigorous trees which yield higher, mature earlier and are more disease resistant.

Cocoa does best where the rainfall is 1200–1800 mm per annum and is well distributed. A mean temperature of around 29°C is best for cocoa growth and for good yields. Cocoa is sensitive to higher temperatures, especially when the plants are young, therefore shade is necessary to establish young cocoa. Traditionally, it was the practice to grow cocoa on superior soils because the crop was grown without fertilisers. With improved crop husbandry a wide range of soils can now be used successfully. Ideally, however, the best soil for cocoa is a fertile clay-loam because of the capacity of such soils to retain sufficient moisture between rains. The soil must be well-drained to avoid waterlogging. The preferred pH is near neutral.

Morphologically, the cocoa tree varies in height from 4.5–8 m. The main trunk is 0.9–1.5 m high and forks into three to five lateral branches referred to as the 'fan' or 'jorquette'. Leaves are borne on the fan branches in two ranks arranged alternately. Below the fan, new leading shoots called 'chupons' are produced. The chupons carry leaves in a spiral arrangement. In plantations it is usual to remove the chupons as they form.

Planting

Propagation of cocoa may be by stem cuttings or seeds. However, to ensure a desirable growth form, the crop is usually planted from seed. Seed for planting is often produced by hand-pollination from selected trees known to produce superior yields. Out-crossing by insects is prevented by pinning a fine insect mesh (20 mesh/cm) around the trunk where the flowers are produced. A seed garden of one hectare would normally provide sufficient seed for about 400 ha/year by about the fifth year, rising to more than 1000 ha after the tenth year. Hand-pollination is, however, very labour-intensive and seed produced by this method is expensive. Another way of producing hybrid seed is to use naturally incompatible parent trees. Thus in such a seed garden, alternate rows of the different parent trees are planted and natural insect cross pollination occurs. The garden must be isolated from other cocoa trees to avoid undesirable crosses.

The best way to grow cocoa seedlings is in polythene bags, 15 × 10 cm in size, filled with rich loamy soil and with one seedling per bag. The seeds should be planted about 1 cm deep and laid flat on their sides to prevent root twisting. The bags may be close-packed in beds and shaded with 50−70% shade. The seedlings are ready for planting out after 6−8 months. Cocoa seeds should be planted as soon as possible after being removed from their pods as they only remain viable for a short period (Figure 12.4).

Some live trees must be left in a new cocoa plantation to provide shade, otherwise some temporary shade must be provided by growing such food crops as plantains, bananas or cocoyams within the rows. Transplanting of seedlings should start as soon as the rains have become established. In the West African region seedlings should not be transplanted later than mid-July. Spacing for unshaded cocoa should be 2−3 m, depending on the variety. For small varieties such as the amelonado type, the closer spacing is recommended, and for larger vigorous types such as the Amazon type, the wider spacing should be used (Figure 12.5).

Fertiliser application

Fertiliser application in the initial stages of the growth of cocoa may not show a marked response,

Figure 12.4 Cacoa seedlings growing in partial shade in a nursury in Nigeria.

Figure 12.5 A cacoa plantation in Camaroon with mature pods ready for harvesting.

especially since the crop is invariably planted on rich soils. However, fertilisers may be applied to older plants at the rate of 100 kg N/ha, 150 kg P_2O_5/ha and 50 kg K_2O/ha per annum.

Weed control

Cocoa forms a close canopy after about five years, thereby suppressing weeds. In the presence of cover crops weeds are suppressed until the cover crops themselves die out due to heavy shading from the cocoa trees. Occasional cutlassing of weeds may also be necessary, especially in places where the cocoa establishment is less than optimum.

Chemical weed control is much used at the present time and is considered to be more economical. Simazine, linuron and diuron are useful for long-term weed control. Dalapon has been recommended for grasses and 2,4-D for broad leaf weeds. Care must be taken to avoid contact of 2,4-D with the cocoa foliage.

Pests and diseases

The most serious pests of cocoa are the capsid bugs, chiefly *Sahlbergella singularis* and *Distantiella theobromae*. These suck sap from young shoots and buds and cause young twigs to die back. The punctures made by the capsids in feeding often become infected with the fungus *Calonectria rigidiuscula*. Control is based on the use of insecticides such as HCH (BHC) and aldrin.

The fungus *Phytophthora palmivora* causes black pod disease. The various symptoms of this disease include pod rotting (black pod), seedling wilt and leaf fall. It can spread from infected pods to invade the flower cushion and stem where it gives rise to a 'canker'. The disease spreads by means of sporangia, which produce zoospores in moist conditions. The disease is controlled by spraying with fungicides such as Perenox and Bordeaux mixture, and by removing infected pods as soon as they are found. Good general field sanitation also helps to check the spread of black pod disease.

Swollen shoot disease is caused by a virus. The symptoms are swelling of the fan and chupon branches and initially a red-vein banding and later, chlorosis along the leaf veins. The virus is spread by mealybugs. The amelonado type of cocoa is particularly susceptible while the Amazon type is more tolerant. Control of the disease is by removing infected trees and burning them.

Harvesting and storage

Harvesting is done all the year round, although in West Africa there are two peak seasons of production: October-February (main crop) and July-August (minor crop). Pods turn yellow or orange when fully ripe. Harvesting is done by means of a sharp knife, either held in the hand or attached to the end of a pole, which is used to cut the pods from the tree. Extra care must be taken not to inflict damage to the pod-bearing cushion. This is necessary because each cushion is expected to bear more fruits for future harvests. Harvested pods are gathered into heaps where they are split and the beans and pulp removed.

The seeds (beans) are collected and fermented in boxes, in baskets or on banana leaves spread on the ground (Figure 12.6). The purpose of fermentation is to kill the embryo so that germination of seeds is arrested and to make the beans more brittle and easier to depulp. Fermentation also makes the beans taste sweeter. After fermentation, the beans are dried, usually by spreading them in the sun or by mechanical driers (Figure 12.7). After drying, the beans are packed in bags and stored in a dry place until they can be marketed.

Tea

Tea, *Camellia sinensis*, belongs to the plant family Theaceae. The family includes about 200 species of trees and shrubs which grow in the wetter tropics. The genus *Camellia* has 45 species of evergreen trees and shrubs all of which are natives of tropical

Figure 12.6 The cacoa beans are removed from the harvested pods and fermented on banana leaves spread on the ground.

Figure 12.7 Cacoa beans drying in the sun in Ghana.

Asia. While some of the species are grown for their decorative foliage, only *C. sinensis* is of any economic importance. Some authorities divide the genus into two; *Camellia*, including those species which have sessile flowers with a deciduous calyx, and *Thea* comprising those species with pedicelled flowers having a persistent calyx. These authorities group the tea plant with the latter subgenus and refer to it as *Thea sinensis*.

Tea is used as a beverage. The stimulating properties of caffeine contained in the leaves and the astringency and taste of the polyphenols and their derivatives are the attributes of tea which have been exploited by man.

Types and ecological adaptation

Tea is grown in a wide range of climates, from the humid tropical lowlands to regions of high altitude. In the tropics cultivation occurs mainly at higher elevations. In the presence of adequate rainfall,

Figure 12.8 Harvesting tea on a tea plantation in West Africa.

268

lowland tea generally produces higher yields, although quality, root growth and longevity are much reduced (Figure 12.8).

Temperature is of such importance to tea plants that leaf expansion only takes place above 21°C, with an optimum range between 25 and 30°C. Above 30°C leaf function is reduced. This infers that at higher altitudes production is seasonal. Trees need adequate and evenly distributed rainfall for high production. The minimum annual rainfall, if evenly distributed, is between 1000—1400 mm. The crop does poorly on hot humid coastal plains.

The tree grows best on acidic soils (pH 4.5—5.5) which are free-draining. Soil pH above 6.0 reduces tree growth. The use of alkaline soils should therefore be avoided. The crop is also sensitive to strong winds and the planting of windbreaks along the ridges or hills has been shown to increase yields.

Despite great variations in the crop, only two main groups with the status of botanical varieties are recognised. These are the assamica (from Assam) and the cambodia (from Indo-China) types. They hybridise freely.

Planting

Tea is generally propagated vegetatively, although seeds may also be used. A clone for vegetative propagation must be able to root easily and establish well from cuttings, preferably without the use of rooting hormones.

Tea bushes selected for propagation are taken out of plucking and allowed to grow freely to produce branches. From healthy selected branches, single leaf cuttings are obtained using a sharp knife to get a clean cut. The cuttings are then planted in a nursery bed. No organic matter should be used in the nursery soil, but rock phosphate should be mixed in at the rate of 500 g per cubic metre. Rooting occurs in 4—6 weeks and the rooted cuttings are then removed to an open nursery and planted in nursery beds or in individual polythene sleeves. Shade is provided heavily at first and then progressively removed in preparation for planting out in the field after about 6—8 months.

Where seeds are preferred as a planting material, they are obtained from high-yielding and good quality tea bushes which are set aside for seed production and which are not pruned. The seeds are tested by floatation in water and the floaters are discarded. The seeds are planted in nursery beds

with soil which has been well worked to a depth of 30 cm or more. The beds are shaded and carefully watered. Nursery seedlings are transplanted to the field when they are about 12—18 months old.

Because of its deep roots, tea requires a site where the soil is deep, well-drained and yet has good water retaining capacity. The minimum soil depth is about 1.8 m. The site should be sheltered from strong winds. For best growth, planting must be done as early as possible in the major rains, so that the young plants become well established before the dry weather sets in. Rows should be 1.5 m apart with the plants spaced 0.75 m apart within the rows. The distances are varied according to the soil, climate and variety of tea.

Being of forest origin, the young plants benefit from being shielded by taller trees, especially if the seedlings have not been adequately hardened in the nursery. In the absence of taller natural vegetation, the plants may be shielded with blackened fronds, pieces of banana leaf or such other suitable material.

Fertiliser application

Tea responds to chemical fertilisers, especially when grown on poor soils. Of the major nutrient elements, nitrogen is the most important. Tea responds to nitrogen application up to 90 kg N/ha in full sun-light, and in general, the level of response to nitrogenous fertilisers is greater in the absence of shade. The soil organic matter is increased by the addition of prunings and green crops. Phosphorus requirements are usually satisfied by annual applications of 20 kg P_2O_5/ha. Potassium fertilisers do not often produce a response in tea, although the crop removes significant quantities of the element. Young tea plants with limited root systems may, however, respond to potassium fertilisers. Micronutrient deficiencies, especially zinc, occasionally occur and are usually corrected by foliar sprays.

Foliar diagnosis is usually used to assess the fertiliser requirements of a tea crop by sampling the two terminal leaves and bud. The levels of each element considered to be optimal for tea are 3.5—4.8% N, 0.25—0.35% P and 1.25—2.0% K. The crop must be fertilised if the level of each plant nutrient is below the lower limits of these defined ranges.

Figure 12.9 A tea processing plant in Kenya.

Weed control

Weed control is usually by hand cultivation, although the use of chemicals for weed control has the advantage of not disturbing the soil. Dalapon and paraquat have been extensively used in tea. When using herbicides for weed control it is generally advised that spraying should be confined to the soil and weed surfaces so as to avoid application to the tea foliage.

Mulching with grass, legumes and other materials also achieves a measure of weed control, in addition to other benefits.

Pests and diseases

Mites and scale insects are the major pests of tea. Control is achieved using systemic insecticides which break down fast enough to avoid contamination of the tea.

The main disease that has caused epidemic losses in tea crops is 'blister-blight' caused by *Exobasidium vexana*. The fungus causes small translucent spots which expand and become concave depressions looking like blisters on the underside of the leaf. The blisters turn white and release spores capable of infecting new leaves. Only young leaves are affected, especially in wet weather. Control is best achieved by using resistant varieties and by confining pruning to drier seasons. Copper sprays such as Bordeaux mixture also control the disease.

Various root diseases, all caused by various fungi, affect tea plants. Most root diseases arise from the stumps of old jungle trees or shade trees which have died and which constitute a source of infection. Removing such stumps and diseased tea bushes and replanting them is the only practicable means of control.

Harvesting

The frequency of harvesting (plucking) varies from every one to two weeks in the tropics during the periods of flushing. In order to ensure high quality tea, especially when harvesting by hand, only the terminal two leaves and a bud are plucked while the lower leaves are left as maintenance leaves which are important for photosynthesis and continued flushing. Pickers use long canes laid across the top of the tea bushes as a guide for picking only top-quality tips.

There are two commercial tea products; the green or unfermented tea used mainly in Asian countries, and the fermented or black tea which forms the bulk of the international tea trade. The difference is that black tea is fermented by enzymes which are present in the tea leaves, while in the green form fermentation is prevented by heating. (Figure 12.9).

Kola

Kola (also known as Cola) is a member of the family Sterculiaceae to which cocoa also belongs. About 500 species of the genus *Cola* are indigenous to the forests of West Africa where several species grow wild, including *C. acuminata* and *C. nitida*. Both of these species are cultivated, but *Cola nitida* (Gbanja kola), which originated in Ghana and the Ivory Coast, is relatively more important than *C. acuminata* (Abata kola) which has a natural distribution from Nigeria to Gabon. Two other species, *C. anomala* (Bamenda kola) and *C. verticillata* (Owe kola), are of less importance and are rarely cultivated. *Cola nitida* is a dicotyledonous plant while the other species have multi-cotyledons. (Figure 12.10).

In addition to the bulk production from West Africa, a small amount of kola is grown in South America. The crop is of much importance in the social lives and religious customs of many ethnic groups in West Africa. The seeds are chewed as a stimulating narcotic. The seeds when ground are also widely used in the beverage industries, particularly in the manufacture of wines and soft drinks.

Types and ecological adaptation

In its natural environment in West Africa, the kola tree lives in the forest zone where the annual rainfall is 1500–2500 mm, unevenly distributed with a marked dry season and temperatures around

Figure 12.10 A flowering branch of a kola tree.

25°C. When they are subjected to severe water stress, especially when the annual dry season is unusually long, kola trees shed some or all of their leaves. Well established and mature trees may survive such harsh conditions, but young trees are invariably killed by drought.

In West Africa, kola cultivation has extended from the forest zone into regions with less rainfall and a longer dry season, especially along rivers and in low-lying areas where there is always some residual moisture. Young kola trees require shade for several years until they are well established after which they can withstand direct exposure to the sun. They prefer fertile soils which are well drained and rich in organic matter.

The fruits of *C. nitida* vary in colour from a creamy white or pink to very dark purple while those of *C. acuminata* are purple or red. There is a strong social preference for white nuts, the selection of which has been associated with small-sized trees. Kola is a cross-pollinated plant, and almost 90% of trees are self-incompatible. Those trees which regularly yield well are chosen for the establishment of new kola plants.

Planting

Kola is generally propagated from seed by first germinating the nuts and then tending the seedlings under shade, in pots or in polythene bags. However, kola may also be vegetatively propagated by marcottage or by rooted cuttings, but this method is only used to a limited extent.

When growing kola from seed, nuts weighing 15 g and over are selected from high-yielding trees. Such seeds should be free from weevils. Seed boxes of convenient sizes are made and filled with a growing medium of moist soil or sawdust, 15 cm deep. The boxes are raised above ground level and the nuts are sown on their sides, 3–5 cm deep, watered carefully and covered with polythene sheeting or wet sacking. Germination takes place several months after sowing because of a long period of dormancy, especially with freshly harvested nuts.

When the shoots of the germinated nuts reach a height of 1.5–2.5 cm they are gently removed from the seed boxes and planted in polythene bags which are watered and kept under shade. Watering should be done daily during dry weather and the shade reduced gradually until the seedlings are ready to be transplanted into the field.

Kola does best under forest conditions. Savanna areas recently cleared of forest (derived savanna) are suitable, provided there is adequate soil moisture in the rooting region and the soil is high in organic matter content. Kola roots do not stand flooding and so good drainage is essential. A few trees may be left during land clearing to provide overhead shade for the kola crop, otherwise shade trees have to be planted for early seedling growth.

Seedlings should be transplanted early in the wet season when they are about 15–25 cm high and 6–8 months old. Seedlings raised in polythene bags can withstand transplanting shock better than those grown in the ground. This infers that seedlings in polythene bags may be transplanted a little later in the rainy season.

Nuts for planting are treated in various ways to enhance germination. It is common practice to leave these nuts on the trees for a much longer time than those intended for consumption. Nuts intended as seed may also be allowed to lie on the ground for sometime before being used. Alternatively the nuts may be stored before planting in the nursery. The principle behind all these measures is to break or reduce the dormancy period. Chemical treatments, including the use of thiourea or thiourea dioxide and heat have been found to induce early germination.

The recommended spacing for kola is 7.5 × 7.5 m, but a closer spacing of 6 × 6 m in sometimes adopted. In either case, the spaces between the rows are cropped for one or two years with such crops as cassava, plantain or other catch crops.

Fertiliser application

In areas where kola is produced, it is not routinely fertilised. However, the crop is always planted on rich forest soils where it derives nutrients from decomposed vegetative matter and leaf litter.

Weed control

The base of the trees should be cleanly ring-weeded as far as the canopy spread, and the intervening bushes should be kept down. Weeding around the base of the plants facilitates the identification of fallen pods when picking. Presently no herbicides are used for weed control in kola plantations, but it is likely that the herbicides used for cocoa would work well for kola.

Pests and diseases

Many insect pests have been reported on kola, among which are:

Kola stem borer. The adult borer of this beetle feeds and lays its eggs on young growth. The larvae then bore into the twigs and towards the thicker woody parts of the plant, retarding growth and causing die-back.

No effective control methods have been established. In less serious infestations, mechanical destruction of the larvae and pupae may be undertaken by poking the tunnels made by the insects with long wires. The adults may be collected and destroyed. Trees that are seriously affected by the kola stem borer should be cut back and allowed to regenerate, with the cut parts being burned. The young regrowth may be protected chemically from reinfestation by spraying with 85 g of Sevin in 27.5 litres of water per hectare.

Kola pod borer, *Characoma strictigrapta*, is the larva of a moth which attacks kola pods causing them to drop. The insect may be controlled by a suitable insecticide such as endrin applied as a fortnightly spray to the tree canopy during the fruiting season.

Kola fruit fly, *Ceratitis colae*. This insect bores into and oviposits (lays her eggs) in mature soft pods and the larvae develop within the pods. The exit holes of the mature larvae provide entry points for the highly destructive kola weevil. For effective control prompt harvesting should be adopted, before the pods become over-ripe and the pod walls decay. Aldrex-T may be used to effect control, but then the nuts can only be used for planting and should not to be used for eating as Aldrex-T is harmful if consumed.

Kola weevils, *Balanogastris kolae* and *Sophrohinus insperatus*. These are two kola weevils known to cause serious losses of kola nuts. They are important pests prior to harvest as well as in storage. Control lies in good crop sanitation. The weevils are unable to penetrate kola pods to reach the nuts except through exit holes left by the fruit fly *Ceratitis colae* or when over-ripe pods fall from the trees and become broken. Therefore, pods should be harvested regularly and all fallen pods picked up promptly. Nuts for storage should be inspected and all infested nuts removed before they are stored. This exercise should be done regularly for nuts in storage.

Kola is also attacked by a number of diseases, including nut spotting caused by *Botryodiplodia theobromae*. This fungus infests the follicles (kola fruits) causing a black rot to develop which subsequently spreads to the nuts. Affected nuts first show rusty brown spots which later turn black and become hard and dry. Where the disease has been found in the field, affected pods should be treated promptly with a suitable fungicide such as Bordeaux mixture to prevent its spread.

Many fungi cause nut diseases, including species of *Fusarium*, *Diplodia*, *Gliocadium*, *Penicillium*, *Pleurotus* and *Schizophyllum*. Control consists of employing good curing and storage methods for the nuts. Fungicides should not be used particularly where the nuts are to be chewed raw.

Kola trees are often attacked by root diseases caused by species of *Fomes* and *Ganoderma*. Control may be achieved by cutting off affected roots and burying them. The choice of a well-drained site is also important for controlling the incidence of these diseases.

Harvesting and storage

The first nuts are produced 6–8 years after planting, and after 12–15 years in the field the tree begins to yield an economic crop. From the age of 20 years the tree is considered to be in full production and may continue to bear until 70–100 years old. However, the mortality rate of trees is high from the age of 25 years onwards.

Harvesting is done by either allowing the nuts to fall from the tree then collecting them or by using a long pole to which a knife is attached to harvest them from the tree. In West Africa the harvesting season extends over a long period, opening in October and gradually increasing to a peak in December. Towards the end of March, the main crop has been harvested in most areas, but a further small mid-season crop is harvested in late May and June. Yields are variable and average about 1000 nuts per tree per year. It is estimated that one hectare of naturally propagated kola in a pure stand could produce 25 000 nuts per year.

After collecting the pods from the trees they are cracked open to release the nuts. The nuts are then soaked in water, buried in moist sand or made into lightly watered heaps for 24 hours to make skinning easy. The skinned nuts are washed and placed in unlined baskets, covered lightly with banana leaves and left for about five days to 'sweat'.

This process reduces the water content of the nuts. The nuts are checked for weevil damage and are then ready to be sold fresh or put into storage, either by the farmer or the trader.

Kola nuts are stored in baskets made from raphia palms which are lined with the broad leaves of *Marantochloa cuspidata* or *Phyrnium ramosissimum*. The nuts are placed in the baskets, covered with leaves and stored in well-shaded places. Often, water is sprinkled over the baskets to keep the nuts from drying out. The leaves used to line the baskets should be changed occasionally. Routine checking of the nuts every three to four weeks is essential to elimate weevil-damaged and decaying nuts. Well-cured and well-stored nuts will keep for several months without spoilage. Exported kola nuts are usually sun-dried.

Tobacco

Tobacco, *Nicotiana tabacum*, is one of the most widely grown commercial non-food plants in the world. It is a member of the family Solanaceae. Although tobacco is a tropical crop originating in South America, the bulk of world production comes from outside the tropics or from areas of high altitude within the tropics. It is one of only a few crop plants which have become commodities of international commerce entirely on the basis of their foliage. Because of its word-wide economic and social significance, despite its possible danger to health as a habit-forming narcotic, attempts to stop or curtail its production or even ban its use entirely have not met with success. Indeed, the consumption of tobacco, particularly by smoking, has continued to increase as evidenced by world production statistics (Table 12.1). Smoked tobacco is normally in the form of cigarettes, cigars or pipes. In addition to smoking, tobacco may also be chewed or taken as snuff. The alkaloid nicotine may be extracted from tobacco waste and used as an insecticide.

Types and ecological adaptation

Of the over 50 species of the genus *Nicotiana*, only two are cultivated; *N tabacum* and *N. rustica*. *N. rustica* is cultivated on a much smaller scale, mainly in Asia and a few places in East and West Africa. The type of leaf produced depends as much

Table 12.1. Annual production of tobacco, 1982–84

Country	Year	Area (million ha)	Yield (kg/ha)	Total production (million tonnes)
China	1982	1.13F	1946	2.21F
	1983	0.78F	1809	1.40F
	1984	0.78F	1960	1.51F
India	1982	0.44	1171	0.52
	1983	0.50	1157	0.58
	1984	0.44	1130	0.50
USA	1982	0.37	2449	0.91
	1983	0.32	2029	0.65
	1984	0.32	2452	0.80
Brazil	1982	0.32	1325	0.42
	1983	0.32	1252	0.40
	1984	0.29	1454	0.42
Turkey	1982	0.21	1008	0.21
	1983	0.23	994	0.23
	1984	0.21	1010	0.21
Africa	1982	0.30	943	0.28
	1983	0.32	953	0.30
	1984	0.31	1044	0.32
Europe	1982	0.50	1605	0.79
	1983	0.50	1425	0.71
	1984	0.51	1526	0.77
World	1982	4.53	1521	6.90
total	1983	4.25	1403	5.97
	1984	4.15	1493	6.21

Souce: FAO production yearbook 1984.
F = FAO estimate.

on the varietal origin as on other factors such as soil, climate, fertiliser and manuring regimes and curing procedures. On the basis of leaf characteristics arising from the curing process, tobacco is normally classified into four types; flue-cured, air-cured, sun-cured and fire-cured.

The correct combination of variety and environment is necessary to achieve the ultimate tobacco product. In general, the tobacco crop requires moderately fertile, medium to heavy, well-drained, slightly acid (pH 6.5–6.8) loamy soils. High temperatures, moderate but well distributed rainfall and soils rich in organic matter are preferred. However, within these generalisations considerable differences in requirements do exist between the various types.

Flue-cured tobacco normally prefers freely drained medium coarse sands, sandy loams, fine sandy loams or sandy clay. The ideal pH range is between 5.5 and 6.5. Soils suited to the production of flue-cured tobacco are often those deficient in magnesium, which explains why a deficiency of this element is more commonly reported than deficiencies of other elements. Flue-cured tobacco has become the most widely cultivated type in the world as the bulk of the production goes into cigarette manufacture. Indeed, the unqualified use of the word 'tobacco' is generally taken to refer to the flue-cured type only. Production of flue-cured tobacco is mainly in the tropical southern United States of America, southeast Africa, Brazil and China.

Air-cured tobacco is a relatively high nitrogen and low soluble carbohydrate type which, although it may be found growing in various soil types, invariably prefers fertile, well-drained sandy loams, clay loams or silty clay loams. The maintenance of a good soil structure is essential. There are two grades of air-cured tobacco; the light air-cured and the dark air-cured. The difference between the two grades is largely a function of the variety of tobacco as well as the soils on which it is grown. Thus the terms 'light' (bright) and 'heavy' (dark) have simultaneous application to both the tobacco and the soil. However, specific varieties have been developed to give the light air-cured leaf. Among common air-cured tobaccos are the well known 'Burley' tobacco produced mainly in the southern USA, Mexico and, to a lesser extent, Central and Southern Africa. 'Maryland' tobacco is of local importance in the USA. Major producers of cigar-tobacco include Cuba, Indonesia and the USA.

The best known sun-cured tobacco is Turkish or Oriental tobacco which is characterised by its aromatic oil content. The crop requires hot, dry and sunny weather in which to mature. The typical Mediterranean climate provides an ideal environment. The shallow, eroded calcareous soils of the slopes of Greek and Turkish mountains are ideal soils. The traditional home of Oriental tobacco is, the eastern Mediterranean and Black Sea coastal areas. While not strictly a tropical crop, very acceptable Oriental tobacco has been produced in Central and East Africa by utilising flue-cured tobacco type soils and planting late in the season. In general, soils with low fertility, particularly those low in nitrogen, give the best sun-cured tobacco.

Soils which would normally support most other annual crops are generally suitable for the cultivation of fire-cured tobacco. Ideally, the soils should be fertile, medium to heavy textured loams, clay loams or sandy clay loams, rich in organic matter. Major areas of production of fire-cured tobacco include southern USA, Italy and East Africa. However, the world demand for this type of tobacco appears to be in a steady decline.

Planting

Tobacco seeds are very small so in some parts of the tropics they are sown in the nursery mixed with sieved wood ash or fine sand. One gram of tobacco seed should normally produce 3000–4000 seedlings, sufficient to plant 0.13 ha, including replacement of missing stands. Germination takes place within 6–7 days; seedlings are ready for transplanting into the field when they are 6–7 weeks old. The seedlings should be sprayed with appropriate fungicides and insecticides to take care of pests and diseases.

Seedlings are transplanted 70 cm apart on 90 cm ridges, preferably after heavy rain. The crop should be fertilised within about four days, if this was not done before planting. It is usual practice for the phosphate and potash where required to be broadcast before the ridges are moulded, or applied at planting together with the nitrogen fertiliser.

The time from planting seeds in the nursery to transplanting takes 7–8 weeks while the period between transplanting and maturity is only another 8 weeks. In general, harvest should be completed within 12 weeks of the crop being planted out in the field.

After a time, the crop has to be 'topped'. This practice varies with the different types; for flue-curing and bright (light) air-curing, topping is achieved by allowing the flower to emerge completely and then breaking it off with about six of the small top leaves. Topping causes suckers to develop in the axils of the leaves and 'suckering' should be done about once a week until harvest, to remove them. 'Priming', the removal of the small leaves at the base of the plant, should be carried out before and at topping (Figure 12.11).

Fertiliser application

The rates of the various recommended fertilisers for tobacco depend on the soil type. The rates of nitrogenous fertilisers used in various parts of the

Figure 12.11 A field of tobacco before 'topping'.

world are in the range 80–140 kg N/ha, particularly for air and sun-cured tobaccos. About 20–25 metric tonnes/ha of farmyard manure may be substituted for conventional fertilisers. The application of phosphate and potash fertilisers is often based on soil tests, but 120–180 kg K/ha and 25–50 kg P/ha have been suggested. Care must be exercised in the use of chlorine-containing fertilisers since a high chloride content in the leaves lowers the quality of the tobacco. Therefore the use of sulphate of potash to supply the potassium requirements of the crop is often preferred to the use of the chloride form. As a rule, the chloride content of any form of fertiliser applied to tobacco should not be more than 3%.

For flue-cured tobacco, which is a low nitrogen product, only the minimum amount of nitrogenous fertiliser needed for growth should be supplied. In order to ensure a high quality flue-cured product,

only very low uptake of nitrogen should be allowed in the latter part of the cropping season. The application of 25–50 kg N/ha to fire-cured tobacco, in the form of ammonium sulphate or urea, is generally adequate for most soils. The fertiliser should be worked into the soil just before planting.

Crop rotation is an important aspect of soil management in the successful production of tobacco. Not only does the practice help to restore soil fertility, but it also helps to check erosion, and pest and disease build-up. It has been shown in some parts of Africa that the yield and the quality of the tobacco are enhanced when production is preceeded by groundnuts.

Weed control

It is essential that the tobacco crop be kept free of weeds throughout the 8–9 weeks growing period in the field. This can be effected by reshaping the ridges either by tractor or by ox-drawn equipment. In Africa, however, weeding is still done manually in many areas.

Pests and diseases

Damping off caused by *Rhizoctonia solani* is a seedbed fungal disease which attacks both the roots and stems of tobacco at soil level causing the seedlings to fall over and die. Excessive watering, poor ventilation and overcrowding of seedlings all tend to encourage the disease. Control measures include proper regulation of the watering regime, good drainage, raising the nursery shades to give adequate ventilation and spraying the affectd plants with the appropriate fungicides.

Frog-eye or barn spot is a fungus disease caused by *Cercospora nicotianae*. The term 'frog-eye' refers to the nursery and field infection seen as brown circular lesions with a central white spot, while 'barn spot' refers to the brown or black spots and lesions which develop during curing, caused by the organisms in the leaf tissue. Excessive nitrogen application, overcrowding of seedlings in the nursery and inadequate ventilation and temperature control during curing are some of the predisposing factors. Control measures include spraying with fungicides in the nursery, avoiding overcrowding of seedlings, the provision of adequate ventilation, timely harvesting of leaves, avoiding overloading of barns and ensuring adequate ventilation and temperature control in the barns.

The leaf-curl virus, *Nicotiana* virus, is transmitted by white flies which are usually found on the under-surface of the leaves. Mosaic virus is transmitted by contact. With both types of virus disease, the infected plants should be uprooted and burned or buried. The *Nicotiana* virus has several alternate hosts, including pepper, cassava and melon. It is advisable that where the disease has occurred, these alternate hosts should not be allowed in the vicinity.

Root rot and stem rot are fungal diseases caused by *Sclerotium* sp. and *Thielaviopsis* sp., respectively. Control measures include good drainage both in the nursery and in the field, the use of resistant varieties and the burning of infected plants.

Root knot is caused by soil nematodes which feed on the roots of growing plants causing distortion and gall formation, resulting in destruction of the root system. Crop rotation and the use of resistant varieties provide the most satisfactory control measures.

Harvesting

The method of harvesting tobacco is dependent on the method of curing used. For flue-curing, the leaves are picked individually as they mature. A leaf is considered mature when it shows a general clear change to a yellow colour. Picking naturally starts with the lower leaves and progresses upwards and the harvest is usually completed after 4–5 pickings. For flue-cured tobacco, the leaves must be shaded as soon as they are collected as even a short period of exposure to hot sun can cause them to darken on curing. Similarly, green and freshly picked leaves should not be stacked in containers, on tables or on the floor for long periods. With Oriental or Turkish tobacco, the operation must be carried out even more carefully with the leaves from different positions on the plant being kept separately as they give different qualities of tobacco.

For air-cured tobacco the whole plant can be cut when mature. Since the leaves of tobacco ripen from the bottom upwards, the cutting has to be timed so that most of the leaves are at or near their optimum stage. This is usually achieved when the middle leaves are fully ripe and the lower or bottom leaves are beginning to fall off. However, for Burley tobacco, it is important to harvest only when fully ripe. The choice therefore is between losing some bottom leaves and the risk of lowering the quality of the leaves at the top. The best compromise is to prime the lower leaves as they ripen and cut the stalk when the top leaves will cure best.

For fire-cured tobacco, the full-ripe stage required in flue-cured tobacco is not necessary. Leaves are ready for harvest when they become thick, brittle and with a mild to moderate degree of yellow mottling, but still possessing a darkish-green background colour. The leaves may be primed as they ripen or they may be stalk-cured. Priming is the common practice in Africa but stalk-curing is more commonly practised in the USA and Canada.

Curing and storage

The curing of tobacco is a complex operation which largely dictates the quality of the final product.

Flue-curing

Flue-curing is a relatively rapid process using smokeless heat. It allows the shortest period of time after the leaves are picked from the plant and before they are killed by desiccation. The process demands a well-built barn through which flues run, and heat from an outside furnace. The barn is essentially an airtight building with controllable sources of heat and ventilation. Humidity and temperature are fully regulated and the height in relation to the floor area of most barns gives them the appearance of chimneys. Sound construction is needed both to bear the weight of the freshly loaded green tobacco and to ensure that air intake is confined to the ventilator openings. Various construction materials are used, including wood, mud, burnt brick and cement blocks. The harvested leaves are tied to sticks about 1.5 m long and these sticks are hung on tiered poles. A barn of $6 \times 6 \times 8$ m tiers normally contains about 1600 sticks of tobacco with a green weight of 450 kg and a cured weight of about 300–350 kg. Barns measuring 6×6 m, 5×6 m and 5×5 m are capable of handling about 3, 2.5 and 1.5 hectares of tobacco, respectively. The curing barns have wood or coal-fired furnaces. The hot flue-pipe passes round the inside of the barn but the smoke goes through a chimney. The temperature of the barn can be regulated according to ventilation and the amount of heat generated in the furnace. In summary, the curing process goes through the following stages:

(a) The yellowing stage (with a temperature of 35–46°C), during which the leaves turn from greenish yellow to yellow. The top and bottom ventilators are closed to conserve humidity.
(b) The fixing of the colour stage, during which the yellow colour obtained in (a) is fixed at a temperature of 46–52°C, while the top and bottom ventilators are kept open.
(c) The drying stage, during which the temperature is raised gradually from 52 to 77°C with the top and bottom ventilators open, so that the leaves become completely dry and brittle. The fire is then withdrawn and the door and ventilators are left open overnight so that the leaves absorb some moisture. This conditions them for grading and baling ready for sale. The average curing period is about six days and during this period, the process must be carefully supervised.

The curing process causes the oxidation of chlorophyll, and hence the disappearance of the green colour, and the breakdown of starch into sugars. After this, the leaf is killed so quickly by desiccation that the sugars cannot hydrolyse nor can browning occur due to the oxidation of phenolic compounds. This ensures a high ratio of sugars to nitrogen and increases the stability of the pale yellow colour.

Air-curing

Air-curing entails exposure of the harvested leaves to natural drying by suspending them in some sort of shaded structure. The structure may allow manipulation of the ventilation or even the provision of modest heat. In the air-curing process, the harvested leaves are piled against the wall with their butt ends on the floor, covered with hessian bags and left to wilt and yellow. The leaves are then strung 4–6 cm apart and the strings tied about 30 cm apart in the barn. Although air-curing is a natural process, it is essential that the drying should not be too fast in dry climates and not too slow in humid conditions. However, in practice, must local air-cured tobacco is left to the mercy of the weather. In some countries, the tobacco leaves are primed and hung in well-built, roofed but open-sided structures. Sometimes, they are partially dried in the open and the curing is completed in built structures. In some parts of West Africa, the leaves are placed in roofed but open-sided structures and dried by the dry Harmattan air and the curing period lasts 2–4 weeks. In these regions, the air-cured tobacco is light (or bright) to deep orange in colour whereas the air-cured tobacco produced in the more humid zones is light to dark mahogany in colour.

Fire-curing

Fire-curing is a much longer process than flue-curing, lasting about 4–7 weeks. The leaves are killed relatively slowly and so the phenolic compounds are oxidised, giving a brown colour, and the sugars are largely hydrolysed giving a characteristically low ratio of sugars to nitrogen. The intensity of smoking adds greatly to the brown colour and to the flavour. In most parts of Africa, fire-curing is done in thatched barns which contain only two tiers and pits on their floors for the fires. The leaves are first yellowed, either by hanging them in the barns without a fire for 4–7 days, or by piling them in a heap in the shade before placing them in the barn. After yellowing, fires are lit and kept going by day for several weeks. The fires should give much smoke but relatively little heat. Fire-curing is used mainly for producing plug and twist tobacco for chewing and pipe tobacco, as well as for cigar or dark cigarette tobacco.

Sun-curing

In sun-curing the harvested leaves are exposed to the full rays of the sun for most of the curing period. The drying is faster than for air-curing. The sugar content is higher than in air-cured leaves, but nitrogen and nicotine contents are lower. The process is used mainly for Oriental or Turkish tobacco. The leaves are normally strung on the same day they are picked, and not later than 24 hours after. The leaves, having been strung, are wilted in the shade until yellowing is fairly pronounced. This normally takes 24 hours although in cool weather such as in Central Africa, this may take up to 72 hours. To complete the curing, the strings are exposed to the sun in different ways, ranging from the use of sophisticated frames which are wheeled in and out of sheds to laying the strings on the ground. The main concern is the occurrence of dew during the drying process, particularly in Africa where the leaves are covered every night. There are various approaches to doing this, but most commonly the racks are covered with paper or a plastic material. If the latter is used, it must be so arranged as to prevent water from dripping onto the leaves. Sun-curing is used extensively for tobacco intended for local consumption and for chewing tobacco.

After curing, the tobacco leaves are too dry to be handled and must be conditioned to increase their moisture content. This may be achieved by leaving them in a cool barn with water sprinkled on the floor or, in dry countries, by placing the leaves in pits or cellars. After they are fully conditioned, they are sorted into different grades. The graded leaves are then tied in bundles of 12–30 leaves, depending on the country. The small-scale or peasant farmer would normally sell his crop at this stage. The estate grower will bale it. Tobacco bales vary in weight for different qualities from 25–75 kg.

Pyrethrum

Pyrethrum, *Chrysanthemum cinerariifolium*, is a native of Yugoslavia. It is a small, perennial crop plant belonging to the family Compositae (Asteraceae) which is grown for the pyrethrins contained in its achenes (developing fruits). Pyrethrins are the six intecticidal constituents extracted after the flowers have been dried. 'Pyrethrum' is also the name of a contact insecticide manufactured from the extract. Pyrethrum insecticides continue to be in demand, despite their high price compared with synthetic insecticides. This is because of their low toxicity to mammals, allowing them to be used safely near humans and cattle. Pyrethrum insecticides are used in aerosols for household use because of their rapid paralytic action, as well as in mosquito coils, in grain storage powders and in dips for dried fish and meat, to protect them against beetle and blowfly infestations.

Types and ecological adaptation

As pyrethrum is basically a temperate crop it requires a cool climate. In particular, flower initiation takes place at low temperatures, usually less than 15–16°C. An inverse relationship exists between pyrethrum content and temperature, therefore the higher the altitude (lower temperatures) the higher the pyrethrins content. Consequently in the tropics the pyrethrum crop is grown in high altitude areas, particularly in East Africa. The optimum altitudes for flower initiation and high pyrethrins content are 2400–2500 m. Vegetative growth is greatest in warm conditions at lower altitudes, but few flowers are produced.

Pyrethrum grows well in areas with a well-distributed annual rainfall of 1000–1250 mm. However, increased flower production in the following rainy season results from a short dry spell of 1–2 months. The crop thrives in deep, fertile well-drained soils with good structure. As the pyrethrum crop is harvested over a period of 3 years the soil becomes increasingly compressed from weeding and harvesting activities. In these circumstances, soils with poor structure are prone to reduced water infiltration, waterlogging and increased soil erosion. The soil pH should be above 5.6, as the crop does not grow well on highly acidic soils. Where the pH is below this, liming should be carried out.

Planting

The pyrethrum crop is generally propagated by clone selection and vegetative propagation using splits. This is done in an effort to maintain high yields and a high pyrethrins content. The varieties used for propagation by seed in the past produced crops of great variability, especially in their pyrethrins content. As seed propagation is an easier method of propagation than the production of splits, research is presently being carried out in an effort to produce better high-yielding varieties with a consistently high pyrethrins content in the flowers.

The production of pyrethrum splits usually takes place in specialised nurseries. As vegetative growth increases with the warm temperatures at lower altitudes, such areas are particularly suitable for the establishment of nurseries. At warm temperatures, young plants can be maintained in a vigorously growing vegetative condition, allowing them to be divided into many splits without damage to their roots. Older plants become woody and can only be divided into a few splits with a higher tendency to root injury.

The soil in the nursery beds should be cultivated deeply to promote root growth. The beds should be about 1.5 m wide and reasonably flat. The recommended spacing for plants is 25 × 25 cm or 30 × 30 cm. Both nitrogen and phosphate fertilisers should be dug into the soil before splits from a recommended clone are planted at a depth of about 15 cm.

For the first few weeks of plant growth in a nursery the soil should be kept moist, using irrigation water where possible. The provision of shade may be required in areas with low cloud cover. One essential but laborious task in the nursery is the removal of flower buds. This should be done as soon as they appear, to encourage vegetative growth.

The plants produced in nurseries are ready for splitting 3–4 months after planting. They are divided into 4 or 5 large splits, to ensure their survival, when producing material for planting out in field conditions. However, when producing plants for further propagation in nursery conditions they can be divided into as many as 12 small splits. After splitting, any stems or woody parts should be removed, leaving only young vegetative growth. The roots should be trimmed to a length equal to the depth of planting holes, usually about 15 cm, as longer roots curl upwards resulting in poor growth.

Wherever possible splitting and replanting should take place on the same day. If left for longer periods, splits often die from desiccation. The splits can be left overnight before planting if they are spread out, as they tend to heat up when they are left in a pile.

The dipping of splits in a fungicide solution is common practice, to prevent the spread of soil borne fungal infections in young plants. The planting depth for splits is critical, as either soil covering the growing points leading to rotting, or exposed roots leading to desiccation, can cause death. The height of the soil around the split should be identical to the soil level around the parent plant and the soil around each split should be well firmed.

Planting on ridges is recommended for the pyrethrum crop as a means of reducing waterlogging and soil compaction around the plants. The spacing of plants within the row should be 30 cm and the inter-row spacing should be 70–90 cm. To avoid poor plant establishment and low yields in the first year the planting time should be as soon as possible after the beginning of the rainy season. When splits or young plants die, new splits should be planted in the spaces during the first growing seasons, but not thereafter as the exercise becomes uneconomical.

A pyrethrum crop usually gives economic yields for 3 years. After this, yields decline significantly and it is advisable to rotate the crop with a grass ley, a pulse crop or a bush fallow for 3 years. This practice helps to restore soil structure and to prevent the spread of soil-borne pests and diseases.

Fertiliser application

Pyrethrum responds well to phosphate fertilisers and the recommended rate for P_2O_5 is 70–90 kg/ha. The application of nitrogenous fertilisers results in excessive vegetative growth at the expense of flower production.

Weed control

Because the pyrethrum crop grows in areas of high rainfall and the plants have an open growth habit, weeds flourish between the plants. About one weeding a month is necessary and this is usually done by hand (Figure 12.12). At regular intervals

Figure 12.12 Hoeing a pyrethrum crop in Kenya.

weeding should be accompanied by earthing-up of the ridges to further effect weed control. Care should be taken during weeding not to disturb the plant roots. Most herbicides either damage the crop or cause a reduction in pyrethrin levels. Paraquat has been used for weed control with limited success in East Africa.

Pests and diseases

The pyrethrum thrip, *Thrips nigropilosus*, causes damage to the leaves during the dry season. For heavy infestations, control is effected by dusting with dimethoate and 1% HCH (BHC) at 80 kg/ha.

The onion thrip, *Thrips tabaci*, feeds on the flowers but does not cause a reduction in yields, even at fairly high population densities. Root knot nematodes, *Meloidogyne* spp. can cause extensive damage to the root systems and can only be effectively controlled by introducing a crop rotation.

Various soil-borne fungi belonging to the genus *Sclerotium* can cause root rot in the pyrethrum crop. The visible symptoms are leaf wilt and discolouration. The most effective control measure is to adopt a crop rotation.

Harvesting

The pyrethrins content of the flowers is highest when the outer 3 rows of disc florets are open and the ray florets are in a horizontal position. During the flowering season picking should be carried out every 2−3 weeks to ensure maximum pyrethrin levels in the harvested crop. Flower production commences 4 months after the planting of splits, and in the tropics continues for up to 10 months of the year in the presence of adequate moisture. By harvesting the crop regularly and removing all flowers with horizontal ray florets, flower initiation is stimulated.

The technique used for picking is to detach the flower head only, leaving the stem behind. On a dry weight basis the percentage yield of pyrethrins is considerably reduced if stems and buds are included, because of their low pyrethrins content. Flowers should be harvested in open baskets to promote air circulation and prevent fermentation. Fermentation causes a marked reduction in pyrethrins content and can be caused by harvesting wet flowers or by leaving the harvested flowers in heaps before spreading out to dry. The average yield of dried flowers obtained by farmers is 500 kg/ha, although 900–1200 kg/ha can be obtained with good management. At the end of each picking season the pyrethrum plants should be pruned to a height of 15–20 cm and all the old stems should be removed.

Drying

The flowers have to be dried to a moisture content of 10–12% before they can be sent to factories for extraction of the pyrethrins. Inadequate drying can lead to fermentation and a subsequent reduction in pyrethrins content. The flowers are either sun dried or dried in artificial driers. Sun drying usually takes 7–14 days, depending on the weather, and is the method used by most small-scale producers. Artificial driers are frequently used by large-scale growers. In this method the moisture content of flowers is reduced using forced hot air, with temperatures not exceeding 80°C. The dried flowers are then stored in bales, and transported to extraction plants as soon as possible, to prevent reductions in pyrethrins content.

CHAPTER 13

Management of forage crops

Grasses are by far the most extensively grown of all the forage crops, and in the tropical regions they may cover up to two-thirds of the land area. This is particularly the case in regions of broad-leaved woodland, thorny woodland and subdesert scrub where grass, rather than forest, tends to be the natural vegetation.

Forage grasses

The grass family Gramineae (Poaceae) includes about three quarters of the cultivated forage crops and all the cereals. Of the approximately 6000 known species in the family, a large proportion are grown in the tropics.

Grasses are botanically characterised by having alternate leaves with parallel veins and jointed, usually hollow, cylindrical stems, with cross partitions at the nodes. The basal portion of the leaf (sheath) encloses the stem. The sheath is open on the opposite side to the blade. The majority of grasses are cross-pollinated, mostly by wind, the few exceptions including wheat, barley and oats, are normally self-pollinated. The roots of all grasses are slender and have few branches.

It is not possible to list in this volume all the grass species and strains which occur in cultivated pastures in various parts of the tropics. Some of the important grasses are listed in Table 13.1. A selection of these which are extensively used for both permanent and temporary pastures in different environmental conditions are discussed with respect to their agronomic characters. Although most of these grasses yield high amounts of dry matter, their content of crude protein is invariably low relative to forage legumes. When it is desirable to combine high yields of dry matter with high levels of crude protein from a forage crop, only two options are available; either to apply adequate amounts of nitrogenous fertilisers, or to grow the grasses in mixtures with leguminous forage crops. Some of the low-yielding natural grasslands, such as those in the sahel, have relatively high crude protein contents (Figure 13.1).

Figure 13.1 An Ayreshire cow grazing on cultivated pasture in Tanzania.

283

Table 13.1. Some important forage grasses grown in the tropics

Botanical name	Common name
Andropogon gayanus	Gamba grass
Brachiaria brizantha	Signal or Palisade grass
Brachiaria decumbens	—
Brachiaria mutica	Para grass
Cenchrus ciliaris	Buffel grass
Cynodon dactylon	Star, Bahama or Bermuda grass
Chloris gayana	Rhodes grass
Digitaria decumbens	Pangola grass
Echinochloa pyramidalis	Antelope grass
Eragrostis curvula	Weeping love grass
Eriochloa polystachyus	Carib grass
Hyparrhenia rufa	Jaragua grass or Thatch grass
Ischaemum aristatum	Batiki blue grass
Melinis minutiflora	Molasses grass
Panicum maximum	Guinea grass
Paspalum notatum	Bahia grass
Pennisetum purpureum	Elephant or Napier grass
Pennisetum clandestinum	Kikuyu grass
Pennisetum pedicellatum	Kyasuwa grass
Setaria sphacelata	Golden Timothy
Stenotaphrum secundatum	Buffalo or Crab grass
Themeda triandra	Red oat grass
Tripsacum laxum	Guatemala grass

Gamba grass

Gamba grass, *Andropogon gayanus*, is a widely distributed perennial grass which grows to a height of 2–3 m. It is tufted, relatively drought and fire tolerant and grows well even in regions with an annual rainfall of less than 600 mm. It is a commom species in the dry savanna zone of West Africa and had been successfully introduced into Brazil and tropical Australia. Gamba grass is able to survive in conditions of low soil fertility and is widely used in the control of erosion. The crop sets seed profusely.

Signal grass

Signal grass, *Brachiaria brizantha*, which is also referred to as palisade grass, is a tall rhizomatous perennial which originated in tropical Africa. It is an important pasture grass in many tropical areas, including Sri Lanka, northeast Australia and the West Indies. The plant is leafy, drought resistant, tolerant of shade and more productive and palatable than *Andropogon gayanus*. Signal grass remains green into the dry season and is therefore good for grazing. Propagation is mostly by seed, although it can also be by division of root stocks.

Star grass

Star grass, *Cynodon dactylon*, also variously referred to as Bermuda grass, Bahama grass or Dhub grass, is a perennial grass with rapidly growing stolons which quickly cover bare ground, forming a dense turf. *C. dactylon* is considered by some authorities to be a distinct species from *C. plectostachyum*. However, in this volume they are treated together.

Star grass originated in East Africa, but is now widely grown throughout the tropics and subtropics. It does well at altitudes of up to 2500 m and in areas with an annual rainfall of more than 500 mm. The plant grows to a height of 1.2 m. It withstands grazing well, indeed close grazing of star grass is essential for the maintenance of a good pasture. The grass responds well to nitrogenous fertilisers which are applied to limit the development of stolons. It remains relatively green even in the dry season and is therefore a valuable species for grazing. In areas of low humidity some varieties produce viable seed. However, propagation is generally by vegetative means. In general, the rhizomatous strains are used for permanent pastures whilst the stoloniferous varieties are used in leys.

Rhodes grass

Rhodes grass, *Chloris gayana*, is a leafy perennial grass of Central and East African origin. It is well adapted to many areas in the tropics with an annual rainfall range of 750–1500 mm.

The plant is a leafy perennial which grows to a height of 0.3–1.2 m when ungrazed. It spreads to some extent by creeping stems (stolons) which form roots at the nodes to give the appearance of a tufted sward when mature. The swards are sufficiently open to permit the mixing of other forage grasses or legumes. The pasture is used for both grazing and for hay or silage making. It is always advantageous to allow new stands of Rhodes grass to flower and set seed before being grazed or cut. Propagation is normally by seed. Among high-yielding cultivars of the species are Pioneer, Katambora, Samford and Callide.

Jaragua grass

Jaragua grass, *Hyparrhenia rufa*, is a tufted perennial that grows up to a height of 3 m. In many ways it is similar to *Andropogon gayanus*, and in Nigeria it is the second most important forage grass. Jaragua grass is widely distributed throughout tropical Africa, and Central and South America. It has poor persistence with close grazing. Propagation is mostly by seed, although root stocks can also be used.

Guinea grass

Guinea grass, *Panicum maximum*, is an erect perennial grass which is a native of Africa, but is now widely distributed in all parts of the tropics and subtropics, from sea level to an altitude of 2000 m. It is best adapted to warm and humid regions with over 900 mm of rainfall each year. The plant grows up to a height of 3 m, remains tufted but does not form a sward. As a result of its deep root system, the species is capable of surviving relatively long droughts. It is shade-tolerant and competes reasonably well with other tall grasses and legumes. However, with continuous grazing Guinea grass dies out rapidly.

The plant sets seed readily, but the seeds are invariably of low viability. Propagation is either by seed or by divided stolons. Among high-yielding cultivars of Guinea grass are Hamil, Coloniao Guinea and Gatton Penic.

Elephant grass

Elephant grass, *Pennisetum purpureum*, also referred to as Napier grass, is distributed throughout tropical Africa. The plant is a tufted perennial which grows to a height of 4.5 m. It is deeply rooted, with thick cane-like stems. It spreads by short and stout rhizomes to form large clumps or stools, up to 1 m across. Elephant grass grows best in warm conditions, especially where the annual rainfall exceeds 1000 mm. It is strongly drought-tolerant and highly responsive to irrigation. The grass grows in a wide range of soils, but is most productive on deep, loamy and free-draining soils of moderate to high fertility.

Elephant grass has good persistence under intensive grazing and has also been used for restoration of soil fertility. In addition to giving a palatable feed for grazing, it is an excellent grass for silage. Propagation is from shoots, stem cuttings, stolons or crown splits.

Pangola grass

Pangola grass, *Digitaria decumbens*, is a low-growing, creeping perennial which is a native of southern Africa. It is now widely distributed in the Caribbean countries, the southern USA, Australia and other tropical regions. It flourishes with abundant rainfall, requiring more than 1000 mm yearly. Growth virtually ceases at below 11°C and during the dry season in the savanna region of West Africa little growth occurs. It tolerates a wide range of soils, including those with a slight waterlogging problem as well as those prone to drought. The plant spreads by long-rooting stolons and forms an open turf. Seed production is generally poor and as a result propagation is invariably by plant division (tillers).

Buffel grass

Buffel grass, *Cenchrus ciliaris*, also referred to as the African foxtail, is a native of India and Indonesia. It is cultivated for permanent pastures in Africa, Australia and other parts of the tropics. The grass is a drought resistant perennial, which thrives well in areas with a rainfall of 300–1000 mm. The plants remain relatively green in the dry season, although the mature stems become wiry. Most varieties grow well on light-textured, deep and free-draining soils, but the taller and more rhizomatous types thrive best on heavy clays, provided calcium is not deficient. Buffel grass tolerates flooding better than most other grasses, and soils with poor drainage.

The widely cultivated and high-yielding varieties of Buffel grass are Biloela, Nunbank, Boorara, Tarewinnabar, Molopo and Lawes, all of which grow to about 1.5 m. Gayndah, American and Cloncurry are all medium in height at 1.05 m and Western Australia is a variety which averages only 0.75 m. Buffel grass is established mainly from seed, although vegetative propagation from tuft splits is occasionally used. Storage, moist-chilling or pelleting with superphosphate improves the seed germination rate.

Forage legumes

With respect to the composition of forage crops, the family Leguminosae (Fabaceae) is second in importance to the grass family. A distinguishing botanical characteristic of this family is that the roots bear swellings called 'nodules' which are caused by the activity of the bacteria *Rhizobium* spp. Each nodule contains a number of bacteria which are able to assimilate atmospheric nitrogen. The activity of these organisms accounts for the great agricultural importance of legumes in increasing soil nitrogen levels. Different leguminous crops vary in their ability to fix nitrogen from the air. About 250 other species of plants are able to fix atmospheric nitrogen, but only the legumes are important in agriculture.

A great number of leguminous crops have been successfully used as forage for the feeding of animals. The important species are listed in Table 13.2 and a few are briefly discussed with respect to their agronomic characteristics.

Forage legumes are rarely grown in pure stands, primarily because of their low yield and unsuitability as silage material. However, in areas with the problem of seasonal rainfall, a valuable use of irrigated land is the production of protein-rich leguminous green fodder. To a limited extent, legumes are also grown in pure stands without irrigation for the production of hay. As an example, in South Africa, Zimbabwe and many parts of South America, *Medicago sativa* (alfalfa or lucerne) is the most widely used species for hay without irrigation. In several parts of Africa *Glycine max*, *Vigna unguiculata*, *Stizolobium deeringianum* and *Dolichos lablab* are grown in pure stands for hay in the absence of irrigation (Figure 13.2).

Stylo

A bushy or partially erect herbaceous legume, stylo *Stylosanthes guyanensis* (formerlly *S. gracilis*), is a perennial herb which is believed to have originated in Central and South America. It is a popular pasture species in Hawaii, Kenya, Uganda, Brazil and Australia. There are many members of the genus, but apart from *S. guyanensis* only two others are of

Figure 13.2 A protein rich leguminous green fodder crop.

Table 13.2. Some important forage legumes grown in the tropics

Botanical name	Common name
Alysicarpus vaginalis	Alyce clover
Alysicarpus glumaceus	
Alysicarpus rogosus	
Arachis hypogaea	Groundnut
Calapogonium mucunoides	
Centrosema pubescens	Centro
Clitoria ternata	
Desmodium spp.	
Dolichos spp.	Hyacinth bean
Glycine javanica	
Glycine max	Soyabean
Indigofera endecaphylla	
Leucaena leucocephala	
Leucaena glauca	
Lotononis bainesii	
Medicago sativa	Lucerne, alfalfa
Mucuna stizolobium	
Phaseolus lathyroides	
Pueraria phaseoloides	Tropical kudzu
Stizolobium deeringianum	Velvet bean
Stylosanthes guyanensis	Stylo
Trifolium spp.	Clover
Vigna spp.	Cowpea

major importance as pasture legumes; *S. humilis* (Townsville stylo) and *S. hamata* (Caribbean stylo).

Stylo grows well in a wide range of environmental conditions. It is more tolerant to cool weather than most tropical forage legumes and can survive frost. It is not shade tolerant and for best yields cultivation should not be at altitudes above 1800 m. The desired rainfall range should be between 600–2500 mm. The crop is able to survive long dry periods, and compared with other cultivated legumes, it produces more dry matter during the dry season. The production of dry matter increases with increasing daylength. The crop tolerates low soil fertility, acidic soils, poor drainage, shallow soils, heavy clays and a limited degree of flooding. From the point of view of usage, stylo is palatable when young, but this reduces with age as the stems are inclined to become woody.

Propagation is normally by seed. Due to the nature of the seed which has a hard seedcoat, the germination rate is usually low, around 5–10%. However, when the seed is mechanically scarified or treated with hot water or acid, the germination rate is increased to almost 100%. This process is therefore essential for good crop establishment.

Centro

Centrosema or centro, *Centrosema pubescens*, is a prostrate, trailing, leafy perennial. It produces pods which are dark-brown, about 12 cm in length, containing up to 20 brownish-black seeds. The plant is specific in its requirements for adapted strains of nodule-forming bacteria.

A short-day plant, centro prefers humid conditions, where annual rainfall is over 1000 mm. It prefers warm temperatures, although it survives frosty conditions. Because it is tolerant of shade, it is commonly used as a cover crop in plantations as well as in pasture mixtures. Even though it is tolerant of shade, centro is able to compete strongly with tall forage grasses. Propagation is by seed which establishes easily. The plant also regenerates easily and freely when grazed.

Grass–legume mixtures

Although pure stands of fodder grasses are easier to establish and manage than grasses grown in mixtures with legumes, one of the most important problems of management for high forage grass production is the continued provision of adequate amounts of nitrogen at frequencies and levels that ensure high nutritive values. By growing grasses and legumes in mixtures, not only is the quality of the forage improved and the need for nitrogen fertilisers reduced, but also the practice helps in the improvement and maintenance of soil structure and fertility (Table 13.3).

Unfortunately, it is not easy to identify mixtures in which a satisfactory balance is maintained between grass and legume species, while still yielding as well as pure grass stands. Success depends not only on finding compatible perennial legumes and grasses, but also on determining the best sowing time, spacing and grazing procedures to enable a balanced mixture to be maintained. While some legumes show excellent qualities as forage crops, there are problems of combining them satisfactorily with tall fodder grasses. In Zimbabwe, a mixture involving *Glycine javanica* and *Pennisetum purpureum* (Elephant grass) has proved satisfactory for silage and some grazing in high-rainfall areas, but

Table 13.3. Yield and protein content of grasses grown alone and in association with clover, Kenya

	Grass grown alone	Grass in association with clover
Dry matter yield (tonnes/ha) 1957	2.04	2.81
Dry matter yield (tonnes/ha) 1958	1.94	3.06
Crude protein content (%) 1957	7.7	9.6

Source: Strange, 1961.

has not been successful elsewhere in the same country. This and other similar examples indicate that combinations of forage grass−legume mixtures in which a satisfactory balance is maintained can only be suggested for specific areas and growth conditions.

Examples of some popular grass−legume mixtures as practised in different parts of the tropics include the following:

(a) *Chloris gayana* grown in association with *Centrosema pubescens, Clitoria ternata, Desmodium uncinatum, Glycine wightii, Stylosanthes* spp. *Trifolium* spp. *Lotononis bainesii* or *Medicago sativa.*

(b) *Pennisetum purpureum* grown in association with *Centrosema pubescens, Glycine javanica, Clitoria* spp., *Cajanus indicus, Desmodium uncinatun, Indigofera endecaphylla, Pueraria phaseoloides* or *Trifolium* spp.

(c) *Panicum maximum* grown in association with *Centrosema pubescens, Desmodium salicifolium, Glycine javanica, Indigofera endecaphylla, Leucaena glauca, Pueraria phaseoloides* or *Stylosanthes guyanensis.*

(d) *Paspalum dilatatum* grown in association with *Alysicarpus vaginalis, Desmodium salicifolium, Indigofera endecaphylla, Lotononis bainesii, Pueraria phaseoloides* or *Trifolium* spp.

(e) *Melinis minutiflora* grown in association with *Pueraria javanica, Pueraria phaseoloides, Stylosanthes guyanensis, Phaseolus lathyroides, Centrosema pubescens* or *Glycine javanica.*

(f) *Brachiaria mutica* grown in association with *Stylosanthes guyanensis, Centrosema pubescens* or *Trifolium* spp.

(g) *Brachiaria decumbens* in association with *Stylosanthes guyanensis.*

(h) *Brachiaria distachya* in association with *Alysicarpus vaginalis.*

(i) *Cynodon dactylon* in association with *Centrosema pubescens.*

(j) *Cynodon plectostachyum* in association with *Centrosema pubescens.*

(k) *Digitaria decumbens* in association with *Lotononis bainesii.*

(l) *Hyparrhenia rufa* in association with *Pueraria phaseoloides.*

(m) *Bromus ciliatus* in association with *Trifolium* spp.

(n) *Phalaris tuberosa* in association with *Trifolium* spp.

(o) *Andropogon gayanus* in association with *Centrosema pubescens.*

(p) *Cenchrus ciliaris* in association with *Stylosanthes guyanensis.*

(q) *Axonopus compressus* in association with *Desmodium canum.*

(r) *Dichanthium cariosum* in association with *Stylosanthes hamata.*

(s) *Setaria sphacelata* in association with *Glycine javanica.*

(t) *Stenotaphrum secundatum* in assocation with *Pueraria phaseoloides.*

(u) *Heteropogon contortus* in association with *Stylosanthes guyanensis.*

Fodder from trees and shrubs

In most tropical areas where there are distinct dry and rainy seasons, trees and shrubs make a considerable contribution to animal feed during the dry season, particularly if it is prolonged. Goats, cattle and sheep, especially in the dry tropics, obtain an appreciable proportion of their feed from leaves, flowers, pods, seeds, twigs and bark of a large number of plants, including *Acacia* spp., *Cajanus cajan* (pigeon pea) and *Leucaema* spp. It is not uncommon, especially during drought periods, for herdsmen to depend almost entirely on breaking off leaves and branches from trees and shrubs, to feed to their animals.

Most of the trees and shrubs browsed by animals come into leaf well before the rains and the edible parts retain their fairly high nutritive value well into the dry season.

Cultivation practices

Land preparation

Wherever possible, the site for the production of forage crops, whether legumes or grasses, as pure stands or in mixtures and for permanent or temporary pasture (ley), should be well-drained. The first stage in the establishment of sown pastures involves under-brush felling and stumping of trees. If land clearing is done in the dry season, fire plays a prominent role. Some of the cost for land clearing may be recovered by growing an early season crop prior to sowing the pasture.

The degree of success achieved in the establishment of pasture very much depends on the extent of tillage prior to sowing. A satisfactory seedbed can be achieved by disc-ploughing or sometimes by cross ploughing, following clearing and stumping. This may be followed by harrowing and firming. The aim is to compact the soil almost to the top and yet leave a shallow surface mulch by harrowing, to ensure soil and seed contact for enhanced germination, seedling emergence and development. In rainfed conditions care must be taken not to cultivate the soil excessively, to avoid the danger of moisture depletion by evaporation.

Where propagation is by stem cuttings, stolons, plant division, side shoots (tillers) or other vegetative means, the demand in terms of seedbed preparation is relatively less rigorous, although a uniform and weed-free seedbed must be ensured.

Planting

Seeds and cuttings should be sown at the beginning of the rains. Dry seeding of forage grasses or legumes is advisable only where the time of the onset of rains is reliably predictable. Since many tropical legumes yield a high percentage of hard seed, their scarification by hot water or acid treatment may be advantageous or sometimes vital. Where feasible, an inoculum of *Rhizobium* should be used in the establishment of legumes, unless appropriate strains of *Rhizobium* are known to exist in the particular soil and crop sequence.

The seed rate is dependent on the variety, the location and the method of seeding. In general, a greater amount of seed is required if it is sown broadcast rather than by drilling in rows. In a grass and legume mixture, the seed rate of each species is approximately halved. There are no exact rules regarding how the two crops should be arranged in the field, but the best mixtures usually result when the grass is drilled in rows and the legume is grown between the rows. Where propagation is by cuttings, especially with grasses, the number of nodes on each cutting should not be more than three. The cuttings should be planted end-to-end in furrows about 6 cm deep. Cuttings that are longer than three nodes tend to bend upwards and they become exposed.

Fertiliser application

The response of pure stands of forage grasses to most mineral fertilisers is small, except where fairly heavy dressings of nitrogen are being applied. Grasses generally have a lower requirement than legumes for phosphorus and lime and are efficient in extracting soil potassium. Where productivity is maintained at a high level by large dressings of nitrogen, both phophorus and potassium fertilisers may also be required. This is certainly the case on soils which are deficient in these nutrients and where the forage is grazed by high-yielding animals.

The successful establishment of forage grasses in association with legumes often reduces the need for nitrogen, although it increases the requirement for other minerals, especially phosphorus, potassium and calcium. Liming materials may have to be applied to raise the pH of acid soils to a level more suited to some of the legumes. In some areas it may be necessary to supply certain of the micronutrients which may be deficient. In general, the application of potassium is not necessary in the early years of well-established mixed pastures because the element is returned to the soil in a readily available form in the urine of grazing animals. However, in later years of the mixture, potassium would need to be applied, especially on light soils.

Weeds, pests and diseases

It is important to control weed growth during the first four to six weeks of the establishment of sown pastures. This may be achieved by hand-pulling of weed plants or by the use of appropriate herbicides. Once the pasture is well established weeds are mostly smothered and subsequent weeding may only be necessary for broad-leaved species, which can be controlled by spraying appropriate herbicides. The type of herbicide to use, and the rate and method of application are necessarily dictated by the crop species, the cultivars in use, the prevalent weed species, the soil type and the climatic conditions. In addition, the choice of herbicide should ensure that there can be no harmful residual effects when the forage is fed to animals. The control of broad-leaved weeds using herbicides is complicated by the presence of legumes in the mixture.

Forage grasses and legumes in the tropics are relatively free from serious pests and diseases. However, occasional leaf-spots and virus infections may be observed, especially during wet weather, but the effects are generally negligible. The control of virus attacks invariably requires the use of resistant cultivars.

Grazing of sown forage

Although maximum yields of dry matter are obtained when forage crops are harvested at or near maturity, the nutritive value and digestibility of the forage at this advanced stage are low. Young plant growth, with a high proportion of leaf to stem, is the highest quality in terms of maximum crude protein and minimum crude fibre content. Management should be regulated to prevent most forage plants from flowering. If this cannot be achieved by grazing, the fodder should be cut and conserved as silage or dried as hay.

Good grazing management provides for periods of rest to allow the forage to recover from grazing. In order to allow for adequate periods of rest and to preserve a favourable balance of crop species, it is necessary to carefully control the grazing. The utilisation of a sown pasture is often grouped into five approaches as follows:

Continuous grazing

This involves leaving the animals to feed on the same pasture area for prolonged periods. This system of grazing is the most prevalent on range lands and tropical grasslands where fencing is economically impracticable. The system is particularly suited to a situation where the number of animals is relatively low. The animals invariably exist through the dry season on sub-maintenance rations, as a result of reduced plant growth and availability of forage. Continuous grazing in the tropics has the disadvantage of creating a build-up of tick and nematode infections.

Rotational grazing

The system of rotational grazing is invariably practised on improved permanent or ley pastures. The system overcomes the disadvantage of under- and over-grazing which is inherent in the continuous system. In rotational grazing the grazing area is sub-divided into a number of paddocks and the animals are moved systematically from one to another, in rotation. The stocking rate of each paddock is high. By the time the last paddock in the sequence has been grazed, the first should be ready for grazing for the second time. In this system, pasture is used when it is young and nutritious and then allowed an adequate period of recovery.

Strip grazing

This is an intensive method of rotational grazing which is based on electric fencing. By the use of a movable electric fence, the amount of herbage offered to the animals is restricted. The system minimises selective grazing by the animals and also checks on wastage. A movable back-fence prevents the premature grazing of regrowth.

Deferred grazing

The system of deferred grazing involves setting aside specific pasture areas or paddocks for later use as standing hay. The system is of particular importance in the dry tropics where range pastures are grazed almost continuously during the dry season. The practice also has the advantage of improving range pastures by allowing self-sown seedlings to become established.

Zero grazing

In zero grazing the animals do not graze the pasture as such, but the forage is cut and fed to the animals, either fresh daily or in the form of hay or silage (Figure 13.3). The system has the advantage of increased animal production due to more efficient forage utilisation. It allows for an increased stocking rate and an intensification of management.

Figure 13.3 The rear view of a tractor-operated baler making hay.

Effects of cutting

When fodder crops are cut regularly, increasing the intervals between cutting results in increased yields and crude fibre levels, but reduced production of crude protein and ash. Therefore, a cutting interval must be selected which gives a satisfactory balance between nutrient value, palatability and amount of dry matter produced. The interval must necessarily vary with the species, cultivar and weather factors. In general, the interval between cuts needs to be longer in the dry season than during the rains. It should also be noted that cutting too frequently weakens the forage, reduces the rate of regrowth and results in reduced dry matter production.

The height of cutting also affects yield, the rate of regrowth and the persistence of the crop. In general, cutting at ground level or too high lowers the rate of regeneration and yield. For most forage species a cutting height of 7.5–10 cm is considered ideal.

Bibliography

Abbot, J.C. and **J.P. Makeham**. (1979). *Agricultural Economics and Marketing in the Tropics*. Longman Group Ltd., London, 168pp.

Acland, J.D. (1971). *East African Crops*. FAO/Longman, 252pp.

Ahn, P.M. (1970). *West African Soils*. Oxford University Press, 332pp.

Akehurst, B.C. (1971). *Tobacco*. Humanities Press, New York, Longman Group Ltd, London, 551pp.

Anderson, H.W., M.D. Hoover and **K.G. Rainhart**, (1976). *Forests and water: effects of forest management on floods, sedimentation and water supply*. USDA Forest Service General Technical Report, **PSW-18**, 115pp.

Anonymous. (1968). *Insect Pests of Nigeria*. Research Bulletin, Univ. of Wisconsin, 83 pp.

Anonymous. (1975). *International Workshop on Grain Legumes*. Proceedings of the International Workshop, ICRISAT Center, India, 350pp.

Anonymous. (1975). *Aviation Weather for Pilots and Flight Operations Personnel*. National Aviation Authority and National Weather Service, US Govt. Printing Office, Washington D.C.

Anonymous. (1980). *Fertilisers and their application to crops in Nigeria*. Fertiliser use series No. 1, Federal Department of Agriculture, Federal Ministry of Agric. & Rural Develop., Lagos, 191pp.

Anonymous (1981). *The winged bean: a high protein crop for the tropics*. National Academy of Sciences, USA, 46pp.

Anonymous. (1984). *Agrometeorology of Sorghum and Millet in the Semi-Arid Tropics*. Proceedings of the International Symposium, ICRISAT Center, India, 322pp.

Anthes, R.A., H.A. Panafsky, J.J. Gahir and **A. Rango**. (1978). *The Atomosphere*. Merrill Publishing Co., Columbus, Ohio.

Are, L.A. and **D.R.C. Gwynne-Jones**. (1973). *Cocoa in West Africa*. Ibadan, Oxford University Press, 146pp.

Bear, F.E. (1965). *Soils in Relation to Crop Growth*. Reinhold Publishing Crop., Chapman and Hall Ltd., London, 303pp.

Cannell, M.G.R. (1983). Coffee. *New Biologist*, **30**(5): 257–263.

Cobley, L.S. and **W.M. Steele**. (1976). *An Introduction to the Botany of Tropical Crops*. (2nd Edn.) Longman, 371pp.

Cohen, I. (1975). From biological to integrated control of citrus pests in Israel. *Citrus*. CIBAGEIGY Technical Monograph, No **4**, Basle: 38–41.

Edmond, J.B., T.L. Senn, F.S. Andrews and **R.G. Halface**. (1975). *Fundamentals of Horticulture*. Hill, 560pp.

Emechebe, A.M. (1975). *Some Aspects of Crop Diseases in Uganda*. Makerere Univ. Printery, Kampala, 43pp.

Emechebe, A.M. (ed.) (1980). *A Checklist of Diseases of Crops in the Savanna and Semi-Arid Areas of Nigeria*. Samaru Miscellaneous Paper, No. 100, Inst. for Agric Research, Ahmadu Bello University, Zaria, 38pp.

Empire Cotton Growing Corporation (1951–56). Progress reports from experimental stations, Namulonge, Uganda, Lake Province, Tanganyika.

Fisher, N.M. (1979). Polyculture and Implications for Appropriate Tillage. In *Proceedings of Appropriate Tillage Workshop*, Zaria, Nigeria, 16–20, Jan., Commonwealth Secretariat, London, 65–72.

Fröhlich, G. and **F.W. Rodewald**. (1970). *Pests and Diseases of Tropical Crops and their Control*. Pergamon Press, Oxford, 371pp.

Grimes R.C. and **R.T. Clarke**. (1962). Continuous arable cropping with the use of manure and fertilizers, *E. Afr. agric. for. J.*, **28**, 74–80.

Hacket, C. and **J. Carolane**. (1982). *Edible Horticultural Crops: A Compendium of Information on Fruit, Vegetable, Spice and Nut Crops*. Academic press, 671pp.

Hartley, C.W.S. (1976). *The oil palm*. Longman, London, Second Edition.

Hartley, C.W.S. (1977). The Oil palm. In *Outlines of perennial crop breeding in the tropics*. Ferwerda, F.P. and F. Wit (Eds.), Misc. Paper, No. 4, Landbouwhogeschool, Wageningen, Netherlands.

Hartmann, H.T. and **D.E. Kester**. (1975). *Plant propagation*. Englewood, Cliffs.

Hayes, W.B. (1945). *Fruit growing in India*. Kitabistan, 283pp.

Herklots, G.A.C. (1972). *Vegetables in South-East Asia*. London, George Allen and Unwin, 525pp.

Herren, G.R. (1981). Biological control in the cassava mealybug. In *Tropical root crops. Research strategies for the 1980s*. Terry, E.R., K.A. Oduro and F. Caveness (eds.). Proceedings, first triennial root crops symposium, Ibadan, Nigeria, 8–12 September 1980.

Hill, P. (1972). *Rural Hauss: A Village and a Setting.* Cambridge University Press, 383pp.

Hill, D.S. (1975). *Agricultural Insect Pests of the Tropics and Their Control.* Cambridge University Press, London, 516pp.

Humphreys, L.R. (1978). *Tropical Pastures and Fodder Crops.* Intermediate Tropical Agriculture Series, Longman, 135pp.

Jones, L.H. (1983). The oil palm and its clonal propagation by tissue culture. *Biologist,* **30(4):** 181–188.

Jones, M.J. and A. Wild. (1975). *Soils of the West African Savanna — Their maintenance and Improvement of Their Fertility.* Comm. Bur. Soils. Tech. Bull. No. 55, 246pp.

Joy, D. and J. Wibberley, (1979). *Tropical Agriculture Handbook,* Cassell Ltd., London and Ibadan, 219pp.

Kowal, J.M. and D.T. Knabe. (1972). *An Agroclimatological Atlas of the Northern States of Nigeria.* Ahmadu Bello University Press, Zaria, Nigeria, 111pp.

Kowal, J.M. and A.H. Kassam. (1978). *Agricultural Ecology of Savanna: A Study of West Africa.* Clarendon Press, Oxford, 403pp.

Kramer, P.J., (1969). *Plant and Soil Water Relationships: A Modern Synthesis.* McGraw Hill Book Co., New York.

Kranz. J.H. Schmutterer, and W. Koch. (eds.) (1978). *Diseases, Pests and Weeds in Tropical Crops.* John Wiley, Chichester. 666pp.

Kulkarni, L.G. (1959). *Castor.* Examiner Press, Fort, Bombay. 107pp.

Leakey, C.L.A. and J.B. Wills. (1977). *Food Crops of the Lowland Tropics.* Oxford University Press, 345pp.

Masefield, G.B. (1949). *A Handbook of Tropical Agriculture.* Oxford University Press London, 196pp.

McIlroy, R.J. (1964). *An Introduction to Tropical Grassland Husbandry.* Oxford University Press, 128pp.

McIlroy, R.J. (1967). *An Introduction to Tropical Cash Crops.* Ibadan University Press, 157pp.

Millar, C.E. (1975). *Soil Fertility.* John Wiley and Sons Inc. New York, 436pp.

Ochse, J.J., M.J. Soule Jr., M.J. Dijkman and C. Wehlburg. (1961). *Tropical and sub-Tropical Agriculture,* Vol. I. The Macmillan Co. Ltd., 760pp.

Ochse, J.J., M.J. Soule Jr., M.J. Dijkman and C. Wehlburg. (1961). *Tropical and sub-Tropical Agriculture,* Vol. II. The Macmillan Co. Ltd., 605pp.

Okigbo, B.N. and D.J. Greenland. (1976). Intercropping Systems in Tropical Africa. In *Multiple Cropping* (Papendick, R.I. *et al*;, eds). African Society.

Olaitan, S.O. and G. Lombin. (1984). *Introduction to tropical soil science.* Macmillan Intermediate Agriculture Series, Macmillan, 126pp.

Olayide, S.O. (1976). *Economic Survey of Nigeria.* Agromolaran Publishing Company Ltd., Ibadan, 203pp.

Onwueme, I.C. (1978). *The Tropical Tuber Crops.* John Wiley and Sons, 234pp.

Onwueme, I.C. (1978). *Crop Science.* Cassell's Tropical Agriculture Series Book 2, Cassell, London, 106pp.

Pearson, L.C. (1967). *Principles of Agronomy.* Reinhold Corporation, New York, 434p.

Philips, T.A. (1977). *An Agricultural Notebook (with special reference to Nigeria).* Longman Group Ltd., London, 312pp.

Purseglove, J.W. (1968). *Tropical Crops — Dicotyledons.* Longman, 719pp.

Purseglove, J.W. (1969). *Tropical Crops — Monocotyledons.* **Vol. I** and **II,** Longman, 607pp.

Quinn, J.C. (1980). *A review of Tomato Cultivar Trials in the Northern States of Nigeria.* Samaru Miscellaneous Paper, No. 84. Inst. for Agric. Research, Ahmadu Bello University, Zaria, 111pp.

Van Rheenen, H.A. (1972). *Major problems of growing Sesame (Sesamum indicum) in Nigeria,* 121pp.

Rosenberg, N.J., (1974). *Microclimate: The biological environment.* John Wiley and Sons, New York.

Ruthenberg, H. (1980). *Farming Systems in the Tropics.* (3rd edn.). Clarendon Press, Oxford, 327pp.

Salisbury, F.B. and C.W. Ross, (1978). *Plant Physiology.* Wadsworth Publishing Co. Inc., Belmont, California.

Samson, J.A. (1970). Rootstock for tropical trees. *Tropical Abstracts,* **25:** 145–151.

Samson, J.A. (1980). *Tropical Fruits.* Tropical Agriculture Series, Longman, 250pp.

Sellers, W.D. (1965). *Physical Climatology.* The University of Chicago Press, 272pp.

Shepherd, G.S. (1958). *Marketing Farm Products— Economics Analysis.* Iowa State Univ. Press, AAmes, Iowa, 532pp.

Simmonds, N.W. (1966). *Banana.* Longman, 512pp.

Singh, S. (1967). *Fruit culture in India.* Indian Council for Agricultural Research, New Delhi, 424pp.

Singh, A. (1980). *Fruit Physiology and Production.* Kslysni Publishers, New Delhi, 513pp.

Stanton, W.R. (1966). *Grain Legumes in Africa.* F.A.O. 184pp.

Strange. R. (1961) Effects of legumes and fertilisers on yields of temporary leys. *E. Afr. agric. for. J.,* **20,** 221–4.

Tempany, H. and D.H. Grist. (1956). *An Introduction to Tropical Agriculture.* Longmans, Green and Co.

Terra, G.J.A. (1966). *Tropical Vegetables.* Publication of the Royal Tropical Institute, Amsterdam, the Netherlands, 107pp.

Thompson, H.C. and W.C. Kelly. (1957). *Vegetable Crops.* McGraw Hill, 611pp.

Tindall, H.D. (1965). *Fruits and Vegetables in W. Africa,* F.A.O. Rome, 269pp.

Tindall, H.D. (1968). *Commercial Vegetable Growing.* Oxford, 300p.

Tindall, H.D. (1971). *Vegetable crop research in anglophone W. Africa*. Seminar on Agric. Res. in West Africa on vegetable crops, University of Ibadan 21pp.

Tindall H.D. (1976) Horticultural training in tropical Africa. *Chronica Horticultura, 16* (13)pp. 21–23.

Tisdale, S.L. and **W.L. Nelson** (1975). *Soil Fertility and Fertilisers*. (3rd Edn.) Macmillan Publishing Company Inc. New York, 694pp.

Truelove, B. (1977). *Research Methods in Weed Science*. Seventh Weed Science Society, 221pp.

Webster, C.C. and **P.N. Wilson.** (1980). *Agriculture in the Tropics*. The English Language Book Society and Longmans, 488pp.

Wellman, F.L. (1961). *Coffee: botany, cultivation and utilisation*, Leonard Hill London.

Westphal, E. (1974). *Pulses in Ethiopia: Their Taxonomy and Agricultural Significance*. Centre for Agricultural Publishing and Documentation, Wageningen, 263pp.

Wheeler, B.E.J. (1969). *An Introduction to Plant Diseases*. John Wiley. Chichester, 374pp.

Williams, C.N. (1975). *The Agronomy of the Major Tropical Crops*. Oxford University Press, 228 pp.

Williams, C.N., W.Y. Chew and **J.H. Rajaratnam.** (1980). *Tree and Field Crops of the Wetter Regions of the Tropics*. Intermediate Tropical Agriculture Series Longman, 262pp.

Wrigley, G. (1971). *Tropical Agriculture. The Development of Production*. Faber and Faber Ltd., London, 376pp.

Appendix 1

Common and trade names of selected agricultural chemicals

Common name	Trade name
Herbicides	
acrolein	Magnacide
alachlor	Lasso
allidochlor (CDAA)	Randox
ametryne	Gesapax
amitrole	Weedazol
asulam	Asulox
atrazine	Gesaprim
barban	Carbyne
benefin	Balan
bensulide	Betasan
bentazone	Basagran
benzadox	Topcide
benzipram	S 18510
bifenox	Modown
bromacil	Duracil 800
bromoxynil	Brominal
	Buctril
butachlor	Machete
butam	S 15544
buthidiazole	Ravage
butralin	Amox 820
butylate	Sutan
cacodylic acid	Phytar 650
	Arsan
carbetamide	Legurame
CDAA (allidochlor)	Randox
CDEC (sulfallate)	Granular Vegedex
chloramben	Amiben
	Vegiben
chlorbromuron	Maloron
chlorfenac	Fenac
chloroxuron	Tenoran
chlorpropham	Zide
CMA	Calcar
cyanazine	Bladex
cycloate	Ro-Neet
cycluron	Alipur
cyperquat	S-21634
cyprazine	S 6115

Common name	Trade name
cyprazole	S-19073
cypromid	Clobber
dalapon	Basfapon
	Dowpon
	Gramevin
dazomet	Mylone 50
DCPA (chlorthal-dimethyl)	Dacthal
desmedipham	Betanex
desmetryne	Semeron
diallate	Avadex
dicamba	Banvel
dichlobenil	Casoron
dichlormate	Rowmate
dichlorprop	Weedone
	2,4-DP
difenzoquat	Avenge
	Finaven
dinitramine	Cobexal
dinoseb	Dow Selective
	Premerge
diphenamid	Enide
dipropetryn	Sancap
diquat	Weedrite
diuron	Karmex
DNOC	Sinox
DSMA	Ansar 8100
	Crab-Erad
endothal	Aquathol
	Endothal Turf Herbicide
	Des-1-cate
EPTC	Eptan G
	Eptam
erbon	Novon
ethalfluralin	Sonalen
ethiolate	Prefox
fenuron	Dybar
fenuron-TCA	Urab
fluchloralin	Basalin
fluometuron	Cotoran
fluorodifen	Preforan
glyphosate	Roundup

295

Common name	Trade name	Common name	Trade name
hexaflurate	TD 480	prometryn	Gesagard
ioxynil	Actril		Gesaten
	Bantrol	pronamide (propyzamide)	Kerb
	Bentrol	propachlor	Ramrod
	Certrol	propanil	Stam-F34
isopropalin	Paarlan	propazine	Gesamil
lenacil	Venzar	propham	Chem-Hoe
linuron	Afalon	prosulfalin	
	Linuron	prynachlor	Basamaize
	Lorox	pyrazon (chloridazon)	Pyramin
MAMA	Ansar 157	secbumeton	Sumitol
MCPA	Agroxone	siduron	Tupersan
MCPB	Tropotox	silvex (fenoprop)	Kuron
mecoprop	Mecopar		Weedone TP
mefluidide	Embark	simazine	Gesatop
	Vistar	simetryn	Gy-bon
metham	Vapam	TCA	Natal
methazole	Probe		NATA
metolachlor	Dual		Sodium TCA
metribuzin	Lexone	tebuthiuron	Spike
	Sencor		Graslan
MH (maleic hydrazide)	MH-30	terbacil	Sinbar
	Slo-Gro	terbuchlor	CP-46358
	Sucker-Stuff	terbuthylazine	Primatol M
molinate	Ordram	terbutol (terbucarb)	Azar
monolinuron	Afesin	terbutryn	Igran
	Aresin	tri-allate	Avadex BW
monuron	Telvar	triclopyr	Garlon
monuron-TCA	Urox	trifluralin	Treflan
MSMA	Ansar	vernolate	Vernam
	Daconate	2,3,6-TBA	Tryben 200
	Weed-Hoe		Benzac
napropamide	Devrinol	2,4-D	2,4-D weedkiller
naptalam	Alanap	2,4-DB	Butoxone
neburon	Kloben		Butyrac
nitralin	Planavin	2,4,5-T	Brushkiller
nitrofen	Tok		
nitrofluorfen		**Avicides**	
norea (noruron)	Herban		
norflurazon	Zorial	Fenthion	Queleatox
oryzalin	Surflan		
oxadiazon	Ronstar	**Insecticides and acaricides**	
oxyfluorfen	Goal	aldrin	Aldrex
paraquat	Gramoxone		Aldrin Dust
PBA	Benzac 354		Aldron
pebulate	Tillam	aluminium phosphide	Phostoxin
pendimethalin	Stomp	BHC — see HCH	
perfluidone	Destun	carbaryl	Carbaryl 25 ULV
phenmedipham	Betanal		Dicarbam
picloram	Tordon		Prosevor
procyazine	Cycle		Sevin
profluralin	Tolban		Vetox
prometon	Pramitol		

Common name	Trade name
coumaphos	Asuntol
	Co-Ral
	Muscatox
	Resitox
cypermethrin	Cymbush
	Ripcord
DDT	Arkotine-DDT
	Didimac
	DDT
DDT 20% + lindane	Cottinex
	Didigam
decamethrin (deltamethrin)	Decis
dieldrin	Alvit
	Antex
	Dieldrex
	Ensoldil
dimethoate	Perfekthion
endosulfan	Thiodan
	Thionex
fenitrothion	Agrothion
	Fenithion
	Folithion
	Sumithion
fenvalerate	Sumicidin
gamma-HCH (lindane or gamma-BHC)	Agrocide
	Chem-Hex
	Gammalin
	Gammamul
	Gammexane
	Gammexine
	Kotol
	Lindane
	Perfekthan
	Sylvogam
malathion	Maladrex
	Malathion
	Malathin

Common name	Trade name
permethrin	Ambush
petroleum oil refined	Albolineum
pirimicarb	Pirimor
pirimiphos-methyl	Actellic
sodium fluoride powder	Flurocid
sulphur powder	Sufran
tetrachlorvinphos	Gardona
	Rabon

Nematicides

Common name	Trade name
carbofuran	Furadan
dibromochloropropane	Nemagon
dichloropropane-dichloropropene mixture	D-D
	Vidden D
ethoprophos	Mocap
fenamiphos	Nemacur
fensulfothion	Dasanit
isazophos	Miral
thionazin	Nemafos

Bacteriostats (B), fungicides (F) and seed dressings (FI)

Common name	Trade name
aldrin + thiram (FI)	Aldrex T
blasticidin-S (F)	Bla-S
bronopol (B)	Bronocot
captafol (F)	Difolatan
	Difosan
cuprous oxide (F)	Caocobre
	Perecot
	Perenox
cupric hydroxide (F)	Kocide 101
fentin acetate (FI)	Brestan
lindane + thiram (FI)	Fernasan 75W
mancozeb (F)	Dithane M-45
phenylmercury acetate (F)	Agrosan
	Leytosan

297

Appendix 2

List of abbreviations

Abbreviation	Full meaning
mm	millimetre
cm	centimetre
m	metre
km	kilometre
g	gram
kg	kilogram
t/ha or tonnes/ha	tonnes per hectare
lit or l	litre
ppm	parts per million
meq	milliequivalent
%	percent
°C	degrees Celsius
°K	degrees Kelvin
cal	calorie
ly	langley
w	watt
sec	second
min	minute

Appendix 3

Miscellaneous conversions

Multiply	By	To obtain
Single superphosphate	0.18	P_2O_5
Single superphosphate	0.14	S
Single superphosphate	0.27	CaO
P_2O_5	0.43	P
P	2.29	P_2O_5
Triple superphosphate	0.45	P_2O_5
Triple superphosphate	0.015	S
Triple superphosphate	0.20	CaO
Phosphate rock	0.27−0.41	P_2O_5
Phosphate rock	0.46	CaO
Ammonium chloride	0.28	N
Ammonium nitrate	0.32	N
Urea	0.46	N
Calcium ammonium nitrate	0.26	N
Calcium ammonium nitrate	0.18	CaO
Ammonium sulphate	0.20	N
Ammonium sulphate	0.23	S
Ammonium phosphate	0.18	N
Ammonium phosphate	0.46	P_2O_5
Potassium nitrate	0.13	N
Potassium nitrate	0.44	K_2O
Potassium chloride	0.60	K_2O
Potassium sulphate	0.50	K_2O
Potassium sulphate	0.17	S
K_2O	0.83	K
K	1.20	K_2O
Gypsum	0.18	S
Gypsum	0.32	CaO
Magnesium sulphate	0.13	S
Magnesium sulphate	0.16	MgO
Manganese sulphate	0.15	S
Manganese sulphate	0.26	Mn
Ferrous sulphate	0.19	S
Ferrous sulphate	0.33	Fe
Copper sulphate	0.13	S
Copper sulphate	0.25	Cu
Zinc sulphate	0.18	S
Zinc sulphate	0.36	Zn

Appendix 4

Conversion factors for imperial and metric units

To convert column 1 into column 2, multiply by	Column 1	Column 2	To convert column 2 into column 1, multiply by
Length			
0.621	kilometre, km	mile, m	1.609
1.094	metre, m	yard, yd	0.914
0.394	centimetre, cm	inch, in	2.540
Area			
0.386	kilometre^2, km^2	mile^2, m^2	2.590
2.47.1	kilometre^2, km^2	acre, acre	0.00405
2.471	hectare, ha (0.01 km^2)	acre, acre	0.405
Volume			
0.00973	metre^3, m^3	acre	102.8
3.532	hectolitre, hl	cubic foot, ft^3	0.2832
2.838	hectolitre, hl	bushel, bu	0.352
1.057	litre	quart (liquid), qt	0.946
Mass			
1.102	tonne (metric)	ton (English)	0.9072
220.5	quintal, q	pound, 1b	0.00454
2.205	kilogram, kg	pound, 1b	0.454
Yield or rate			
0.446	tonne (metric)/hectare	ton (English)/acre	2.242
0.892	kg/ha	1b/acre	1.121
0.892	quintal/hectare	hundredweight/acre	1.121
0.089	litre (1)/hectare	gallons/acre	11.236
Pressure			
14.22	kg/cm^2	1b/inch^2, psi	0.0703
14.50	bar	1b/in^2, psi	0.06895
0.9869	bar	atmosphere, atm *	1.013
0.9678	kg/cm^2	atmosphere, atm *	1.033
14.70	atmosphere, atm *	1b/in^2, psi	0.06805

* An 'atmosphere' may be specified in metric or English units

Temperature			
1.80°C + 32	Celsius, C	Fahrenheit, F	0.555 (°F-32)
°C + 273	Celsius, C	Kelvin, K	°K-273
Light or rate			
0.0929	lux	foot-candle, ft-c	10.764
1.000	langley (ly)/minute	calories (cal.)/cm^2 min	1.000
1.433 × 10^{-3}	Watts (W)/m^2	Cal/cm^2 min	697.837

Appendix 5

Available strains of *Rhizobium,* their host plant and their origin

Strain number	Host plant	Source of innoculum
5005	Prima cowpea	IITA, Ibadan
5008	Groundnut	IITA, Ibadan
5009	Cowpea	IITA, Ibadan
5028	Pale green cowpea	IITA, Ibadan
5029	Winged bean	IITA, Ibadan
5030	*Stylosanthes humilis*	Moor Plantation, Ibadan.
5017	Cowpea	Lagos, soil around *Dolichos* sp.
5000	Cowpea	Samaru, Nigeria
5011	Cowpea	Samaru, Nigeria
5016	Cowpea	Kano Soil, Nigeria
5018	Cowpea	Makerere, Uganda
5200	Pigeon pea	Makerere, Uganda
CB 756	Cowpea	Zimbabwe
CB 1024	Cowpea	ex CSIRO, Brisbane Australia

Index

302

216, 220, 224, 230
maneb, 200
mango (*Mangifera indica*), 76, 248–252
manual labour, 67, 81
marcotting, 77, 244
marketing, transportation and, 95; retail, 98; wholesale, 98
markets, road-side, 97; rural, 96–97; urban, 97
Maruca testulalis (bean borer), 167
Matuga, Kenya, 36
mechanisation, 81
Medicago sativa (lucerne), 286
Melampsora lini (flax rust), 151
Meloidogyne javanica (eelworm), 162, 229
melon (see *Cucumis*)
metobromuron, 200, 209, 210
metolachlor, 110, 134, 156, 160, 194, 209, 216
metribuzin, 209
micro-climate, 56
micro-nutrients, 27, 29, 32–33
micro-organisms, soil, 25, 27, 29
millet (*Pennisetum americanum/ typhoides*), 43, 45, 83, 85, 96, 104, 117–122
mineralisation, 24, 26, 30
mixtures, grass-legume, 287–288
Mokwa, Nigeria, 17, 139
Monilochaetes infuscans (scurf), 198
monocropping (monoculture), 60, 62
monolinuron, 203
Mononychellus tanajoa (spider mite), 188
montmorillonite, 19, 28
monuron, 227
mulch, 63, 72, 218, 227, 229, 241, 262
mutation, 74, 76

neem (*Azadirachta indica*), 152
nematodes, 229, 241
net assimilation rate, 53
New Delhi, India, 162
niger (*Guizotia abyssinica*), 117, 130, 140–141
Nigeria, 117, 234; planting times, 83; rain-fall, 56; ridge cropping, 80; soil erosion, 72; soils, 24; weather, 13, 50–52
nitrates, soil, 29–30
nitrogen, 29–30, 37, 84–85, 88
nitrogen cycle, 29

nitrogen fixation, 30
nitrogenous fertilisers, 87, 88
Nomadacris (red locust), 116
norflurazon, 156, 171, 177
nutrient availability, 19, 26–27, 30, 59, 87
nutrient uptake, 62, 85
nutrients, essential, 29
nutritive value, 153

oats, 104
Oidium heveae, 183
oil palm (*Elaeis guineensis*), 32, 130, 144–147
oilseeds, moisture content, 94
okra (*Abelmoschus esculentus*), 214–216
Oligonychus pretepsis (date mite), 255
Omphisa anastomosalis (vine borer), 198
Ophelia pigmentata (root rot), 256
onion (*Allium*), 211–214
overlap ploughing, 72
oxadiazon, 106, 214
oxisols, 28, 31

Pangola grass (*Digitaria decumbens*), 285
Pantnagar, India, 163
papain, 232–234
paraquat, 146, 182, 200, 227, 233, 245, 251, 270
parathion, 177
Parkia clappertoniana, 257
pasture, sown, 97
pawpaw (*Carica papaya*), 76, 231–234
pea (*Pisum sativum/arvense*), 168
peanut (see groundnut)
pedigree method, 74
pepper (*Piper nigrum*), 218–220
Perenox (cuprous oxide), 246, 266
Perigea capensis (moth), 141
pest control, 90–93, 172; integrated, 93
pesticide applicators, 92
pesticides, 92
Phaeosariopsis (brown spot), 246
Phenacoccus manihot (Cassava mealybug), 92, 188
phosphates, 26–27, 34–35, 84, 87
phosphorus, 29, 30–31, 134
photorespiration, 47
photosynthesis, 41–47, 48; carbon dioxide availability, 43–44; light

absorption, 41, 42; water stress and, 50
photosynthetic rate, 42, 43
Phyllosticta (leaf spot), 151
Phytophthora, 78–79, 178, 183, 220; *P. cinnamomi*, 227, 241; *P. colocasiae*, 203; *P. infestans*, 199, 200; *P. palmivora*, 183, 266; *P. parasitica*, 241
pigeon pea (*Cajanus cajan*), 161–163
pineapple (*Ananas comosus*), 79, 238–241
Piricularia orvzae (rice blast), 107
plant breeding, 74–75
plant growth hormones, 54, 55; mixed cropping and, 61
plant introduction, 73
plantain (*Musa paradisiaca*), 220–226
planting date, 163
Plasmopora viticola (downy mildew), 238
ploughing, 63, 67, 72
Polistes (fruit wasp), 256
polyembryony, 247
Polygonum, 106
polyploidisation, 74
Poncirus trifoliata, 242
potato (*Solanum tuberosum*), 45, 87, 198–201; storage, 95
potash, 84–85; muriate of, 37
potassium, 28, 31, 84–85
potassium sulphate, 35
Pratylenchus (nematode), 195, 198
Prodenia litura (cut worm), 203
prometryne, 164, 203
propagation, sexual, 76–77; vegetative, 77–79; pineapple, 239–240; yams, 191–193
propanil, 106, 122
protein content, grasses, 283
pruning, 63–64; grapevine, 235–237; coffee, 262
Pseudococcus citri (mealybug), 246
Pseudomonas, 178; *P. sesami*, 140; *P. solanacearum*, 134, 229
Pseudotheraptus wayii (coreid bug), 149
Puccinia (rust), 110, 122; *P. carthami*, 143; *P. helianthus*, 143
Pueraria phaseoloides, 182
pulses, moisture content, 94
pure lines, 74
pyrethrins, 279
pyrethroids, 119, 172, 210, 211

306